GENE
CLONING
AND DNA
ANALYSIS

GENE CLONING AND DNA ANALYSIS

An Introduction

T.A. BROWN

University of Manchester
Manchester

Seventh Edition

WILEY Blackwell

Library of Congress Cataloging-in-Publication Data
1007561016
Brown, T.A. (Terence A.), author.
  Gene cloning and DNA analysis : an introduction / T.A. Brown. – Seventh edition.
    p. ; cm.
  Includes bibliographical references and index.
  ISBN 978-1-119-07257-7 (cloth) – ISBN 978-1-119-07256-0 (pbk.)
  I. Title.
  [DNLM: 1. Cloning, Molecular.   2. DNA, Recombinant–analysis.   3. Sequence Analysis,
DNA. QU 450]
  QH442.2
  572.8'633–dc23   2015015462

A catalogue record for this book is available from the British Library.

Wiley also publishes its books in a variety of electronic formats. Some content that appears in print
may not be available in electronic books.

Set in 10/12pt Sabon by Aptara Inc., New Delhi, India
Printed and bound in Singapore by Markono Print Media Pte Ltd

1 2016

Contents in Brief

CONTENTS IN BRIEF

Contents
CONTENTS

Part II The Applications of Gene Cloning and DNA Analysis in Research 173

10 Sequencing Genes and Genomes 175

11 Studying Gene Expression and Function 201

Part III The Applications of Gene Cloning and DNA Analysis in Biotechnology 245

Preface to the Seventh Edition

PREFACE TO THE SEVENTH EDITI

Anyone who works with DNA is well aware of the dramatic changes that have taken place during the past few years in DNA sequencing methodology. To reflect these advances, in this new edition *of Gene Cloning and DNA Analysis: An Introduction* I have completely remodelled the chapter on DNA sequencing to give the new 'next-generation' methods equal prominence alongside the traditional approaches to DNA sequencing, and also to modernize the description of the ways in which genome sequences are generated. Elsewhere, I have stressed the importance of RNA-seq as a means of studying transcriptomes, and ChIP-seq for locating protein-binding sites. These changes correct the major weakness of the Sixth Edition, which was written just before these methods came into mainstream use.

Elsewhere, I have made the usual updates, especially in Part III where I have tried to keep pace with the increasingly rapid developments in the applications of gene cloning and DNA analysis in industry, medicine and agriculture. I have also rewritten the last part of the final chapter, on archaeogenetics, in order to present some of the new information on the human past that has been revealed by the Neanderthal and Denisovan genome sequences. As always, my primary aim is to ensure that *Gene Cloning* remains an introductory text that begins at the beginning and does not assume that the reader has any prior knowledge of the techniques used to study genes and genomes.

For the n-th time I must thank my wife Keri for the unending support that she has given to me in my decision to use up evenings and weekends writing this and other books.

T.A. Brown
University of Manchester

About the companion website

ABOUT THE COMPANION WEBSITE

This book is accompanied by a companion website:

www.wiley.com/go/brown/cloning

The website includes:

- Powerpoints of all figures from the book for downloading
- PDFs of tables from the book

PART I
The Basic Principles of Gene Cloning and DNA Analysis

Chapter 1

Why Gene Cloning and DNA Analysis are Important

Chapter contents

In the middle of the 19th century, Gregor Mendel formulated a set of rules to explain the inheritance of biological characteristics. The basic assumption of these rules is that each heritable property of an organism is controlled by a factor, called a gene, that is a physical particle present somewhere in the cell. The rediscovery of Mendel's laws in 1900 marks the birth of genetics, the science aimed at understanding what these genes are and exactly how they work.

1.1 The early development of genetics

For the first 30 years of its life this new science grew at an astonishing rate. The idea that genes reside on chromosomes was proposed by W. Sutton in 1903, and received experimental backing from T.H. Morgan in 1910. Morgan and his colleagues then developed the techniques for gene mapping, and by 1922 had produced a comprehensive analysis of the relative positions of over 2000 genes on the four chromosomes of the fruit fly, *Drosophila melanogaster*.

Despite the brilliance of these classical genetic studies, there was no real understanding of the molecular nature of the gene until the 1940s. Indeed, it was not until the

Gene Cloning and DNA Analysis: An Introduction, Seventh Edition. T.A. Brown.
© 2016 John Wiley & Sons, Ltd. Published 2016 by John Wiley & Sons, Ltd.
Companion Website: www.wiley.com/go/brown/cloning

experiments of Avery, MacLeod, and McCarty in 1944, and of Hershey and Chase in 1952, that anyone believed that deoxyribonucleic acid (DNA) is the genetic material. Up until then it was widely thought that genes were made of protein. The discovery of the role of DNA was a tremendous stimulus to genetic research, and many famous biologists (Delbrück, Chargaff, Crick, and Monod were among the most influential) contributed to the second great age of genetics. During the 14 years between 1952 and 1966, the structure of DNA was elucidated, the genetic code cracked, and the processes of transcription and translation described.

1.2 The advent of gene cloning and the polymerase chain reaction

These years of activity and discovery were followed by a lull, a period of anticlimax when it seemed to some molecular biologists (as the new generation of geneticists styled themselves) that there was little of fundamental importance that was not understood. In truth, there was a frustration that the experimental techniques of the late 1960s were not sophisticated enough to allow the gene to be studied in any greater detail.

Then, in the years 1971–1973 genetic research was thrown back into gear by what at the time was described as a revolution in experimental biology. A whole new methodology was developed, enabling previously impossible experiments to be planned and carried out, if not with ease, then at least with success. These methods, referred to as recombinant DNA technology or genetic engineering, and having at their core the process of gene cloning, sparked another great age of genetics. They led to rapid and efficient DNA sequencing techniques that enabled the structures of individual genes to be determined, reaching a culmination at the turn of the century with the massive genome sequencing projects, including the human project which was completed in 2000. They led to procedures for studying the regulation of individual genes, which have allowed molecular biologists to understand how aberrations in gene activity can result in human diseases such as cancer. The techniques spawned modern biotechnology, which puts genes to work in production of proteins and other compounds needed in medicine and industrial processes.

During the 1980s, when the excitement engendered by the gene cloning revolution was at its height, it hardly seemed possible that another, equally novel and equally revolutionary process was just around the corner. According to DNA folklore, Kary Mullis invented the polymerase chain reaction (PCR) during a drive along the coast of California one evening in 1985. His brainwave was an exquisitely simple technique that acts as a perfect complement to gene cloning. PCR has made easier many of the techniques that were possible but difficult to carry out when gene cloning was used on its own. It has extended the range of DNA analysis and enabled molecular biology to find new applications in areas of endeavour outside of its traditional range of medicine, agriculture, and biotechnology. Archaeogenetics, molecular ecology, and DNA forensics are just three of the new disciplines that have become possible as a direct consequence of the invention of PCR, enabling molecular biologists to ask questions about human evolution and the impact of environmental change on the biosphere, and to bring their powerful tools to bear in the fight against crime. Forty years have passed since the dawning of the age of gene cloning, but we are still riding the rollercoaster and there is no end to the excitement in sight.

1.3 What is gene cloning?

What exactly is gene cloning? The easiest way to answer this question is to follow through the steps in a gene cloning experiment (Figure 1.1):

1 A fragment of DNA, containing the gene to be cloned, is inserted into a circular DNA molecule called a vector, to produce a recombinant DNA molecule.

2 The vector transports the gene into a host cell, which is usually a bacterium, although other types of living cell can be used.

3 Within the host cell the vector multiplies, producing numerous identical copies, not only of itself but also of the gene that it carries.

4 When the host cell divides, copies of the recombinant DNA molecule are passed to the progeny and further vector replication takes place.

5 After a large number of cell divisions, a colony, or clone, of identical host cells is produced. Each cell in the clone contains one or more copies of the recombinant DNA molecule. The gene carried by the recombinant molecule is now said to be cloned.

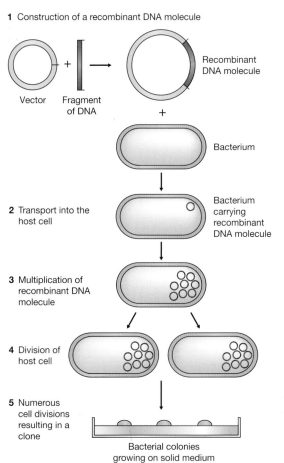

1 Construction of a recombinant DNA molecule

Vector   Fragment of DNA

Recombinant DNA molecule

Bacterium

2 Transport into the host cell

Bacterium carrying recombinant DNA molecule

3 Multiplication of recombinant DNA molecule

4 Division of host cell

5 Numerous cell divisions resulting in a clone

Bacterial colonies growing on solid medium

Figure 1.1

The basic steps in gene cloning.

1.4 What is PCR?

The polymerase chain reaction is very different from gene cloning. Rather than a series of manipulations involving living cells, PCR is carried out in a single test tube simply by mixing DNA with a set of reagents and placing the tube in a thermal cycler, a piece of equipment that enables the mixture to be incubated at a series of temperatures that are varied in a preprogrammed manner. The basic steps in a PCR experiment are as follows (Figure 1.2):

1 The mixture is heated to 94 °C, at which temperature the hydrogen bonds that hold together the two strands of the double-stranded DNA molecule are broken, causing the molecule to denature.

Figure 1.2

The basic steps in the polymerase chain reaction.

Template DNA

1 Denaturation of the template DNA at 94°C

2 Annealing of the oligonucleotide primers at 50–60°C

Primers

3 Synthesis of new DNA at 74°C

4 Repeat the cycle 25–30 times

2 The mixture is cooled down to 50–60 °C. The two strands of each molecule could join back together at this temperature, but most do not because the mixture contains a large excess of short DNA molecules, called oligonucleotides or primers, which anneal to the DNA molecules at specific positions.

3 The temperature is raised to 74 °C. This is a good working temperature for the *Taq* DNA polymerase that is present in the mixture. We will learn more about DNA polymerases on p. 51. All we need to understand at this stage is that the *Taq* DNA polymerase attaches to one end of each primer and synthesizes new strands of DNA, complementary to the template DNA molecules, during this step of the PCR. Now we have four stands of DNA instead of the two that there were to start with.

4 The temperature is increased back to 94 °C. The double-stranded DNA molecules, each of which consists of one strand of the original molecule and one new strand of DNA, denature into single strands. This begins a second cycle of denaturation–annealing–synthesis, at the end of which there are eight DNA strands. By repeating the cycle 30 times the double-stranded molecule that we began with is converted into over 130 million new double-stranded molecules, each one a copy of the region of the starting molecule delineated by the annealing sites of the two primers.

1.5 Why gene cloning and PCR are so important

As can be seen from Figures 1.1 and 1.2, gene cloning and PCR are relatively straight-forward procedures. Why, then, have they assumed such importance in biology? The answer is largely because both techniques can provide a pure sample of an individual gene, separated from all the other genes in the cell.

1.5.1 Obtaining a pure sample of a gene by cloning

To understand exactly how cloning can provide a pure sample of a gene, consider the basic experiment from Figure 1.1, but drawn in a slightly different way (Figure 1.3). In this example, the DNA fragment to be cloned is one member of a mixture of many different fragments, each carrying a different gene or part of a gene. This mixture could indeed be the entire genetic complement of an organism – a human, for instance. Each of these fragments becomes inserted into a different vector molecule to produce a family of recombinant DNA molecules, one of which carries the gene of interest. Usually, only one recombinant DNA molecule is transported into any single host cell, so that although the final set of clones may contain many different recombinant DNA molecules, each individual clone contains multiple copies of just one molecule. The gene is now separated away from all the other genes in the original mixture, and its specific features can be studied in detail.

In practice, the key to the success or failure of a gene cloning experiment is the ability to identify the particular clone of interest from the many different ones that are obtained. If we consider the genome of the bacterium *Escherichia coli*, which contains just over 4000 different genes, we might at first despair of being able to find just one gene among all the possible clones (Figure 1.4). The problem becomes even more overwhelming when we remember that bacteria are relatively simple organisms and that the human genome contains about five times as many genes. As explained in Chapter 8, a variety

Figure 1.3

Cloning allows individual fragments of DNA to be purified.

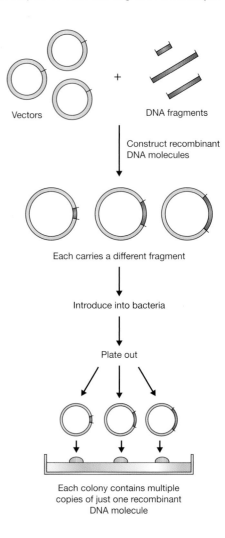

of different strategies can be used to ensure that the correct gene can be obtained at the end of the cloning experiment. Some of these strategies involve modifications to the basic cloning procedure, so that only cells containing the desired recombinant DNA molecule can divide and the clone of interest is automatically selected. Other methods involve techniques that enable the desired clone to be identified from a mixture of lots of different clones.

Once a gene has been cloned there is almost no limit to the information that can be obtained about its structure and expression. The availability of cloned material has stimulated the development of many different analytical methods for studying genes, with new techniques being introduced all the time. Methods for studying the structure and expression of a cloned gene are described in Chapters 10 and 11, respectively.

1.5.2 PCR can also be used to purify a gene

The polymerase chain reaction can also be used to obtain a pure sample of a gene. This is because the region of the starting DNA molecule that is copied during PCR

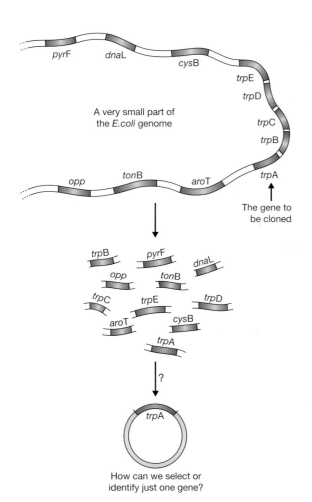

Figure 1.4
The problem of selection.

A very small part of the E.coli genome

The gene to be cloned

How can we select or identify just one gene?

is the segment whose boundaries are marked by the annealing positions of the two oligonucleotide primers. If the primers anneal either side of the gene of interest, many copies of that gene will be synthesized (Figure 1.5). The outcome is the same as with a gene cloning experiment, although the problem of selection does not arise because the desired gene is automatically 'selected' as a result of the positions at which the primers anneal.

A PCR experiment can be completed in a few hours, whereas it takes weeks – if not months – to obtain a gene by cloning. Why then is gene cloning still used? This is because PCR has two limitations:

- In order for the primers to anneal to the correct positions, on either side of the gene of interest, the sequences of these annealing sites must be known. It is easy to synthesize a primer with a predetermined sequence (see p. 149), but if the sequences of the annealing sites are unknown then the appropriate primers cannot be made. This means that PCR cannot be used to isolate genes that have not been studied before – that has to be done by cloning.

Figure 1.5

Gene isolation by PCR.

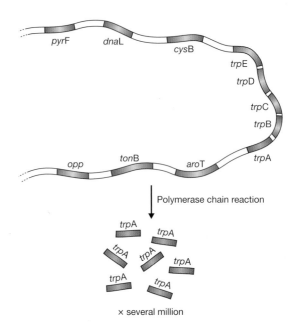

- There is a limit to the length of DNA sequence that can be copied by PCR. Five kilobases (kb) can be copied fairly easily, and segments up to 40 kb can be dealt with by using specialized techniques, but this is shorter than the lengths of many genes, especially those of humans and other vertebrates. Cloning must be used if an intact version of a long gene is required.

Gene cloning is therefore the only way of isolating long genes or those that have never been studied before. But PCR still has many important applications. For example, even if the sequence of a gene is not known, it may still be possible to determine the appropriate sequences for a pair of primers, based on what is known about the sequence of the equivalent gene in a different organism. A gene that has been isolated and sequenced from, say, mouse could therefore be used to design a pair of primers for isolation of the equivalent gene from humans.

In addition, there are many applications where it is necessary to isolate or detect genes whose sequences are already known. A PCR of human globin genes, for example, is used to test for the presence of mutations that might cause the blood disease called thalassaemia. The design of appropriate primers for this PCR is easy because the sequences of the human globin genes are known. After the PCR, the gene copies are sequenced or studied in some other way to determine if any of the thalassaemia mutations are present.

Another clinical application of PCR involves the use of primers specific for the DNA of a disease-causing virus. A positive result indicates that a sample contains the virus and that the person who provided the sample should undergo treatment to prevent onset of the disease. The PCR is tremendously sensitive, with a carefully set up reaction yielding detectable amounts of DNA even if there is just one DNA molecule in the starting mixture. This means that the technique can detect viruses at the earliest stages of an infection, increasing the chances of a treatment being successful. This great sensitivity means that PCR can also be used with DNA from forensic material such as hairs and dried bloodstains, or even from the bones of long-dead humans (see Chapter 16).

1.6 How to find your way through this book

This book explains how gene cloning, PCR and other DNA analysis techniques are carried out, and describes the applications of these techniques in modern biology. The applications are covered in the second and third parts of the book. Part II describes how genes and genomes are studied, while Part III provides accounts of the broader applications of gene cloning and PCR in biotechnology, medicine, agriculture, and forensic science.

In Part I we deal with the basic principles. Most of the nine chapters in Part I are devoted to gene cloning because this technique is more complicated than PCR. When you have understood how cloning is carried out you will have understood many of the basic principles of how DNA is analysed. In Chapter 2 we look at the central component of a gene cloning experiment – the vector – which transports the gene into the host cell and is responsible for its replication. To act as a cloning vector a DNA molecule must be capable of entering a host cell and, once inside, replicating to produce multiple copies of itself. Two naturally occurring types of DNA molecule satisfy these requirements:

- Plasmids, which are small circles of DNA found in bacteria and some other organisms. Plasmids can replicate independently of the host cell chromosome.
- Virus chromosomes, in particular the chromosomes of bacteriophages, which are viruses that specifically infect bacteria. During infection the bacteriophage DNA molecule is injected into the host cell where it undergoes replication.

Chapter 3 describes how DNA is purified from living cells – both the DNA that will be cloned and the vector DNA – and Chapter 4 covers the various techniques for handling purified DNA molecules in the laboratory. There are many such techniques, but two are particularly important in gene cloning. These are the ability first to cut the vector at a specific point, and then to repair it in such a way that the gene is inserted (see Figure 1.1). These and other DNA manipulations were developed as an offshoot of basic research into DNA synthesis and modification in living cells, and most of the manipulations make use of purified enzymes. The properties of these enzymes, and the way they are used in DNA studies, are described in Chapter 4.

Once a recombinant DNA molecule has been constructed, it must be introduced into the host cell so that replication can take place. Transport into the host cell makes use of natural processes for the uptake of plasmid and viral DNA molecules. These processes and the ways they are utilized in gene cloning are described in Chapter 5, while the most important types of cloning vector are introduced, and their uses examined, in Chapters 6 and 7. To conclude the coverage of gene cloning, in Chapter 8 we investigate the problem of selection (see Figure 1.4), before returning in Chapter 9 to a more detailed description of PCR and its related techniques.

Further reading

Blackman, K. (2001) The advent of genetic engineering. *Trends in Biochemical Science*, **26**, 268–270. [An account of the early days of gene cloning.]

Brock, T.D. (1990) *The Emergence of Bacterial Genetics*. Cold Spring Harbor Laboratory Press, New York. [Details the discovery of plasmids and bacteriophages.]

Brown, T.A. (2006) *Genomes*, 3rd edn. Garland Science, Oxford. [An introduction to modern genetics and molecular biology.]

Cherfas, J. (1982) *Man Made Life*. Blackwell, Oxford. [A history of the early years of genetic engineering.]

Judson, H.F. (1979) *The Eighth Day of Creation*. Penguin Science, London. [A very readable account of the development of molecular biology in the years before the gene cloning revolution.]

Mullis, K.B. (1990) The unusual origins of the polymerase chain reaction. *Scientific American*, **262**(4), 56–65. [An entertaining account of how PCR was invented.]

Chapter 2

Vectors for Gene Cloning: Plasmids and Bacteriophages

Chapter contents

A DNA molecule needs to display several features to be able to act as a vector for gene cloning. Most importantly, it must be able to replicate within the host cell, so that numerous copies of the recombinant DNA molecule can be produced and passed to the daughter cells. A cloning vector also needs to be relatively small, ideally less than 10 kb in size, as large molecules tend to break down during purification, and are also more difficult to manipulate. Two kinds of DNA molecule that satisfy these criteria can be found in bacterial cells, namely plasmids and bacteriophage chromosomes.

2.1 Plasmids

Plasmids are circular molecules of DNA that lead an independent existence in the bacterial cell (Figure 2.1). Plasmids almost always carry one or more genes, and often these genes are responsible for a useful characteristic displayed by the host bacterium. For example, the ability to survive in normally toxic concentrations of antibiotics such as chloramphenicol or ampicillin is often due to the presence in the bacterium of a plasmid carrying antibiotic resistance genes. In the laboratory, antibiotic resistance is often used as a selectable marker to ensure that bacteria in a culture contain a particular plasmid (Figure 2.2).

Most plasmids possess at least one DNA sequence that can act as an origin of replication, so they are able to multiply within the cell independently of the main bacterial

Gene Cloning and DNA Analysis: An Introduction, Seventh Edition. T.A. Brown.
© 2016 John Wiley & Sons, Ltd. Published 2016 by John Wiley & Sons, Ltd.
Companion Website: www.wiley.com/go/brown/cloning

Figure 2.1

Plasmids: independent genetic elements found in bacterial cells.

chromosome (Figure 2.3a). The smaller plasmids make use of the host cell's own DNA replicative enzymes in order to make copies of themselves, whereas some of the larger ones carry genes that code for special enzymes that are specific for plasmid replication. A few types of plasmid are also able to replicate by inserting themselves into the bacterial chromosome (Figure 2.3b). These integrative plasmids or episomes may be stably maintained in this form through numerous cell divisions, but they always at some stage exist as independent elements.

2.1.1 Size and copy number

The size and copy number of a plasmid are particularly important as far as cloning is concerned. We have already mentioned the relevance of plasmid size, and stated that less

Figure 2.2

The use of antibiotic resistance as a selectable marker for a plasmid. RP4 (top) carries genes for resistance to ampicillin, tetracycline, and kanamycin. Only those *E. coli* cells that contain RP4 (or a related plasmid) are able to survive and grow in a medium that contains toxic amounts of one or more of these antibiotics.

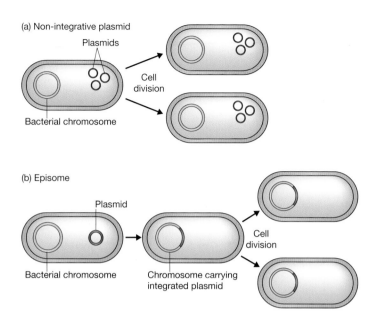

Figure 2.3

Replication strategies for (a) a non-integrative plasmid, and (b) an episome.

than 10 kb is desirable for a cloning vector. Plasmids range from about 1.0 kb for the smallest to over 250 kb for the largest plasmids (Table 2.1), so only a few are useful for cloning purposes. However, as we will see in Chapter 7, larger plasmids can be adapted for cloning under some circumstances.

The copy number refers to the number of molecules of an individual plasmid that are normally found in a single bacterial cell. The factors that control copy number are not well understood. Some plasmids, especially the larger ones, are stringent and have a low copy number of perhaps just one or two per cell; others, called relaxed plasmids, are present in multiple copies of 50 or more per cell. Generally speaking, a useful cloning vector needs to be present in the cell in multiple copies so that large quantities of the recombinant DNA molecule can be obtained.

Table 2.1

Sizes of representative plasmids.

| | SIZE | | |
PLASMID	NUCLEOTIDE LENGTH (kb)	MOLECULAR MASS (MDa)	ORGANISM
pUC8	2.1	1.8	*E. coli*
ColE1	6.4	4.2	*E. coli*
RP4	54	36	*Pseudomonas* and others
F	95	63	*E. coli*
TOL	117	78	*Pseudomonas putida*
pTiAch5	213	142	*Agrobacterium tumefaciens*

Figure 2.4

Plasmid transfer by conjugation between bacterial cells. The donor and recipient cells attach to each other by a pilus, a hollow appendage present on the surface of the donor cell. A copy of the plasmid is then passed to the recipient cell. Transfer is thought to occur through the pilus, but this has not been proven and transfer by some other means (e.g., directly across the bacterial cell walls) remains a possibility.

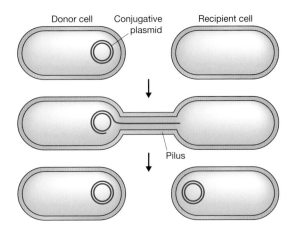

2.1.2 Conjugation and compatibility

Plasmids fall into two groups: conjugative and non-conjugative. Conjugative plasmids are characterized by the ability to promote sexual conjugation between bacterial cells (Figure 2.4), a process that can result in a conjugative plasmid spreading from one cell to all the other cells in a bacterial culture. Conjugation and plasmid transfer are controlled by a set of transfer or *tra* genes, which are present on conjugative plasmids but absent from the non-conjugative type. A non-conjugative plasmid may, under some circumstances, be cotransferred along with a conjugative plasmid when both are present in the same cell.

Several different kinds of plasmid may be found in a single cell, including more than one different conjugative plasmid at any one time. In fact, cells of *E. coli* have been known to contain up to seven different plasmids at once. To be able to coexist in the same cell, different plasmids must be compatible. If two plasmids are incompatible, then one or the other will be rapidly lost from the cell. Different types of plasmid can therefore be assigned to different incompatibility groups on the basis of whether or not they can coexist, and plasmids from a single incompatibility group are often related to each other in various ways. The basis of incompatibility is not well understood, but events during plasmid replication are thought to underlie the phenomenon.

2.1.3 Plasmid classification

The most useful classification of naturally occurring plasmids is based on the main characteristic coded by the plasmid genes. The five major types of plasmid according to this classification are as follows:

- Fertility or F plasmids carry only *tra* genes and have no characteristic beyond the ability to promote conjugal transfer of plasmids. A well-known example is the F plasmid of *E. coli*.
- Resistance or R plasmids carry genes conferring on the host bacterium resistance to one or more antibacterial agents, such as chloramphenicol, ampicillin, and mercury. R plasmids are very important in clinical microbiology as their spread through

natural populations can have profound consequences in the treatment of bacterial infections. An example is RP4, which is commonly found in *Pseudomonas*, but also occurs in many other bacteria.

- Col plasmids code for colicins, proteins that kill other bacteria. An example is ColE1 of *E. coli*.
- Degradative plasmids allow the host bacterium to metabolize unusual molecules such as toluene and salicylic acid, an example being TOL of *Pseudomonas putida*.
- Virulence plasmids confer pathogenicity on the host bacterium; these include the Ti plasmids of *Agrobacterium tumefaciens*, which induce crown gall disease on dicotyledonous plants.

2.1.4 Plasmids in organisms other than bacteria

Although plasmids are widespread in bacteria they are by no means as common in other organisms. The best-characterized eukaryotic plasmid is the 2 µm circle that occurs in many strains of the yeast *Saccharomyces cerevisiae*. The discovery of the 2 µm plasmid was very fortuitous as it allowed the construction of cloning vectors for this very important industrial organism (p. 111). The search for plasmids in other eukaryotes (such as filamentous fungi, plants and animals) has proved disappointing, and it is suspected that many higher organisms simply do not harbour plasmids within their cells.

2.2 Bacteriophages

Bacteriophages, or phages as they are commonly known, are viruses that specifically infect bacteria. Like all viruses, phages are very simple in structure, consisting merely of a DNA (or occasionally ribonucleic acid (RNA)) molecule carrying a number of genes, including several for replication of the phage, surrounded by a protective coat or capsid made up of protein molecules (Figure 2.5).

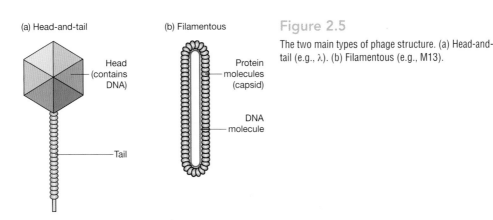

(a) Head-and-tail

Head (contains DNA)

Tail

(b) Filamentous

Protein molecules (capsid)

DNA molecule

Figure 2.5

The two main types of phage structure. (a) Head-and-tail (e.g., λ). (b) Filamentous (e.g., M13).

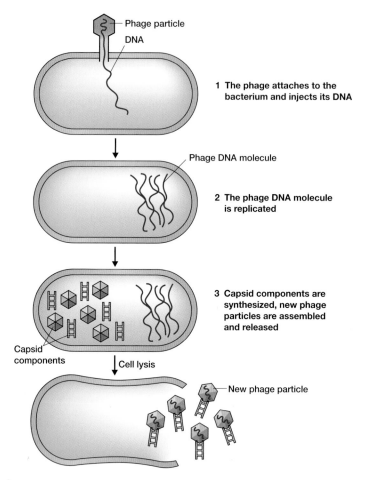

Phage particle

DNA

1 The phage attaches to the bacterium and injects its DNA

Phage DNA molecule

2 The phage DNA molecule is replicated

3 Capsid components are synthesized, new phage particles are assembled and released

Capsid components

Cell lysis

New phage particle

Figure 2.6

The general pattern of infection of a bacterial cell by a bacteriophage.

2.2.1 The phage infection cycle

The general pattern of infection, which is the same for all types of phage, is a three-step process (Figure 2.6):

1 The phage particle attaches to the outside of the bacterium and injects its DNA chromosome into the cell.
2 The phage DNA molecule is replicated, usually by specific phage enzymes coded by genes in the phage chromosome.
3 Other phage genes direct synthesis of the protein components of the capsid, and new phage particles are assembled and released from the bacterium.

With some phage types the entire infection cycle is completed very quickly, possibly in less than 20 minutes. This type of rapid infection is called a lytic cycle, as release of the new phage particles is associated with lysis of the bacterial cell. The characteristic feature of a lytic infection cycle is that phage DNA replication is immediately followed by synthesis of capsid proteins, and the phage DNA molecule is never maintained in a stable condition in the host cell.

2.2.2 Lysogenic phages

In contrast to a lytic cycle, lysogenic infection is characterized by retention of the phage DNA molecule in the host bacterium, possibly for many thousands of cell divisions. With many lysogenic phages the phage DNA is inserted into the bacterial genome, in a manner similar to episomal insertion (see Figure 2.3b). The integrated form of the phage DNA (called the prophage) is quiescent, and a bacterium (referred to as a lysogen) that carries a prophage is usually physiologically indistinguishable from an uninfected cell. The prophage is eventually released from the host genome and the phage reverts to the lytic mode and lyses the cell. The infection cycle of lambda (λ), a typical lysogenic phage of this type, is shown in Figure 2.7.

A limited number of lysogenic phages follow a rather different infection cycle. When M13 or a related phage infects *E. coli*, new phage particles are continuously assembled and released from the cell. The M13 DNA is not integrated into the bacterial genome and does not become quiescent. With these phages, cell lysis never occurs, and the infected bacterium can continue to grow and divide, albeit at a slower rate than uninfected cells. Figure 2.8 shows the M13 infection cycle.

Although there are many different varieties of bacteriophage, only λ and M13 have found a major role as cloning vectors. We will now consider the properties of these two phages in more detail.

Gene organization in the λ DNA molecule

Lambda is a typical example of a head-and-tail phage (see Figure 2.5a). The DNA is contained in the polyhedral head structure and the tail serves to attach the phage to the bacterial surface and to inject the DNA into the cell (see Figure 2.7).

The λ DNA molecule is 49 kb in size and has been intensively studied by the techniques of gene mapping and DNA sequencing. As a result, the positions and identities of all of the genes in the λ DNA molecule are known (Figure 2.9). A feature of the λ genetic map is that genes related in terms of function are clustered together in the genome. For example, all of the genes coding for components of the capsid are grouped together in the left-hand third of the molecule, and genes controlling integration of the prophage into the host genome are clustered in the middle of the molecule. Clustering of related genes is profoundly important for controlling expression of the λ genome, as it allows genes to be switched on and off as a group rather than individually. Clustering is also important in the construction of λ-based cloning vectors, as we will discover when we return to this topic in Chapter 6.

The linear and circular forms of λ DNA

A second feature of λ that turns out to be of importance in the construction of cloning vectors is the conformation of the DNA molecule. The molecule shown in Figure 2.9 is linear, with two free ends, and represents the DNA present in the phage head structure. This linear molecule consists of two complementary strands of DNA, base-paired according to the Watson–Crick rules (that is, double-stranded DNA). At either end of the molecule is a short 12-nucleotide stretch in which the DNA is single-stranded (Figure 2.10a); these two single strands are complementary and so can

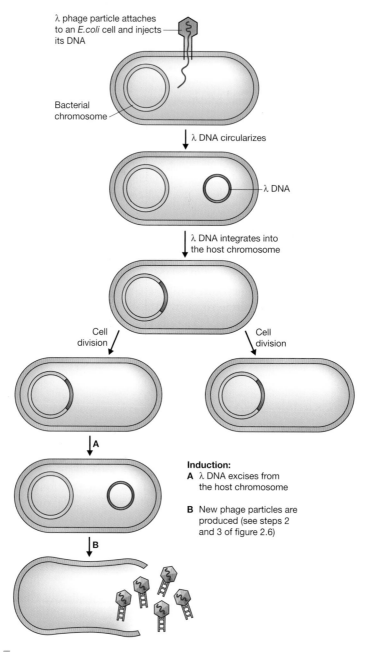

λ phage particle attaches
to an *E.coli* cell and injects
its DNA

Bacterial
chromosome

λ DNA circularizes

λ DNA

λ DNA integrates into
the host chromosome

Cell
division

Cell
division

A

B

Induction:
A λ DNA excises from
   the host chromosome

B New phage particles are
   produced (see steps 2
   and 3 of figure 2.6)

Figure 2.7

The lysogenic infection cycle of bacteriophage λ.

base pair with one another to form a circular, completely double-stranded molecule
(Figure 2.10b).

Complementary single strands are often referred to as 'sticky' ends or cohesive ends,
because base pairing between them can 'stick' together the two ends of a DNA molecule
(or the ends of two different DNA molecules). The λ cohesive ends are called the *cos*

Figure 2.8

The infection cycle of bacteriophage M13.

sites and they play two distinct roles during the λ infection cycle. First, they allow the linear DNA molecule that is injected into the cell to be circularized, which is a necessary prerequisite for insertion into the bacterial genome (see Figure 2.7).

The second role of the *cos* sites is rather different, and comes into play after the prophage has excised from the host genome. At this stage a large number of new λ DNA molecules are produced by the rolling circle mechanism of replication (Figure 2.10c), in which a continuous DNA strand is 'rolled off' the template molecule. The result is a catenane consisting of a series of linear λ genomes joined together at the *cos* sites. The role of the *cos* sites is now to act as recognition sequences for an endonuclease that cleaves the catenane at the *cos* sites, producing individual λ genomes. This endonuclease, which is the product of gene *A* on the DNA molecule, creates the single-stranded sticky ends, and also acts in conjunction with other proteins to package each λ genome into a phage head structure. The cleavage and packaging processes recognize just the

Figure 2.9

The λ genetic map, showing the positions of the important genes and the functions of the gene clusters.

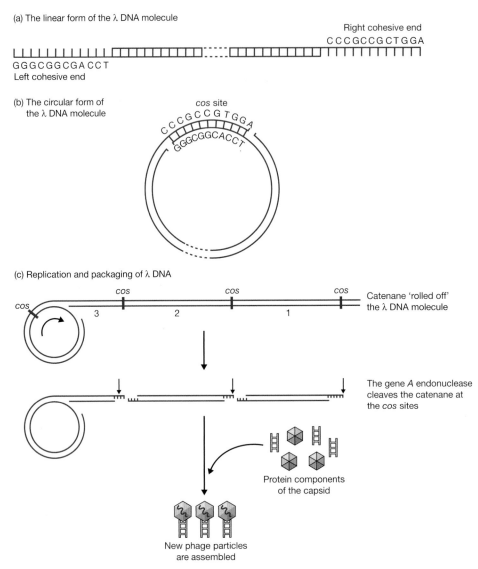

Figure 2.10

The linear and circular forms of λ DNA. (a) The linear form, showing the left and right cohesive ends. (b) Base pairing between the cohesive ends results in the circular form of the molecule. (c) Rolling circle replication produces a catenane of new linear λ DNA molecules, which are individually packaged into phage heads as new λ particles are assembled.

cos sites and the DNA sequences to either side of them, so changing the structure of the internal regions of the λ genome, for example by inserting new genes, has no effect on these events so long as the overall length of the λ genome is not altered too greatly.

M13 – a filamentous phage

M13 is an example of a filamentous phage (see Figure 2.5b) and is completely different in structure from λ. Furthermore, the M13 DNA molecule is much smaller than the λ

Figure 2.11

The M13 infection cycle, showing the different types of DNA replication that occur. (a) After infection, the single-stranded M13 DNA molecule is converted into the double-stranded replicative form (RF). (b) The RF replicates to produce multiple copies of itself. (c) Single-stranded molecules are synthesized by rolling circle replication and used in the assembly of new M13 particles.

(a) Injection of single-stranded DNA into the host cell, followed by synthesis of the second strand

M13 particle injects DNA into cell

Pilus

Single-stranded DNA

Double-stranded DNA - replicative form (RF)

(b) Replication of the RF to produce new double-stranded molecules

(c) Mature M13 phage are continuously produced

RF replicates by rolling circle mechanism to produce linear single-stranded DNA

Circularized DNA

Mature phage particles

genome, being only 6407 nucleotides in length. It is circular and is unusual in that it consists entirely of single-stranded DNA.

The smaller size of the M13 DNA molecule means that it has room for fewer genes than the λ genome. This is possible because the M13 capsid is constructed from multiple copies of just three proteins (requiring only three genes), whereas synthesis of the λ head-and-tail structure involves over 15 different proteins. In addition, M13 follows a simpler infection cycle than λ, and does not need genes for insertion into the host genome.

Injection of an M13 DNA molecule into an *E. coli* cell occurs via the pilus, the structure that connects two cells during sexual conjugation (see Figure 2.4). Once inside the cell the single-stranded molecule acts as the template for synthesis of a complementary strand, resulting in normal double-stranded DNA (Figure 2.11a). This molecule is not inserted into the bacterial genome, but instead replicates until over 100 copies are present in the cell (Figure 2.11b). When the bacterium divides, each daughter cell receives copies of the phage genome, which continues to replicate, thereby maintaining its overall numbers per cell. As shown in Figure 2.11c, new phage particles are continuously assembled and released, with about 1000 new phages being produced during each generation of an infected cell.

Several features of M13 make this phage attractive as a cloning vector. The genome is less than 10 kb in size, well within the range desirable for a potential vector. In addition, the double-stranded replicative form (RF) of the M13 genome behaves very much like a plasmid, and can be treated as such for experimental purposes. It is easily prepared from a culture of infected *E. coli* cells (p. 43) and can be reintroduced by transfection (p. 83). Most importantly, genes cloned with an M13-based vector can be obtained in the form of single-stranded DNA. Single-stranded versions of cloned genes are useful for several techniques, such as *in vitro* mutagenesis (p. 216). Cloning in an M13 vector is an easy and reliable way of obtaining single-stranded DNA for this type of work. M13 vectors are also used in phage display, a technique for identifying pairs of genes whose protein products interact with one another (p. 241).

2.2.3 Viruses as cloning vectors for other organisms

Most living organisms are infected by viruses, and it is not surprising that there has been great interest in the possibility that viruses might be used as cloning vectors for higher organisms. This is especially important when it is remembered that plasmids are not commonly found in organisms other than bacteria and yeast. Several eukaryotic viruses have been employed as cloning vectors for specialized applications. For example, human adenoviruses are used in gene therapy (p. 286), baculoviruses are used to synthesize important pharmaceutical proteins in insect cells (p. 262), and caulimoviruses and geminiviruses have been used for cloning in plants (p. 127). These vectors are discussed more fully in Chapter 7.

Further reading

Dale, J.W. and Park, S.T. (2010) *Molecular Genetics of Bacteria*, 5th edn, Wiley-Blackwell, Chichester. [Provides a detailed description of plasmids and bacteriophages.]
Willey, J., Sherwood, L., and Woolverton, C. (2013) *Prescott's Microbiology*, 9th edn, McGraw-Hill Higher Education, Maidenhead. [A good introduction to microbiology, including plasmids and phages.]

Chapter 3

Purification of DNA from Living Cells

Chapter contents

The genetic engineer will, at different times, need to prepare at least three distinct kinds of DNA. First, total cell DNA will often be required as a source of material from which to obtain genes to be cloned. Total cell DNA may be DNA from a culture of bacteria, from a plant, from animal cells, or from any other type of organism that is being studied. It consists of the genomic DNA of the organism along with any additional DNA molecules, such as plasmids, that are present.

The second type of DNA that will be required is pure plasmid DNA. The preparation of plasmid DNA from a culture of bacteria follows the same basic steps as purification of total cell DNA, with the crucial difference that at some stage the plasmid DNA must be separated from the main bulk of chromosomal DNA also present in the cell.

Finally, phage DNA will be needed if a phage cloning vector is to be used. Phage DNA is generally prepared from bacteriophage particles rather than from infected cells, so there is no problem with contaminating bacterial DNA, although special techniques are needed to remove the phage capsid. An exception is the double-stranded replicative form of M13, which is prepared from *E. coli* cells in the same way as a bacterial plasmid.

3.1 Preparation of total cell DNA

The fundamentals of DNA preparation are most easily understood by first considering the simplest type of DNA purification procedure, that where the entire DNA complement of a bacterial cell is required. The modifications needed for plasmid and phage DNA preparation can then be described later.

Gene Cloning and DNA Analysis: An Introduction, Seventh Edition. T.A. Brown.
© 2016 John Wiley & Sons, Ltd. Published 2016 by John Wiley & Sons, Ltd.
Companion Website: www.wiley.com/go/brown/cloning

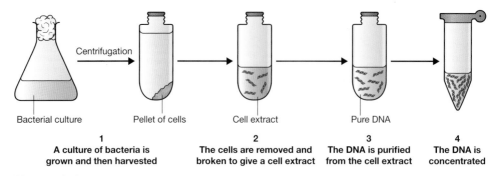

Figure 3.1
The basic steps in preparation of total cell DNA from a culture of bacteria.

The procedure for total DNA preparation from a culture of bacterial cells can be divided into four stages (Figure 3.1):

1 A culture of bacteria is grown and then harvested.
2 The cells are broken open to release their contents.
3 This cell extract is treated to remove all components except the DNA.
4 The resulting DNA solution is concentrated.

3.1.1 Growing and harvesting a bacterial culture

Most bacteria can be grown without too much difficulty in a liquid medium (broth culture). The culture medium must provide a balanced mixture of the essential nutrients at concentrations that will allow the bacteria to grow and divide efficiently. Two typical growth media are detailed in Table 3.1.

M9 is an example of a defined medium in which all the components are known. This medium contains a mixture of inorganic nutrients to provide essential elements such as nitrogen, magnesium and calcium, as well as glucose to supply carbon and energy. In practice, additional growth factors such as trace elements and vitamins must be added

Table 3.1
The composition of two typical media for the growth of bacterial cultures.

MEDIUM	COMPONENT	g l^{-1} OF MEDIUM
M9 medium	Na_2HPO_4	6.0
	KH_2PO_4	3.0
	NaCl	0.5
	NH_4Cl	1.0
	$MgSO_4$	0.5
	Glucose	2.0
	$CaCl_2$	0.015
LB (Luria–Bertani medium)	Tryptone	10.0
	Yeast extract	5.0
	NaCl	10.0

to M9 before it will support bacterial growth. Precisely which supplements are needed depends on the species concerned.

The second medium described in Table 3.1 is rather different. Luria-Bertani (LB) is a complex or undefined medium, meaning that the precise identity and quantity of its components are not known. This is because two of the ingredients – tryptone and yeast extract – are complicated mixtures of unknown chemical compounds. Tryptone in fact supplies amino acids and small peptides, while yeast extract (a dried preparation of partially digested yeast cells) provides the nitrogen requirements, along with sugars and inorganic and organic nutrients. Complex media such as LB need no further supplementation and support the growth of a wide range of bacterial species.

Defined media must be used when the bacterial culture has to be grown under precisely controlled conditions. This is not necessary when the culture is being grown simply as a source of DNA, and under these circumstances a complex medium is appropriate. In LB medium at 37 °C, aerated by shaking at 150–250 rpm on a rotary platform, *E. coli* cells divide once every 20 min or so until the culture reaches a maximum density of about 2–3×10^9 cells ml^{-1}. The growth of the culture can be monitored by reading the optical density (OD) at 600 nm (Figure 3.2), at which wavelength 1 OD unit corresponds to about 0.8×10^9 cells ml^{-1}.

In order to prepare a cell extract, the bacteria must be obtained in as small a volume as possible. Harvesting is therefore performed by spinning the culture in a centrifuge (Figure 3.3). Fairly low centrifugation speeds will pellet the bacteria at the bottom of the centrifuge tube, allowing the culture medium to be poured off. Bacteria from a 1000 ml culture at maximum cell density can then be resuspended into a volume of 10 ml, or less.

(a) Measurement of optical density

(b) Estimation of cell number from a calibration curve

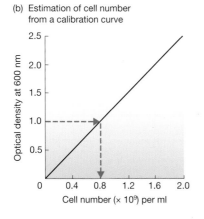

Figure 3.2

Estimation of bacterial cell number by measurement of optical density (OD). (a) A sample of the culture is placed in a glass cuvette and light with a wavelength of 600 nm shone through. The amount of light that passes through the culture is measured and the OD (also called the absorbance) calculated as:

$$1 \text{ OD unit} = \log_{10} \frac{\text{intensity of transmitted light}}{\text{intensity of incident light}}$$

The operation is performed with a spectrophotometer. (b) The cell number corresponding to the OD reading is calculated from a calibration curve. This curve is plotted from the OD values of a series of cultures of known cell density. For *E. coli*, 1 OD unit = 0.8×10^9 cells ml^{-1}.

Figure 3.3
Harvesting bacteria by centrifugation.

3.1.2 Preparation of a cell extract

The bacterial cell is enclosed in a cytoplasmic membrane and surrounded by a rigid cell wall. With some species, including *E. coli*, the cell wall may itself be enveloped by a second, outer membrane. All of these barriers have to be disrupted to release the cell components.

Techniques for breaking open bacterial cells can be divided into physical methods, in which the cells are disrupted by mechanical forces, and chemical methods, where cell lysis is brought about by exposure to chemical agents that affect the integrity of the cell barriers. Chemical methods are most commonly used with bacterial cells when the object is DNA preparation.

Chemical lysis generally involves one agent attacking the cell wall and another disrupting the cell membrane (Figure 3.4a). The chemicals that are used depend on the species of bacterium involved, but with *E. coli* and related organisms, weakening of the cell wall is usually brought about by lysozyme, ethylenediamine tetraacetate (EDTA), or a combination of both. Lysozyme is an enzyme that is present in egg white and in secretions such as tears and saliva, and which digests the polymeric compounds that give the cell wall its rigidity. EDTA removes magnesium ions that are essential for preserving the overall structure of the cell envelope, and also inhibits cellular enzymes that could degrade DNA. Under some conditions, weakening the cell wall with lysozyme or EDTA is sufficient to cause bacterial cells to burst, but usually a detergent such as sodium dodecyl sulphate (SDS) is also added. Detergents aid the process of lysis by removing lipid molecules and thereby cause disruption of the cell membranes.

Figure 3.4
Preparation of a cell extract. (a) Cell lysis. (b) Centrifugation of the cell extract to remove insoluble debris.

Having lysed the cells, the final step in the preparation of a cell extract is the removal of insoluble cell debris. Components such as partially digested cell wall fractions can be pelleted by centrifugation (Figure 3.4b), leaving the cell extract as a reasonably clear supernatant.

3.1.3 Purification of DNA from a cell extract

In addition to DNA, a bacterial cell extract contains significant quantities of protein and RNA. A variety of methods can be used to purify the DNA from this mixture. One approach is to treat the mixture with reagents which degrade the contaminants, leaving a pure solution of DNA (Figure 3.5a). Other methods use chromatography or some other fractionation process to separate the mixture into its various components, so the DNA is removed from the proteins and RNA in the extract (Figure 3.5b).

Removing contaminants by organic extraction and enzyme digestion

The standard way to deproteinize a cell extract is to add phenol or a 1 : 1 mixture of phenol and chloroform. These organic solvents precipitate proteins but leave the nucleic acids (DNA and RNA) in aqueous solution. The result is that if the cell extract is mixed gently with the solvent, and the layers are then separated by centrifugation, precipitated protein molecules are left as a white coagulated mass at the interface between the aqueous and organic layers (Figure 3.6). The aqueous solution of nucleic acids can then be removed using a pipette.

With some cell extracts the protein content is so great that a single phenol extraction is not sufficient to completely purify the nucleic acids. This problem could be solved by carrying out several phenol extractions one after the other, but this is undesirable as each mixing and centrifugation step results in a certain amount of breakage of the DNA molecules. The answer is to treat the cell extract with a protease such as pronase

(a) Degradation of contaminants

DNA   Contaminants   Pure DNA

(b) Separation of DNA from contaminants

Pure DNA   Contaminants

Figure 3.5

Two approaches to DNA purification. (a) Treating the mixture with reagents which degrade the contaminants, leaving a pure solution of DNA. (b) Separating the mixture into different fractions, one of which is pure DNA.

Figure 3.6

Removal of protein contaminants by phenol extraction.

Cell extract   Mix with phenol   Separate layers by centrifugation

Aqueous layer (DNA + RNA)

Interface (coagulated proteins)

Phenol

or proteinase K before phenol extraction. These enzymes break polypeptides down into smaller units, which are more easily removed using phenol.

Some RNA molecules, especially messenger RNA (mRNA), are removed by phenol treatment, but most remain with the DNA in the aqueous layer. The only effective way to remove the RNA is with the enzyme ribonuclease, which rapidly degrades these molecules into ribonucleotide subunits.

Using ion-exchange chromatography to purify DNA from a cell extract

Biochemists have devised various methods for using differences in electrical charge to separate mixtures of chemicals into their individual components. One of these methods is ion-exchange chromatography, which separates molecules according to how tightly they bind to electrically charged particles present in a chromatographic matrix or resin. DNA and RNA are both negatively charged, as are some proteins, and so bind to a positively charged resin. The electrostatic attachment is disrupted by salt (Figure 3.7a), and the removal of more tightly bound molecules requires higher concentrations of salt. By gradually increasing the salt concentration, different types of molecule can be detached from the resin, one after another.

The simplest way to carry out ion-exchange chromatography is to place the resin in a glass or plastic column and then to add the cell extract at the top (Figure 3.7b). The extract passes through the column, but because the extract contains very little salt all of the negatively charged molecules bind to the resin and are retained in the column. If a salt solution of gradually increasing concentration is now passed through the column, the different types of molecule will be eluted (i.e., become unbound) in the sequence protein, RNA, and finally DNA. Such careful separation is usually not needed and just two salt solutions are used, one whose concentration is sufficient to elute the protein and RNA, leaving just the DNA bound, followed by a second solution of a higher concentration which elutes the DNA, which is now free from protein and RNA contaminants.

Using silica to purify DNA from a cell extract

A second type of fractionation method for DNA purification makes use of silica particles. In the presence of guanidinium thiocyanate, DNA binds tightly to silica particles, providing an easy way of recovering the DNA from a cell extract. As with the ion-exchange method, silica purification is usually carried out in a chromatography column, where the silica is placed in the column and the cell extract, to which guanidinium thiocyanate has been added, is passed through (Figure 3.8). DNA binds to the silica and is retained in the column, whereas the other biochemicals are immediately eluted. After washing away the last contaminants with guanidinium thiocyanate solution, the

(a) Attachment of DNA to ion-exchange particles

Ion-exchange
particle

DNA
molecule

Figure 3.7

DNA purification by ion-exchange chromatography. (a) Attachment of DNA to ion-exchange particles. (b) DNA is purified by column chromatography. The solutions passing through the column can be collected by gravity flow or by the **spin column** method, in which the column is placed in a low-speed centrifuge.

(b) DNA purification by ion-exchange chromatography

Cell extract

Ion-exchange
resin

Salt

More
salt

Protein
+ RNA

DNA

Discard

DNA is recovered by adding water, which destabilizes the interactions between the DNA molecules and the silica.

Alternatively, the silica and guanidinium thiocyanate can be added directly to the cell extract and, after DNA binding, the sample is centrifuged to collect the silica particles (Figure 3.9a). The pellet is then resuspended in water to release the DNA, and a second centrifugation is applied to re-pellet the silica, leaving the DNA in solution. A clever modification of this makes use of magnetic beads that are coated with silica where, rather than centrifugation, the beads are collected at the bottom of the test tube using a magnet (Figure 3.9b). Magnetic collection has two advantages compared to the other types of fractionation method used to purify DNA. First, it is quicker, as no time is lost waiting for solutions to pass through a chromatography column or waiting for the centrifuge to complete its run. Second, magnetic collection is easier to automate.

Figure 3.8

DNA purification by the guanidinium thiocyanate and silica method.

By using special equipment many DNA purifications can be carried out in parallel (96 in the simplest format, but much higher numbers with more sophisticated systems). This is an important consideration in applied settings such as clinical microbiology laboratories, where a high throughput is needed so that many DNA samples can be processed as quickly as possible.

3.1.4 Concentration of DNA samples

Organic extraction often results in a very thick solution of DNA that does not need to be concentrated any further. Other purification methods provide more dilute solutions and it is therefore important to consider methods for increasing the DNA concentration.

The most frequently used method of concentration is ethanol precipitation. In the presence of a salt (strictly speaking, monovalent cations such as sodium ions, Na^+) and at a temperature of $-20\,°C$ or less, absolute ethanol efficiently precipitates polymeric nucleic acids. With a thick solution of DNA the ethanol can be layered on top of the sample, causing molecules to precipitate at the interface. A spectacular trick is to push a glass rod through the ethanol into the DNA solution. When the rod is removed, DNA molecules adhere and can be pulled out of the solution in the form of a long fibre (Figure 3.10a). Alternatively, if ethanol is mixed with a dilute DNA solution, the precipitate can be collected by centrifugation (Figure 3.10b) and then redissolved in an appropriate volume of water. Ethanol precipitation has the added advantage of leaving short-chain and monomeric nucleic acid components in solution. Ribonucleotides produced by ribonuclease treatment are therefore lost at this stage.

(a) Silica particles added directly to the cell extract

(b) Silica-coated magnetic beads

Figure 3.9

Alternative methods of using silica to purify DNA from a cell extract. (a) Silica particles are added directly to a cell extract and, after DNA binding, collected by centrifugation. (b) Silica-coated magnetic beads are collected with a magnet.

3.1.5 Measurement of DNA concentration

It is crucial to know exactly how much DNA is present in a solution when carrying out a gene cloning experiment. Fortunately, DNA concentrations can be accurately measured using ultraviolet (UV) absorbance spectrophotometry. The amount of UV

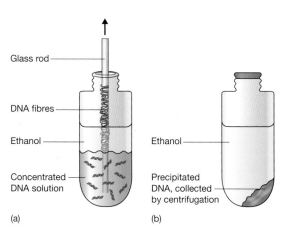

Figure 3.10

Collecting DNA by ethanol precipitation. (a) Absolute ethanol is layered on top of a concentrated solution of DNA. Fibres of DNA can be withdrawn with a glass rod. (b) For less concentrated solutions, ethanol is added (at a ratio of 2.5 volumes of absolute ethanol to 1 volume of DNA solution) and the precipitated DNA collected by centrifugation.

radiation absorbed by a solution of DNA is directly proportional to the amount of DNA in the sample. Usually, absorbance is measured at 260 nm, at which wavelength an absorbance (A_{260}) of 1.0 corresponds to 50 mg of double-stranded DNA per ml. Measurements of as little as 1 µl of a DNA solution can be carried out in spectrophotometers designed especially for this purpose.

Ultraviolet absorbance can also be used to check the purity of a DNA preparation. With a pure sample of DNA, the ratio of the absorbances at 260 and 280 nm (A_{260}/A_{280}) is 1.8. Ratios of less than 1.8 indicate that the preparation is contaminated, either with protein or with phenol.

3.1.6 Other methods for the preparation of total cell DNA

Bacteria are not the only organisms from which DNA may be required. Total cell DNA from, for example, plants or animals will be needed if the aim of the genetic engineering project is to clone genes from these organisms. Although the basic steps in DNA purification are the same whatever the organism, some modifications may have to be introduced to take account of the special features of the cells being used.

Obviously, the growth of cells in a liquid medium is appropriate only for bacteria, other microorganisms, and plant and animal cell cultures. The major modifications, however, are likely to be needed at the cell breakage stage. The chemicals used for disrupting bacterial cells do not usually work with other organisms: lysozyme, for example, has no effect on plant cells. Specific degradative enzymes are available for most cell wall types, but often physical techniques (e.g., grinding frozen material with a mortar and pestle) are more efficient. In contrast, most animal cells have no cell wall at all and can be disrupted simply by treating with detergent.

Animal tissue is also disrupted by guanidinium thiocyanate, the compound used in the silica method for DNA purification. Guanidinium thiocyanate is a chaotropic agent, which means that it interferes with the hydrogen bonding that normally holds water molecules together. This in turn destabilizes membranes and causes proteins and other biochemicals to be denatured. Guanidinium thiocyanate can therefore be used to release DNA from any type of cell or tissue that is not enclosed in a protective cell wall. The silica binding procedure, possibly in its automated magnetic bead format, is therefore a very popular method for DNA purification from clinical material such as blood samples.

Another important consideration is the biochemical content of the cells from which DNA is being extracted. With most bacteria, the main biochemicals present in a cell extract are protein, DNA and RNA, and consequently a phenol extraction and/or protease treatment, followed by the removal of RNA with ribonuclease, leaves a pure DNA sample. These treatments may not be sufficient to produce pure DNA if the cells also contain significant quantities of other biochemicals. Plant tissues are particularly difficult in this respect as they often contain large amounts of carbohydrates that are not removed by phenol extraction. Instead, a different approach must be used. One method employs the detergent cetyltrimethylammonium bromide (CTAB), which forms an insoluble complex with nucleic acids. When CTAB is added to a plant cell extract the nucleic acid–CTAB complex precipitates, leaving carbohydrate, protein and other contaminants in the supernatant (Figure 3.11). The precipitate is then collected by centrifugation and resuspended in 1 M sodium chloride, which causes the complex to break down. The

Figure 3.11

The CTAB method for the purification of plant DNA.

nucleic acids can then be concentrated by ethanol precipitation and the RNA removed by ribonuclease treatment.

3.2 Preparation of plasmid DNA

The purification of plasmids from a culture of bacteria involves the same general strategy as the preparation of total cell DNA. A culture of cells, containing plasmids, is grown in liquid medium, harvested, and a cell extract prepared; the protein and RNA are then removed and the DNA probably concentrated by ethanol precipitation. However, there is an important distinction between plasmid purification and the preparation of total cell DNA. In a plasmid preparation it is always necessary to separate the plasmid DNA from the large amount of bacterial chromosomal DNA that is also present in the cells.

Separating the two types of DNA can be very difficult, but is nonetheless essential if the plasmids are to be used as cloning vectors. The presence of the smallest amount of contaminating bacterial DNA in a gene cloning experiment can easily lead to undesirable results. Fortunately, several methods are available for removal of bacterial DNA during plasmid purification, and the use of these – either individually or in combination – can result in the isolation of very pure plasmid DNA.

The methods are based on the several physical differences between plasmid DNA and bacterial DNA, the most obvious of which is size. The largest plasmids are only 8% of the size of the *E. coli* chromosome, and most are much smaller than this. Techniques that can separate small DNA molecules from large ones should therefore effectively purify plasmid DNA.

In addition to size, plasmids and bacterial DNA differ in conformation. When applied to a polymer such as DNA, the term conformation refers to the overall spatial configuration of the molecule, with the two simplest conformations being linear and circular. Plasmids and the bacterial chromosome are circular, but during preparation of the cell extract the chromosome is always broken to produce linear fragments. A method for separating circular from linear molecules would therefore result in pure plasmids.

3.2.1 Separation on the basis of size

The usual stage at which size fractionation is performed is during preparation of the cell extract. If the cells are lysed under very carefully controlled conditions, only a minimal

Figure 3.12

Preparation of a cleared lysate.

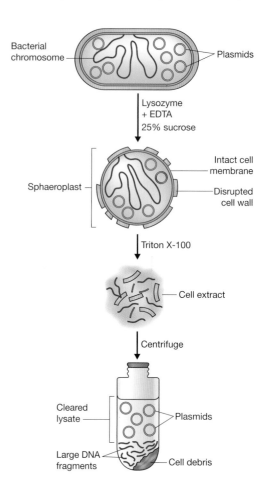

amount of chromosomal DNA breakage occurs. The resulting DNA fragments are still very large – much larger than the plasmids – and can be removed with the cell debris by centrifugation. This process is aided by the fact that the bacterial chromosome is physically attached to the cell envelope, so fragments of the chromosome sediment with the cell debris if these attachments are not broken.

Cell disruption must therefore be carried out very gently to prevent wholesale breakage of the bacterial DNA. For *E. coli* and related species, controlled lysis is performed as shown in Figure 3.12. Treatment with EDTA and lysozyme is carried out in the presence of sucrose, which prevents the cells from bursting immediately. Instead, sphaeroplasts – cells with partially degraded cell walls that retain an intact cytoplasmic membrane – are formed. Cell lysis is then induced by adding a non-ionic detergent such as Triton X-100 (ionic detergents, such as SDS, cause chromosomal breakage). As this method causes very little breakage of the bacterial DNA, centrifugation results in a cleared lysate consisting almost entirely of plasmid DNA.

A cleared lysate will, however, invariably retain some chromosomal DNA. Furthermore, if the plasmids themselves are large molecules they may also sediment with the cell debris. Size fractionation is therefore rarely sufficient in its own right, and alternative ways of removing the bacterial DNA contaminants must be considered.

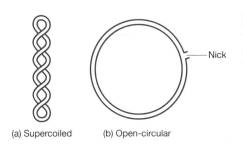

(a) Supercoiled   (b) Open-circular

Figure 3.13

Two conformations of circular double-stranded DNA.
(a) Supercoiled: both strands are intact.
(b) Open-circular: one or both strands are nicked.

3.2.2 Separation on the basis of conformation

Before considering the ways in which conformational differences between plasmids and bacterial DNA can be used to separate the two types of DNA, we must look more closely at the overall structure of plasmid DNA. It is not strictly correct to say that plasmids have a circular conformation, because double-stranded DNA circles can take up one of two quite distinct configurations. Most plasmids exist in the cell as supercoiled molecules (Figure 3.13a). Supercoiling occurs because the double helix of the plasmid DNA is partially unwound during the plasmid replication process by enzymes called topoisomerases (p. 71). The supercoiled conformation can be maintained only if both polynucleotide strands are intact, hence the more technical name of covalently closed-circular (ccc) DNA. If one of the polynucleotide strands is broken, the double helix reverts to its normal relaxed state and the plasmid takes on the alternative conformation, called open-circular (oc) (Figure 3.13b).

Supercoiling is important in plasmid preparation because supercoiled molecules can be fairly easily separated from non-supercoiled DNA. Two different methods are commonly used; both can purify plasmid DNA from crude cell extracts although in practice best results are obtained if a cleared lysate is first prepared.

Alkaline denaturation

The basis of this technique is that there is a narrow pH range at which non-supercoiled DNA is denatured, whereas supercoiled plasmids are not. If sodium hydroxide is added to a cell extract or cleared lysate, so that the pH is adjusted to 12.0–12.5, then the hydrogen bonding in non-supercoiled DNA molecules is broken, causing the double helix to unwind and the two polynucleotide chains to separate (Figure 3.14). If acid is now added, these denatured bacterial DNA strands reaggregate into a tangled mass.

Linear DNA   Supercoiled plasmids   Single-stranded linear DNA   Tangled mass of linear DNA   Supercoiled plasmids   Pellet of linear DNA molecules

pH 12.0 to 12.5   pH 7.0   Centrifuge

Figure 3.14

Plasmid purification by the alkaline denaturation method.

Figure 3.15

Caesium chloride density gradient centrifugation. (a) A CsCl density gradient produced by high-speed centrifugation. (b) Separation of protein, DNA and RNA in a density gradient.

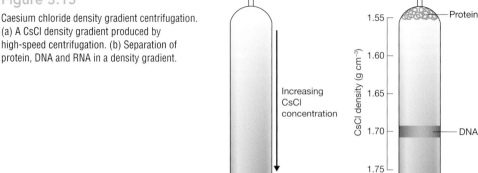

The insoluble network can be pelleted by centrifugation, leaving plasmid DNA in the supernatant. An additional advantage of this procedure is that, under some circumstances (specifically cell lysis by SDS and neutralization with sodium acetate), most of the protein and RNA also becomes insoluble and can be removed by centrifugation. Further purification by organic extraction or column chromatography may therefore not be needed if the alkaline denaturation method is used.

Ethidium bromide–caesium chloride density gradient centrifugation

This is a specialized version of the more general technique of equilibrium or density gradient centrifugation. A density gradient is produced by centrifuging a solution of caesium chloride (CsCl) at very high speed (Figure 3.15a). Any macromolecules present in the CsCl solution when it is centrifuged form bands at distinct points in the gradient (Figure 3.15b), and exactly where a particular molecule bands will depend on its buoyant density. DNA has a buoyant density of about 1.70 g cm^{-3}, and therefore it will migrate to the point in the gradient where the CsCl density is also 1.70 g cm^{-3}. In contrast, protein molecules have much lower buoyant densities and so float at the top of the tube, whereas RNA forms a pellet at the bottom (Figure 3.15b). Density gradient centrifugation can therefore separate DNA, RNA and protein, and is an alternative to organic extraction or column chromatography for DNA purification.

More importantly, density gradient centrifugation in the presence of ethidium bromide (EtBr) can be used to separate supercoiled DNA from non-supercoiled molecules. Ethidium bromide binds to DNA molecules by intercalating between adjacent base pairs, causing a partial unwinding of the double helix (Figure 3.16). This unwinding results in a decrease in the buoyant density, by as much as 0.125 g cm^{-3} for linear DNA. Supercoiled DNA, with no free ends, has very little freedom to unwind and can only bind a limited amount of EtBr. The decrease in buoyant density of a supercoiled molecule is therefore much less, only about 0.085 g cm^{-3}. As a consequence, supercoiled molecules will form a band in an EtBr–CsCl gradient at a different position to linear and open-circular DNA (Figure 3.17a).

Ethidium bromide–caesium chloride density gradient centrifugation is a very efficient method for obtaining pure plasmid DNA. When a cleared lysate is subjected to this procedure, plasmids band at a distinct point, separated from the linear bacterial DNA, with the protein floating on the top of the gradient and RNA pelleted at the

Figure 3.16

Partial unwinding of the DNA double helix by EtBr intercalation between adjacent base pairs. The normal DNA molecule shown on the left is partially unwound by taking up four EtBr molecules, resulting in the 'stretched' structure on the right.

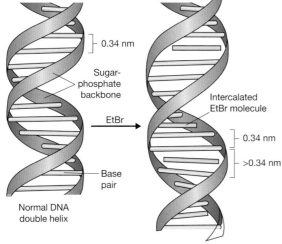

bottom. The position of the DNA bands can be seen by shining UV radiation on the tube, which causes the bound EtBr to fluoresce. The pure plasmid DNA is then removed by puncturing the side of the tube and withdrawing a sample with a syringe (Figure 3.17b). The EtBr bound to the plasmid DNA is extracted with *n*-butanol (Figure 3.17c) and the CsCl removed by dialysis (Figure 3.17d). The resulting plasmid preparation is virtually 100% pure and ready for use as a cloning vector.

3.2.3 Plasmid amplification

The preparation of plasmid DNA can be hindered by the fact that plasmids make up only a small proportion of the total DNA in the bacterial cell. The yield of DNA from a bacterial culture may therefore be disappointingly low. Plasmid amplification offers a means of increasing this yield.

The aim of amplification is to increase the copy number of a plasmid. Some multi-copy plasmids (those with copy numbers of 20 or more) have the useful property of being able to replicate in the absence of protein synthesis. This contrasts with the main bacterial chromosome, which cannot replicate under these conditions. This property can be utilized during the growth of a bacterial culture for plasmid DNA purification. After a satisfactory cell density has been reached, an inhibitor of protein synthesis (e.g., chloramphenicol) is added and the culture incubated for a further 12 h. During this

(a) An EtBr–CsCl
     density gradient

(b) Removal of the
     DNA band

(c) Extraction of the
     EtBr with *n*-butanol

(d) Removal of the
     CsCl by dialysis

Figure 3.17

Purification of plasmid DNA by EtBr–CsCl density gradient centrifugation.

time the plasmid molecules continue to replicate, even though chromosome replication and cell division are blocked (Figure 3.18). The result is that plasmid copy numbers of several thousand may be attained. Amplification is therefore a very efficient way of increasing the yield of multicopy plasmids.

3.3 Preparation of bacteriophage DNA

The key difference between phage DNA purification and the preparation of either total cell DNA or plasmid DNA is that, for phages, the starting material is not normally a cell extract. This is because bacteriophage particles can be obtained in large numbers from the extracellular medium of an infected bacterial culture. When such a culture is centrifuged the bacteria are pelleted, leaving the phage particles in suspension (Figure 3.19). The phage particles are then collected from the suspension and their DNA extracted by a single deproteinization step to remove the phage capsid.

This overall process is more straightforward than the procedure used to prepare total cell or plasmid DNA. Nevertheless, successful purification of significant quantities of

Figure 3.18

Plasmid amplification.

Figure 3.19

Preparation of a phage suspension from an infected culture of bacteria.

phage DNA is subject to several pitfalls. The main difficulty, especially with λ, is growing an infected culture in such a way that the extracellular phage titre (the number of phage particles per ml of culture) is sufficiently high. In practical terms, the maximum titre that can reasonably be expected for λ is 10^{10} ml^{-1}, yet 10^{10} λ particles will yield only 500 ng of DNA. Large culture volumes, in the range of 500–1000 ml, are therefore needed if substantial quantities of λ DNA are to be obtained.

3.3.1 Growth of cultures to obtain a high λ titre

Growing a large-volume culture is no problem (bacterial cultures of 100 l and more are common in biotechnology), but obtaining the maximum phage titre requires a certain amount of skill. The naturally occurring λ phage is lysogenic (p. 19), and an infected culture consists mainly of cells carrying the prophage integrated into the bacterial DNA (see Figure 2.7). The extracellular λ titre is extremely low under these circumstances.

In order to obtain a high yield of extracellular λ, the culture must be induced, so that all the bacteria enter the lytic phase of the infection cycle, resulting in cell death and the release of λ particles into the medium. Although induction is normally very difficult to control, most laboratory strains of λ carry a temperature-sensitive (ts) mutation in the *cI* gene. This is one of the genes responsible for maintaining the phage in an integrated state. If inactivated by a mutation the *cI* gene no longer functions correctly and the switch to lysis occurs. In the *cIts* mutation, the *cI* gene is functional at 30 °C, at which temperature normal lysogeny can occur, but at 42 °C the *cIts* gene product does not function properly and lysogeny cannot be maintained. A culture of *E. coli* infected with a λ phage carrying the *cIts* mutation can therefore be induced to produce extracellular phages by transferring from 30 °C to 42 °C (Figure 3.20).

3.3.2 Preparation of non-lysogenic λ phages

Although most λ strains are lysogenic, many cloning vectors derived from λ are modified, by deletions of the *cI* and other genes, so that lysogeny never occurs. These phages cannot integrate into the bacterial genome and can infect cells only by a lytic cycle (p. 18).

With these phages, the key to obtaining a high titre lies in the way in which the culture is grown, and in particular the stage at which the cells are infected by adding phage particles. If phages are added before the cells are dividing at their maximal rate, then all the cells are lysed very quickly, resulting in a low titre (Figure 3.21a). If the cell density is too high when the phages are added, the culture will never be completely lysed and again the phage titre will be low (Figure 3.21b). The ideal situation is when

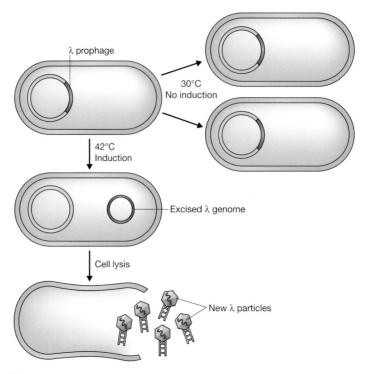

Figure 3.20

Induction of a λ *cIts* lysogen by transferring from 30 °C to 42 °C.

Figure 3.21

Achieving the correct balance between culture age and inoculum size when preparing a sample of a non-lysogenic phage.

Figure 3.22

Collection of phage particles by polyethylene glycol (PEG) precipitation.

the age of the culture and the size of the phage inoculum are balanced such that the culture continues to grow, but eventually all the cells are infected and lysed (Figure 3.21c). As can be imagined, skill and experience are needed to judge the matter to perfection.

3.3.3 Collection of phages from an infected culture

The remains of lysed bacterial cells, along with any intact cells that are inadvertently left over, can be removed from an infected culture by centrifugation, leaving the phage particles in suspension (see Figure 3.19). The problem now is to reduce the size of the suspension to 5 ml or less, which is a manageable size for DNA extraction.

Phage particles are so small that they are pelleted only by very high-speed centrifugation. Hence, the collection of phages is usually achieved by precipitation with polyethylene glycol (PEG) a long-chain polymeric compound which, in the presence of salt, absorbs water, thereby causing macromolecular assemblies such as phage particles to precipitate. The precipitate can then be collected by centrifugation, and redissolved in a suitably small volume (Figure 3.22).

3.3.4 Purification of DNA from λ phage particles

Deproteinization of the redissolved PEG precipitate with a protease enzyme, followed by fractionation of the DNA by ion-exchange chromatography, is often sufficient to extract the DNA from the phage particles. Alternatively, the phage capsids can be disrupted with guanidinium thiocyanate and the DNA recovered by silica binding. However, for the highest quality preparations the λ phages are subjected to an intermediate purification step. This is necessary because the PEG precipitate also contains a certain amount of bacterial debris, possibly including unwanted cellular DNA. These contaminants can be separated from the λ particles by CsCl density gradient centrifugation. The λ particles band in a CsCl gradient at 1.45–1.50 g cm^{-3} (Figure 3.23) and can be withdrawn from the gradient as described previously for DNA bands (see Figure 3.17). The removal of CsCl by dialysis then leaves a phage preparation from which high-purity DNA can be extracted.

3.3.5 Purification of M13 DNA causes few problems

Most of the differences between the M13 and λ infection cycles are to the advantage of the molecular biologist wishing to prepare M13 DNA. First, the double-stranded

Figure 3.23

Purification of λ phage particles by CsCl density gradient centrifugation.

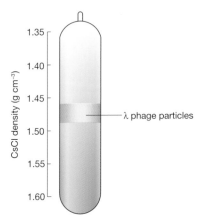

replicative form of M13 (p. 23), which behaves like a high-copy number plasmid, is very easily purified by the standard procedures for plasmid preparation. A cell extract is prepared from cells infected with M13, and the replicative form separated from bacterial DNA by alkaline denaturation or EtBr–CsCl density gradient centrifugation.

The single-stranded form of the M13 genome, contained in the extracellular phage particles, is also frequently required. In this respect the major advantage compared with λ is that high titres of M13 are very easy to obtain. As infected cells continually secrete M13 particles into the medium (see Figure 2.8), with lysis never occurring, a high M13 titre is achieved simply by growing the infected culture to a high cell density. In fact, titres of 10^{12} ml^{-1} and above are quite easy to obtain without any special tricks being used. Such high titres mean that significant amounts of single-stranded M13 DNA can be prepared from cultures of small volume (5 ml or less). Furthermore, as the infected cells are not lysed there is no problem with cell debris contaminating the phage suspension.

In summary, a typical method for single-stranded M13 DNA preparation involves the growth of a small volume of infected culture, centrifugation to pellet the bacteria, precipitation of the phage particles with PEG, phenol extraction to remove the phage protein coats, and ethanol precipitation to concentrate the resulting DNA (Figure 3.24).

Figure 3.24

Preparation of single-stranded M13 DNA from an infected culture of bacteria.

Further reading

Berensmeier, S. (2006) Magnetic particles for the separation and purification of nucleic acids. *Applied Microbiology and Biotechnology*, **73**, 495–504.

Birnboim, H.C. and Doly, J. (1979) A rapid alkaline extraction procedure for screening recombinant plasmid DNA. *Nucleic Acids Research*, **7**, 1513–1523. [A method for preparing plasmid DNA.]

Boom, R., Sol, C.J.A., Salimans, M.M.M., Jansen, C.L., Wertheim van Dillen, P.M.E., and van der Noordaa, J. (1990) Rapid and simple method for purification of nucleic acids. *Journal of Clinical Microbiology*, **28**, 495–503. [The guanidinium thiocyanate and silica method for DNA purification.]

Clewell, D.B. (1972) Nature of ColE1 plasmid replication in *Escherichia coli* in the presence of chloramphenicol. *Journal of Bacteriology*, **110**, 667–676. [The biological basis of plasmid amplification.]

Marmur, J. (1961) A procedure for the isolation of deoxyribonucleic acid from microorganisms. *Journal of Molecular Biology*, **3**, 208–218. [Genomic DNA preparation.]

Radloff, R., Bauer, W., and Vinograd, J. (1967) A dye-buoyant-density method for the detection and isolation of closed-circular duplex DNA. *Proceedings of the National Academy of Sciences of the USA*, **57**, 1514–1521. [The original description of ethidium bromide density gradient centrifugation.]

Rogers, S.O. and Bendich, A.J. (1985) Extraction of DNA from milligram amounts of fresh, herbarium and mummified plant tissues. *Plant Molecular Biology*, **5**, 69–76. [The CTAB method.]

Yamamoto, K.R., Alberts, B.M., Benzinger, R., Lawhorne, L., and Trieber, G. (1970) Rapid bacteriophage sedimentation in the presence of polyethylene glycol and its application to large scale virus preparation. *Virology*, **40**, 734–744. [Preparation of λ DNA.]

Zinder, N.D. and Boeke, J.D. (1982) The filamentous phage (Ff) as vectors for recombinant DNA. *Gene*, **19**, 1–10. [Methods for M13 phage growth and DNA preparation.]

Chapter 4

Manipulation of Purified DNA

Chapter contents

Once pure samples of DNA have been prepared, the next step in a gene cloning experiment is construction of the recombinant DNA molecule (see Figure 1.1). To produce this recombinant molecule, the vector, as well as the DNA to be cloned, must be cut at specific points and then joined together in a controlled manner. Cutting and joining are two examples of DNA manipulative techniques, a wide variety of which have been developed over the past few years. As well as being cut and joined, DNA molecules can be shortened, lengthened, copied into RNA or into new DNA molecules, and modified by the addition or removal of specific chemical groups. These manipulations, all of which can be carried out in the test tube, provide the foundation not only for gene cloning but also for studies of DNA biochemistry, genome organization, gene structure, and the control of gene expression.

Almost all DNA manipulative techniques make use of purified enzymes. Within the cell, these enzymes participate in essential processes such as DNA replication and transcription, breakdown of unwanted or foreign DNA (e.g., invading virus DNA), repair of mutated DNA, and recombination between different DNA molecules. After purification from cell extracts, many of these enzymes can be persuaded to carry out their natural reactions, or something closely related to them, under artificial conditions. Although these enzymatic reactions are often straightforward, most are absolutely impossible to perform by standard chemical methods. Purified enzymes are therefore crucial to genetic engineering and an important industry has sprung up around their preparation, characterization, and marketing. Commercial suppliers of high-purity enzymes provide an essential service to the molecular biologist.

Gene Cloning and DNA Analysis: An Introduction, Seventh Edition. T.A. Brown.
© 2016 John Wiley & Sons, Ltd. Published 2016 by John Wiley & Sons, Ltd.
Companion Website: www.wiley.com/go/brown/cloning

The cutting and joining manipulations that underlie gene cloning are carried out by enzymes termed restriction endonucleases (for cutting) and ligases (for joining). Most of this chapter will be concerned with the ways in which these two types of enzyme are used. First, we must consider the whole range of DNA manipulative enzymes, to see exactly what types of reaction can be performed. Many of these enzymes will be mentioned in later chapters when procedures that make use of them are described.

4.1 The range of DNA manipulative enzymes

DNA manipulative enzymes can be grouped into four broad classes, depending on the type of reaction that they catalyse:

- Nucleases are enzymes that cut, shorten, or degrade nucleic acid molecules.
- Ligases join nucleic acid molecules together.
- Polymerases make copies of molecules.
- Modifying enzymes remove or add chemical groups.

Before considering in detail each of these classes of enzyme, two points should be made. The first point is that, although most enzymes can be assigned to a particular class, a few display multiple activities that span two or more classes. Most importantly, many polymerases combine their ability to make new DNA molecules with an associated DNA degradative (i.e. nuclease) activity.

Second, it should be appreciated that, as well as the DNA manipulative enzymes, many similar enzymes able to act on RNA are known. The ribonuclease used to remove contaminating RNA from DNA preparations (p. 30) is an example of such an enzyme. Although some RNA manipulative enzymes have applications in gene cloning and will be mentioned in later chapters, we will in general restrict our thoughts to those enzymes that act on DNA.

4.1.1 Nucleases

Nucleases degrade DNA molecules by breaking the phosphodiester bonds that link one nucleotide to the next in a DNA strand. There are two different kinds of nuclease (Figure 4.1):

- Exonucleases remove nucleotides one at a time from the end of a DNA molecule.
- Endonucleases are able to break internal phosphodiester bonds within a DNA molecule.

The main distinction between different exonucleases lies in the number of strands that are degraded when a double-stranded molecule is attacked. The enzyme called Bal31 (purified from the bacterium *Alteromonas espejiana*) is an example of an exonuclease that removes nucleotides from both strands of a double-stranded molecule (Figure 4.2a). The greater the length of time that Bal31 is allowed to act on a group of DNA molecules, the shorter the resulting DNA fragments will be. In contrast, enzymes such as *E. coli* exonuclease III degrade just one strand of a double-stranded molecule, leaving single-stranded DNA as the product (Figure 4.2b).

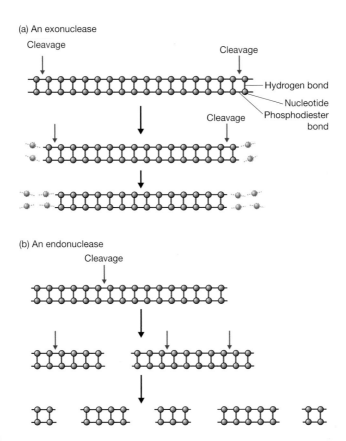

Figure 4.1

The reactions catalysed by the two different types of nuclease. (a) An exonuclease, which removes nucleotides from the end of a DNA molecule. (b) An endonuclease, which breaks internal phosphodiester bonds.

Figure 4.2

The reactions catalysed by different types of exonuclease. (a) Bal31, which removes nucleotides from both strands of a double-stranded molecule. (b) Exonuclease III, which removes nucleotides only from the 3′ terminus.

Figure 4.3

The reactions catalysed by different types of endonuclease. (a) S1 nuclease, which cleaves only single-stranded DNA, including single-stranded nicks in mainly double-stranded molecules. (b) DNase I, which cleaves both single- and double-stranded DNA. (c) A restriction endonuclease, which cleaves double-stranded DNA, but only at a limited number of sites.

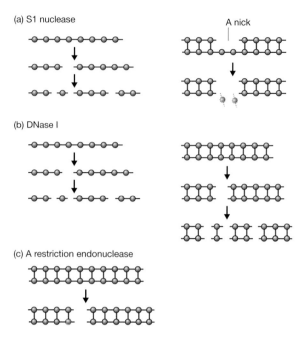

(a) S1 nuclease

A nick

(b) DNase I

(c) A restriction endonuclease

The same criterion can be used to classify endonucleases. S1 endonuclease (from the fungus *Aspergillus oryzae*) only cleaves single strands (Figure 4.3a), whereas deoxyribonuclease I (DNase I), which is prepared from cow pancreas, cuts both single- and double-stranded molecules (Figure 4.3b). DNase I is non-specific in that it attacks DNA at any internal phosphodiester bond, so the end result of prolonged DNase I action is a mixture of mononucleotides and very short oligonucleotides. On the other hand, the special group of enzymes called restriction endonucleases cleave double-stranded DNA only at a limited number of specific recognition sites (Figure 4.3c). These important enzymes are described in detail on p. 53.

4.1.2 Ligases

Within the cell, the function of DNA ligase is to repair single-stranded breaks ('discontinuities') that arise in double-stranded DNA molecules during, for example, DNA replication. DNA ligases from most organisms can also join together two individual fragments of double-stranded DNA (Figure 4.4). The role of these enzymes in construction of recombinant DNA molecules is described on p. 66.

Figure 4.4

The two reactions catalysed by DNA ligase. (a) Repair of a discontinuity – a missing phosphodiester bond in one strand of a double-stranded molecule. (b) Joining two molecules together.

(a) Discontinuity repair

Discontinuity

DNA ligase

(b) Joining two molecules

DNA ligase

4.1.3 Polymerases

DNA polymerases are enzymes that synthesize a new strand of DNA complementary to an existing DNA or RNA template (Figure 4.5a). Most polymerases can function only if the template possesses a double-stranded region that acts as a primer for the initiation of polymerization.

Four types of DNA polymerase are used routinely in genetic engineering. The first is DNA polymerase I, which is usually prepared from *E. coli*. This enzyme attaches to a short, single-stranded region (or nick) in a mainly double-stranded DNA molecule, and then synthesizes a completely new strand, degrading the existing strand as it proceeds (Figure 4.5b). DNA polymerase I is therefore an example of an enzyme with a dual activity, namely DNA polymerization and DNA degradation.

The polymerase and nuclease activities of DNA polymerase I are controlled by different parts of the enzyme molecule. The nuclease activity is contained in the first 323 amino acids of the polypeptide, so removal of this segment leaves a modified enzyme that retains the polymerase function but is unable to degrade DNA. This modified enzyme, called the Klenow fragment, can still synthesize a complementary DNA strand on a single-stranded template, but as it has no nuclease activity it cannot continue the synthesis once the nick is filled in (Figure 4.5c). Several other enzymes – natural polymerases and modified versions – have similar properties to the Klenow fragment. For many years, these polymerases were used in DNA sequencing but have now been largely

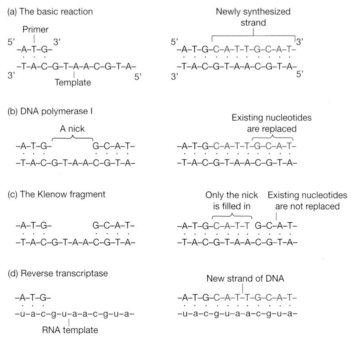

Figure 4.5

The reactions catalysed by DNA polymerases. (a) The basic reaction: a new DNA strand is synthesized in the 5′ to 3′ direction. (b) DNA polymerase I, which initially fills in nicks but then continues to synthesize a new strand, degrading the existing one as it proceeds. (c) The Klenow fragment, which only fills in nicks. (d) Reverse transcriptase, which uses a template of RNA.

superseded in this role, although they still have important applications, for example in DNA labelling (p. 145).

The *Taq* DNA polymerase used in the polymerase chain reaction (PCR) (see Figure 1.2) is the DNA polymerase I enzyme of the bacterium *Thermus aquaticus*. This organism lives in hot springs, and many of its enzymes – including the *Taq* DNA polymerase – are thermostable, which means that they are resistant to denaturation by heat treatment. This is the special feature of *Taq* DNA polymerase that makes it suitable for PCR, because if it was not thermostable it would be inactivated when the temperature of the reaction is raised to 94 °C to denature the DNA.

The final type of DNA polymerase that is important in genetic engineering is reverse transcriptase, an enzyme involved in the replication of several kinds of virus. Reverse transcriptase is unique in that it uses as a template not DNA but RNA (Figure 4.5d). The ability of this enzyme to synthesize a DNA strand complementary to an RNA template is central to the technique called complementary DNA (cDNA) synthesis (p. 142).

4.1.4 DNA-modifying enzymes

Numerous enzymes exist that modify DNA molecules by the addition or removal of specific chemical groups. The most important are as follows:

- Alkaline phosphatase (from *E. coli*, calf intestinal tissue, or arctic shrimp), which removes the phosphate group present at the 5′ terminus of a DNA molecule (Figure 4.6a).

Figure 4.6

The reactions catalysed by DNA-modifying enzymes. (a) Alkaline phosphatase, which removes 5′-phosphate groups. (b) Polynucleotide kinase, which attaches 5′-phosphate groups. (c) Terminal deoxynucleotidyl transferase, which attaches deoxyribonucleotides to the 3′ termini of polynucleotides in either single- or double-stranded molecules.

- Polynucleotide kinase (from *E. coli* infected with T4 phage), which has the reverse effect to alkaline phosphatase, adding phosphate groups onto free 5′ termini (Figure 4.6b).
- Terminal deoxynucleotidyl transferase (from calf thymus tissue), which adds one or more deoxyribonucleotides onto the 3′ terminus of a DNA molecule (Figure 4.6c).

4.2 Enzymes for cutting DNA: Restriction endonucleases

Gene cloning requires that DNA molecules be cut in a very precise and reproducible fashion. This is illustrated by the way in which the vector is cut during the construction of a recombinant DNA molecule (Figure 4.7a). Each vector molecule must be cleaved at a single position, to open up the circle so that new DNA can be inserted. A molecule that is cut more than once will be broken into two or more separate fragments and will be of no use as a cloning vector. Furthermore, each vector molecule must be cut at exactly the same position on the circle; as will become apparent in later chapters, random cleavage is not satisfactory. It should be clear that a very special type of nuclease is needed to carry out this manipulation.

Often, it is also necessary to cleave the DNA that is to be cloned (Figure 4.7b). There are two reasons for this. First, if the aim is to clone a single gene, which may consist of only 2 or 3 kb of DNA, then that gene will have to be cut out of the large (often greater than 80 kb) DNA molecules produced by skilful use of the preparative

(a) Vector molecules

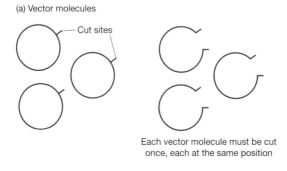

Each vector molecule must be cut once, each at the same position

Figure 4.7

The need for very precise cutting manipulations in a gene cloning experiment.

(b) The DNA molecule containing the gene to be cloned

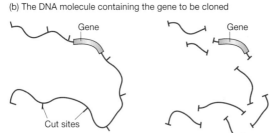

Large DNA molecule

Fragments small enough to be cloned

techniques described in Chapter 3. Second, large DNA molecules may have to be broken down simply to produce fragments small enough to be carried by the vector. Most cloning vectors exhibit a preference for DNA fragments that fall into a particular size range: most plasmid-based vectors, for example, are very inefficient at cloning DNA molecules more than 8 kb in length.

Purified restriction endonucleases allow the molecular biologist to cut DNA molecules in the precise, reproducible manner required for gene cloning. The discovery of these enzymes, which led to Nobel Prizes for W. Arber, H. Smith, and D. Nathans in 1978, was one of the key breakthroughs in the development of genetic engineering.

4.2.1 The discovery and function of restriction endonucleases

The initial observation that led to the eventual discovery of restriction endonucleases was made during the early 1950s, when some strains of bacteria were shown to be immune to bacteriophage infection, a phenomenon referred to as host-controlled restriction.

The mechanism of restriction is not very complicated, even though it took over 20 years to be fully understood. Restriction occurs because the bacterium produces an enzyme that degrades the phage DNA before it has time to replicate and direct the synthesis of new phage particles (Figure 4.8a). The bacterium's own DNA – the destruction

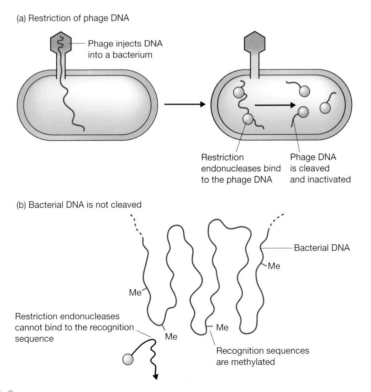

Figure 4.8

The function of a restriction endonuclease in a bacterial cell. (a) Phage DNA is cleaved, but (b) bacterial DNA is not.

of which would of course be lethal – is protected from attack because it carries additional methyl groups that block the degradative enzyme action (Figure 4.8b).

These degradative enzymes are termed restriction endonucleases and are synthesized by many, perhaps all, species of bacteria. To date, almost 4000 different enzymes have been isolated, and more than 600 are currently available for use in the laboratory. Three different classes of restriction endonuclease are recognized, each distinguished by a slightly different mode of action. Types I and III are rather complex and have only a limited role in genetic engineering, whereas type II restriction endonucleases are the cutting enzymes that are so important in gene cloning.

4.2.2 Type II restriction endonucleases cut DNA at specific nucleotide sequences

The central feature of type II restriction endonucleases (which will be referred to simply as 'restriction endonucleases' from now on) is that each enzyme has a specific recognition sequence at which it cuts a DNA molecule. A particular enzyme cleaves DNA at the recognition sequence and nowhere else. For example, the restriction endonuclease called *Pvu*I (isolated from *Proteus vulgaris*) cuts DNA only at the hexanucleotide CGATCG; in contrast, a second enzyme from the same bacterium, called *Pvu*II, cuts at a different hexanucleotide, in this case CAGCTG.

Many restriction endonucleases recognize hexanucleotide target sites, but others cut at four, five, eight, or even longer nucleotide sequences. *Sau*3A (from *Staphylococcus aureus* strain 3A) recognizes GATC, while *Alu*I (*Arthrobacter luteus*) cuts at AGCT. There are also examples of restriction endonucleases with degenerate recognition sequences, which means that they cut DNA at any one of a family of related sites. *Hinf*I (*Haemophilus influenzae* strain R$_f$), for instance, recognizes GANTC, and so cuts at GAATC, GATTC, GAGTC, and GACTC. The recognition sequences for some of the most frequently used restriction endonucleases are listed in Table 4.1.

4.2.3 Blunt ends and sticky ends

The exact nature of the cut produced by a restriction endonuclease is of considerable importance in the design of a gene cloning experiment. Many restriction endonucleases make a simple double-stranded cut in the middle of the recognition sequence (Figure 4.9a), which results in a blunt end or a flush end. *Pvu*II and *Alu*I are examples of blunt end cutters.

Other restriction endonucleases cut DNA in a slightly different way. With these enzymes, the two DNA strands are not cut at exactly the same position but instead the cleavage is staggered, usually by two or four nucleotides, so that the resulting DNA fragments have short single-stranded overhangs at each end (Figure 4.9b). These are called sticky ends or cohesive ends, as base pairing between them can stick the DNA molecule back together again (recall that sticky ends were encountered on p. 20 during the description of λ phage replication). One important feature of sticky end enzymes is that restriction endonucleases with different recognition sequences may produce the same sticky ends. *Bam*HI (recognition sequence GGATCC) and *Bgl*II (AGATCT) are examples, both of which produce GATC sticky ends (Figure 4.9c). The same sticky end is also produced by *Sau*3A, which recognizes only the tetranucleotide GATC. Fragments

Table 4.1

The recognition sequences for some of the most frequently used restriction endonucleases.

ENZYME	ORGANISM	RECOGNITION SEQUENCE[a]	BLUNT OR STICKY END
EcoRI	Escherichia coli	GAATTC	Sticky
BamHI	Bacillus amyloliquefaciens	GGATCC	Sticky
BglII	Bacillus globigii	AGATCT	Sticky
PvuI	Proteus vulgaris	CGATCG	Sticky
PvuII	Proteus vulgaris	CAGCTG	Blunt
HindIII	Haemophilus influenzae R_d	AAGCTT	Sticky
HinfI	Haemophilus influenzae R_f	GANTC	Sticky
Sau3A	Staphylococcus aureus	GATC	Sticky
AluI	Arthrobacter luteus	AGCT	Blunt
TaqI	Thermus aquaticus	TCGA	Sticky
HaeIII	Haemophilus aegyptius	GGCC	Blunt
NotI	Nocardia otitidis-caviarum	GCGGCCGC	Sticky
SfiI	Streptomyces fimbriatus	GGCCNNNNNGGCC	Sticky

The sequence shown is that of one strand, given in the 5′ to 3′ direction. 'N' indicates any nucleotide. Note that almost all recognition sequences are palindromes: when both strands are considered they read the same in each direction, for example:

```
        5′—GAATTC—3′
EcoRI      ||||||
        3′—CTTAAG—5′
```

(a) Production of blunt ends

```
–N–N–A–G–C–T–N–N–      AluI      –N–N–A–G      C–T–N–N–
 · · · · · · · ·     ——————>      · · · ·      · · · ·
–N–N–T–C–G–A–N–N–                –N–N–T–C      G–A–N–N–

 'N' = A, G, C, or T                              Blunt ends
```

(b) Production of sticky ends

```
–N–N–G–A–A–T–T–C–N–N–   EcoRI    –N–N–G        A–A–T–T–C–N–N–
 · · · · · · · · · ·   ——————>    · · ·        · · · · · · ·
–N–N–C–T–T–A–A–G–N–N–             –N–N–C–T–T–A–A        G–N–N–
                                                   Sticky ends
```

(c) The same sticky ends produced by
different restriction endonucleases

```
BamHI     –N–N–G                G–A–T–C–C–N–N–
           · · ·                    · · · ·
          –N–N–C–C–T–A–G         G–N–N–

BglII     –N–N–A                G–A–T–C–T–N–N–
           · · ·                    · · · ·
          –N–N–T–C–T–A–G         A–N–N–

Sau3A     –N–N–N                G–A–T–C–N–N–N–
           · · ·                    · · · ·
          –N–N–N–C–T–A–G         N–N–N–
```

Figure 4.9

The ends produced by cleavage of DNA with different restriction endonucleases. (a) A blunt end produced by AluI. (b) A sticky end produced by EcoRI. (c) The same sticky ends produced by BamHI, BglII and Sau3A.

of DNA produced by cleavage with either of these enzymes can be joined to each other, as each fragment carries a complementary sticky end.

4.2.4 The frequency of recognition sequences in a DNA molecule

The number of recognition sequences for a particular restriction endonuclease in a DNA molecule of known length can be calculated mathematically. A tetranucleotide sequence (e.g., GATC) should occur once every $4^4 = 256$ nucleotides, and a hexanucleotide (e.g., GGATCC) once every $4^6 = 4096$ nucleotides. These calculations assume that the nucleotides are ordered in a random fashion, and that the four different nucleotides are present in equal proportions (i.e., the GC content = 50%). In practice, neither of these assumptions is entirely valid. For example, the λ DNA molecule, at 49 kb, should contain about 12 sites for a restriction endonuclease with a hexanucleotide recognition sequence. In fact, many of these recognition sites occur less frequently (e.g., six for *Bgl*II, five for *Bam*HI, and only two for *Sal*I), a reflection of the fact that the GC content for λ is rather less than 50% (Figure 4.10a).

Furthermore, restriction sites are generally not evenly spaced out along a DNA molecule. If they were, then digestion with a particular restriction endonuclease would produce fragments of roughly equal sizes. Figure 4.10b shows the fragments produced by cutting λ DNA with *Bgl*II, *Bam*HI, and *Sal*I; in each case there is a considerable

Figure 4.10

Restriction of the λ DNA molecule. (a) The positions of the recognition sequences for *Bgl*II, *Bam*HI, and *Sal*I. (b) The fragments produced by cleavage with each of these restriction endonucleases. The numbers are the fragment sizes in base pairs.

spread of fragment sizes, indicating that in λ DNA the nucleotides are not randomly ordered.

The lesson to be learned from Figure 4.10 is that although mathematics may give an idea of how many restriction sites to expect in a given DNA molecule, only experimental analysis can provide the true picture. We must therefore move on to consider how restriction endonucleases are used in the laboratory.

4.2.5 Performing a restriction digest in the laboratory

As an example, we will consider how to digest a sample of λ DNA (concentration 125 μg ml^{-1}) with BglII.

First, the required amount of DNA must be pipetted into a test tube. The amount of DNA that will be restricted depends on the nature of the experiment. In this case, we will digest 2 μg of λ DNA, which is contained in 16 μl of the sample (Figure 4.11a). Very accurate micropipettes will therefore be needed. The other main component in the reaction will be the restriction endonuclease, obtained from a commercial supplier as a pure solution of known concentration. Before adding the enzyme, the solution containing the DNA must be adjusted to provide the correct conditions to ensure maximal activity of the enzyme. Most restriction endonucleases function adequately at pH 7.4, though different enzymes vary in their requirements for ionic strength, usually provided by sodium chloride (NaCl), and magnesium (Mg^{2+}) concentration (all type II restriction endonucleases require Mg^{2+} in order to function). It is also advisable to add a reducing agent, such as dithiothreitol (DTT), which stabilizes the enzyme and prevents its inactivation. Providing the correct conditions for the enzyme is very important. Incorrect NaCl or Mg^{2+} concentrations will not only decrease the activity of the restriction endonuclease, they might also cause changes in the specificity of the enzyme, so that DNA cleavage occurs at additional, non-standard recognition sequences.

The composition of a suitable buffer for BglII is shown in Table 4.2. This buffer is ten times the working concentration, and is diluted by being added to the reaction mixture. In this example, a suitable final volume for the reaction mixture would be 20 μl, so 2 μl of 10× BglII buffer is added to the 16 μl of DNA already present (Figure 4.11b).

The restriction endonuclease can now be added. By convention, 1 unit of enzyme is defined as the quantity needed to cut 1 μg of DNA in 1 h, so 2 units of BglII are needed to cut 2 μg of λ DNA. BglII is frequently obtained at a concentration of 4 units per μl, so 0.5 μl is sufficient to cleave the DNA. The final ingredients in the reaction mixture are therefore 0.5 μl BglII + 1.5 μl water, giving a final volume of 20 μl (Figure 4.11c).

Figure 4.11
Performing a restriction digest in the laboratory.

Table 4.2

A 10× buffer suitable for restriction of DNA with *Bgl*I.

COMPONENT	CONCENTRATION (mM)
Tris–HCl, pH 7.4	500
MgCl$_2$	100
NaCl	500
Dithiothreitol	10

The last factor to consider is incubation temperature. Most restriction endonucleases, including *Bgl*II, function best at 37 °C, although a few have different requirements. *Taq*I, for example, is a restriction enzyme from *Thermus aquaticus* and, like *Taq* DNA polymerase, has a high working temperature. Restriction digests with *Taq*I must be incubated at 65 °C to obtain maximum enzyme activity.

After 1 h, the restriction should be complete (Figure 4.11d). If the DNA fragments produced by restriction are to be used in cloning experiments, the enzyme must somehow be destroyed so that it does not accidentally digest other DNA molecules that may be added at a later stage. There are several ways of 'killing' the enzyme. In many cases a short incubation at 70 °C is sufficient, but for others a phenol extraction or the addition of EDTA, which binds Mg^{2+} ions preventing restriction endonuclease action, is used (Figure 4.11e).

4.2.6 Analysing the result of restriction endonuclease cleavage

A restriction digest results in a number of DNA fragments, the sizes of which depend on the exact positions of the recognition sequences for the endonuclease in the original molecule (see Figure 4.10). A method of determining the number and sizes of the fragments is needed if restriction endonucleases are to be of use in gene cloning. Whether or not a DNA molecule is cut at all can be determined fairly easily by testing the viscosity of the solution. As larger DNA molecules result in a more viscous solution than smaller molecules, cleavage is associated with a decrease in viscosity. Determining the number and sizes of the individual cleavage products is more difficult, and in fact for several years this was one of the most tedious aspects of experiments involving DNA. Eventually, these problems were solved during the early 1970s when the technique of gel electrophoresis was developed.

Separation of molecules by gel electrophoresis

Electrophoresis, like ion-exchange chromatography (see p. 30), is a technique that employs the differences in electrical charge to separate the molecules in a mixture. DNA molecules have negative charges, and so when placed in an electric field will migrate towards the positive pole (Figure 4.12a). The rate of migration of a molecule depends on two factors, namely its shape and its charge-to-mass ratio. Unfortunately, most DNA molecules are the same shape and have very similar charge-to-mass ratios. Fragments of different sizes cannot therefore be separated using standard electrophoresis.

The size of the DNA molecule does, however, become a factor if the electrophoresis is performed in a gel. A gel, which is usually made of agarose, polyacrylamide (or a mixture of the two) comprises a complex network of pores through which the DNA

Figure 4.12

(a) Standard electrophoresis does not separate DNA fragments of different sizes, whereas (b) gel electrophoresis does.

(a) Standard electrophoresis

DNA   Buffer

Electrophorese

Buffer

DNA migrates towards the anode, but little separation into size classes occurs

(b) Gel electrophoresis

DNA, loaded into a well cut out of the gel   Gel   Buffer

Electrophorese

DNA separates into bands of different-sized fragments   Smallest

molecules must travel to reach the positive electrode. The smaller the DNA molecule, the faster it can migrate through the gel. Gel electrophoresis therefore separates DNA molecules according to their size (Figure 4.12b).

In practice, the composition of the gel determines the sizes of the DNA molecules that can be separated. A 0.5 cm-thick slab of 0.5% agarose, which has relatively large pores, would be used for molecules in the size range 1 to 30 kb, allowing, for example, molecules of 10 and 12 kb to be clearly distinguished. At the other end of the scale, a very thin (0.3 mm) 40% polyacrylamide gel, with extremely small pores, would be used to separate much smaller DNA molecules, in the range of 1 to 300 bp, and could distinguish molecules that differed in length by just a single nucleotide.

Visualizing DNA molecules in an agarose gel

The easiest way to visualize the results of a gel electrophoresis experiment is to stain the gel with a compound that makes the DNA visible. Ethidium bromide (EtBr), which has already been described on p. 38 as a means of visualizing DNA in caesium chloride gradients, is also used routinely to stain DNA in agarose and polyacrylamide gels (Figure 4.13). Bands showing the positions of the different size classes of DNA fragment are clearly visible under UV irradiation after EtBr staining, as long as sufficient DNA is present. Unfortunately, the procedure is very hazardous because EtBr is a powerful mutagen. Moreover, EtBr staining also has limited sensitivity, and if a band contains less than about 10 ng of DNA then it might not be visible after staining.

Figure 4.13

Visualizing DNA bands in an agarose gel by EtBr staining and ultraviolet (UV) irradiation.

For this reason, non-mutagenic dyes that stain DNA green, red or blue are now used in many laboratories. Most of these dyes can be used either as a post-stain after electrophoresis (as illustrated in Figure 4.13 for EtBr) or, alternatively, because they are non-hazardous they can be included in the buffer solution in which the agarose or polyacrylamide is dissolved when the gel is prepared. Some of these dyes require UV irradiation in order to make the bands visible, but others are visualized by illumination at other wavelengths, for example under blue light. This removes a second hazard as UV radiation can cause severe burns. The most sensitive dyes are able to detect bands that contain less than 1 ng DNA.

4.2.7 Estimation of the sizes of DNA molecules

Gel electrophoresis separates different-sized DNA molecules, with the smallest molecules travelling the greatest distance toward the positive electrode. If several DNA fragments of varying sizes are present (the result of a successful restriction digest, for example), then a series of bands appears in the gel. But how can the sizes of these fragments be determined?

The most accurate method is to make use of the mathematical relationship that links migration rate to molecular mass. The relevant formula is:

$$D = a - b(\log M)$$

where D is the distance moved, M is the molecular mass, and a and b are constants that depend on the electrophoresis conditions.

Because extreme accuracy in estimating DNA fragment sizes is not always necessary, a much simpler – but less precise – method is more generally used. A standard restriction digest, comprising fragments of known size, is usually included in each electrophoresis gel that is run. Restriction digests of λ DNA are often used in this way as size markers. For example, *Hind*III cleaves λ DNA into eight fragments that range in size from 125 bp for the smallest to over 23 kb for the largest. As the sizes of the fragments in this digest

Figure 4.14

Estimation of the sizes of DNA fragments in an agarose gel.
(a) A rough estimate of fragment size can be obtained by eye.
(b) A more accurate measurement of fragment size is gained
by using the mobilities of the *Hind*III–λ fragments to construct
a calibration curve. The sizes of the unknown fragments can
then be determined from the distances they have migrated.

(a) Rough estimation by eye

(b) Accurate graphical estimation

are known, the fragment sizes in the experimental digest can be estimated by comparing
the positions of the bands in the two tracks (Figure 4.14). Special mixtures of DNA
fragments called DNA ladders, whose sizes are multiples of 100 bp or of 1 kb, can
also be used as size markers. Although not precise, size estimation by comparison with
DNA markers can be performed with as little as 5% error, which is satisfactory for most
purposes.

4.2.8 Mapping the positions of different restriction sites in a DNA molecule

So far, we have considered how to determine the number and sizes of the DNA fragments produced by restriction endonuclease cleavage. The next step in restriction analysis is to construct a map showing the relative positions in the DNA molecule of the
recognition sequences for a number of different enzymes. Only when a restriction map is
available can the correct restriction endonucleases be selected for the particular cutting
manipulation that is required (Figure 4.15).

To construct a restriction map, a series of restriction digests must be performed. First,
the number and sizes of the fragments produced by each restriction endonuclease must
be determined by gel electrophoresis, followed by comparison with size markers (Figure 4.16). This information must then be supplemented by a series of double digestions,
in which the DNA is cut by two restriction endonucleases at once. It might be possible
to perform a double digestion in one step if both enzymes have similar requirements
for pH, Mg^{2+} concentration, etc. Alternatively, the two digestions may have to be carried out sequentially, adjusting the reaction mixture after the first digestion to provide
a different set of conditions for the second enzyme.

Figure 4.15

Using a restriction map to determine which restriction endonucleases should be used to obtain DNA fragments containing individual genes.

Comparing the results of single and double digests will allow many, if not all, of the restriction sites to be mapped (Figure 4.16). Ambiguities can usually be resolved by partial digestion, carried out under conditions that result in the cleavage of only a limited number of the restriction sites on any DNA molecule. A partial digestion is usually achieved either by reducing the incubation period, so the enzyme does not have time to cut all the restriction sites, or by incubating at a low temperature (e.g., 4 °C rather than 37 °C), which limits the activity of the enzyme.

The result of a partial digestion is a complex pattern of bands in an electrophoresis gel since, as well as the standard fragments produced by total digestion, additional sizes are seen. These are molecules that comprise two adjacent restriction fragments, separated by a site that has not been cleaved. Their sizes indicate which restriction fragments in the complete digest are next to one another in the uncut molecule (Figure 4.16).

4.2.9 Special gel electrophoresis methods for separating larger molecules

During agarose gel electrophoresis, a DNA fragment will migrate at a rate that is proportional to its size, but this relationship is not a direct one. The formula that links migration rate to molecular mass has a logarithmic component (p. 61), which means that the difference in migration rates become increasingly small for larger molecules (Figure 4.17). In practice, molecules larger than about 50 kb cannot be resolved efficiently using standard gel electrophoresis.

This size limitation is not usually a problem when the restriction fragments being studied have been obtained by cutting the DNA with an enzyme with a tetranucleotide or hexanucleotide recognition sequence. Most, if not all, of the fragments produced

Single and double digestions

Enzyme	Number of fragments	Sizes (kb)
XbaI	2	24.0, 24.5
XhoI	2	15.0, 33.5
KpnI	3	1.5, 17.0, 30.0
XbaI + XhoI	3	9.0, 15.0, 24.5
XbaI + KpnI	4	1.5, 6.0, 17.0, 24.0

Conclusions:

(1) As λ DNA is linear, the number of restriction sites for each enzyme is
XbaI 1, XhoI 1, KpnI 2.

(2) The XbaI and XhoI sites can be mapped:

(3) All the KpnI sites fall in the 24.5 kb XbaI fragment, as the 24.0 kb fragment
is intact after XbaI–KpnI double digestion. The order of the KpnI fragments
can be determined only by partial digestion.

Partial digestion

Enzyme	Fragment sizes (kb)
KpnI – limiting conditions	1.5, 17.0, 18.5, 30.0, 31.5, 48.5

Conclusions:

(1) 48.5 kb fragment = uncut λ.
(2) 1.5, 17.0 and 30.0 kb fragments are products of complete digestion.
(3) 18.5 and 31.5 kb fragments are products of partial digestion.

Figure 4.16

Restriction mapping. This example shows how the positions of the XbaI, XhoI and KpnI sites on the λ DNA molecule
can be determined.

Figure 4.17

The influence of DNA size on migration rate
during conventional gel electrophoresis.

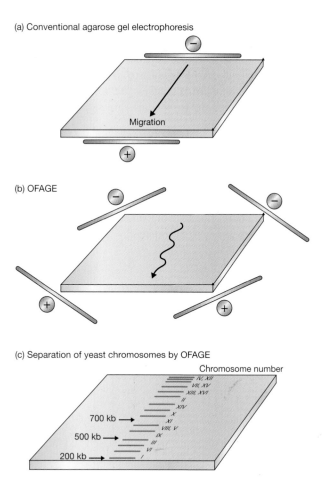

(a) Conventional agarose gel electrophoresis

Migration

(b) OFAGE

(c) Separation of yeast chromosomes by OFAGE

Chromosome number

700 kb →

500 kb →

200 kb →

IV, XII
VII, XV
XIII, XVI
II
XIV
X
XI
VIII, V
IX
III
VI
I

Figure 4.18

The difference between conventional gel electrophoresis and orthogonal field alternation gel electrophoresis (OFAGE).

in this way will be less than 30 kb in length and easily resolved by agarose gel electrophoresis.

Difficulties might arise, however, if an enzyme with a longer recognition sequence is used, such as *Not*I, which cuts at an eight-nucleotide sequence (see Table 4.1). *Not*I would be expected, on average, to cut a DNA molecule once every $4^8 = 65\ 536$ bp. It is therefore unlikely that *Not*I fragments would be separated by standard gel electrophoresis.

The limitations of standard gel electrophoresis can be overcome if a more complex electric field is used. Although several different systems have been designed, the principle is best illustrated by orthogonal field alternation gel electrophoresis (OFAGE). Instead of being applied directly along the length of the gel (as in the standard method; Figure 4.18a), the electric field now alternates between two pairs of electrodes, each pair set at an angle of 45° to the length of the gel (Figure 4.18b). The result is a pulsed electric field, with the DNA molecules in the gel having continually to change direction in accordance with the pulses. As the two fields alternate in a regular fashion, the net movement of the DNA molecules in the gel is still from one end to the other, in more or less a straight line. With every change in field direction each DNA molecule has to realign through 90° before its migration can continue. This is the key point, because a short molecule can realign faster than a long one, allowing the short molecules to progress towards the end of the gel more quickly. This added dimension increases the resolving power of the gel quite dramatically, such that molecules up to several thousand kilobases in length can be separated.

This size range includes not only restriction fragments but also the intact chromosomal molecules of many lower eukaryotes, including yeast, several important filamentous fungi, and protozoans such as the malaria parasite *Plasmodium falciparum*. OFAGE and related techniques such as contour clamped homogeneous electric fields (CHEF) and field inversion gel electrophoresis (FIGE) can therefore be used to prepare gels that show the separated chromosomes of these organisms (Figure 4.18c), enabling the DNA from these individual chromosomes to be purified.

4.3 Ligation: Joining DNA molecules together

The final step in the construction of a recombinant DNA molecule is a joining together of the vector molecule and the DNA to be cloned (Figure 4.19). This process is referred to as ligation, and the enzyme that catalyses the reaction is called DNA ligase.

4.3.1 The mode of action of DNA ligase

All living cells produce DNA ligases, but the enzyme used in genetic engineering is usually purified from *E. coli* bacteria that have been infected with T4 phage. Within the cell, the enzyme carries out the very important function of repairing any discontinuities that may arise in one of the strands of a double-stranded molecule (see Figure 4.4a). A discontinuity is quite simply a position where a phosphodiester bond between adjacent nucleotides is missing (contrast this with a nick, where one or more nucleotides are absent). Although discontinuities may arise by a chance breakage of the cell's DNA molecules, they are also a natural result of processes such as DNA replication and recombination. Ligases therefore play several vital roles in the cell.

In the test tube, purified DNA ligases not only repair single-strand discontinuities but also join together individual DNA molecules, or the two ends of the same molecule. The chemical reaction involved in ligating two molecules is exactly the same as discontinuity repair, except that two phosphodiester bonds must be made, one for each strand (Figure 4.20).

Figure 4.19

Ligation: the final step in construction of a recombinant DNA molecule.

Ligating blunt ends

Figure 4.20

Ligation of blunt-ended DNA molecules by DNA ligase.

4.3.2 Sticky ends increase the efficiency of ligation

The ligation reaction in Figure 4.20 shows two blunt-ended fragments being joined together. Although this reaction can be carried out in the test tube, it is not very efficient because the ligase is unable to 'catch hold' of the molecule to be ligated and so must wait for chance associations to bring the ends together. If possible, blunt end ligation should be performed at high DNA concentrations, to increase the chances of the ends of the molecules coming together in the correct fashion.

In contrast, the ligation of complementary sticky ends is much more efficient. This is because compatible sticky ends can base pair with one another by hydrogen bonding (Figure 4.21), forming a relatively stable structure for the enzyme to work on. If the phosphodiester bonds are not synthesized fairly quickly the sticky ends will fall apart again. Nonetheless, these transient, base-paired structures increase the efficiency of ligation by increasing the period of time that the ends are in contact with one another.

4.3.3 Putting sticky ends on to a blunt-ended molecule

For the reasons detailed in Section 4.3.2, compatible sticky ends are desirable on the DNA molecules to be ligated together in a gene cloning experiment. Often, these sticky ends can be provided by digesting both the vector and the DNA to be cloned with the same restriction endonuclease, or with different enzymes that produce the same sticky end, though it is not always possible to do this. A common situation is where the vector molecule has sticky ends but the DNA fragments to be cloned are blunt-ended. Under these circumstances, one of three methods can be used to put the correct sticky ends onto the DNA fragments.

Figure 4.21

Ligation of sticky-ended molecules.

(a) A typical linker

```
C-G-A-T-G-G-A-T-C-C-A-T-C-C
| | | | | | | | | | | | | |
G-C-T-A-C-C-T-A-G-G-T-A-G-G
```

*Bam*HI site

(b) The use of linkers

Linkers   Blunt-ended molecule

DNA ligase

Linkers attached

*Bam*HI

*Bam*HI sticky end

Cleaved linkers

Figure 4.22

Linkers and their use. (a) The structure of a typical linker. (b) The attachment of linkers to a blunt-ended molecule.

Linkers

Linkers are short pieces of double-stranded DNA of known nucleotide sequence that are synthesized in the test tube. A typical linker, as shown in Figure 4.22a, is blunt-ended but contains a restriction site (*Bam*HI in the example shown). DNA ligase can attach linkers to the ends of larger blunt-ended DNA molecules. Although a blunt end ligation, this particular reaction can be performed very efficiently because synthetic oligonucleotides, such as linkers, can be made in very large amounts and added into the ligation mixture at a high concentration.

More than one linker will attach to each end of the DNA molecule, producing the chain structure shown in Figure 4.22b. Digestion with *Bam*HI cleaves the chains at the recognition sequences, producing a large number of cleaved linkers and the original DNA fragment, now carrying *Bam*HI sticky ends. This modified fragment is ready for ligation into a cloning vector restricted with *Bam*HI.

Adaptors

The use of linkers has one potential drawback. If the blunt-ended molecule shown in Figure 4.22b were to contain one or more *Bam*HI recognition sequences, then the restriction step needed to cleave the linkers and produce the sticky ends would also cleave the blunt-ended molecule (Figure 4.23). The resulting fragments would have the correct

Figure 4.23

A possible problem with the use of linkers. Compare this situation with the desired result of *Bam*HI restriction, as shown in Figure 4.22b.

*Bam*HI

Cleavage due to internal *Bam*HI sites

69

(a) A typical adaptor

```
G–A–T–C–C–C–G–G
        |  |  |  |
        G–G–C–C
```

BamHI sticky end

(b) Adaptors could ligate to one another

```
C–C–G–G  G–A–T–C–C–C–G–G
|  |  |  |         |  |  |  |
G–G–C–C–C–T–A–G  G–G–C–C
```

(c) The new DNA molecule is still blunt-ended

Adaptors   Blunt-ended
           molecule   DNA ligase

Adaptors ligate to themselves

Figure 4.24

Adaptors and the potential problem with their use. (a) A typical adaptor. (b) Two adaptors could ligate to one another to produce a molecule similar to a linker, so that (c) after ligation of adaptors a blunt-ended molecule is still blunt-ended and the restriction step is still needed.

sticky ends, but that would be no consolation if the gene contained in the blunt-ended fragment had now been broken into pieces.

The second method of attaching sticky ends to a blunt-ended molecule is designed to avoid this problem. An adaptor, like a linker, is a short synthetic oligonucleotide but is synthesized so that it already has one sticky end (Figure 4.24a). The idea is of course to ligate the blunt end of the adaptor to the blunt ends of the DNA fragment, to produce a new molecule with sticky ends. This may appear to be a simple method, but in practice a new problem arises. The sticky ends of individual adaptor molecules could base pair with each other to form dimers (Figure 4.24b) so that the new DNA molecule is still blunt-ended (Figure 4.24c). Although the sticky ends could be recreated by digestion with a restriction endonuclease, that would defeat the purpose of using adaptors in the first place.

The answer to this problem lies in the precise chemical structure of the ends of the adaptor molecule. Normally, the two ends of a polynucleotide strand are chemically distinct, a fact that is clear from a careful examination of the polymeric structure of DNA (Figure 4.25a). One end, referred to as the 5′ terminus, carries a phosphate group (5′-P), while the other, the 3′ terminus, has a hydroxyl group (3′-OH). In the double helix the two strands are antiparallel (Figure 4.25b), so each end of a double-stranded molecule consists of one 5′-P terminus and one 3′-OH terminus. Ligation then takes place between the 5′-P and 3′-OH ends (Figure 4.25c).

Adaptor molecules are synthesized so that the blunt end is the same as 'natural' DNA, but the sticky end is different. The 3′-OH terminus of the sticky end is the same as usual, but the 5′-P terminus is modified: it lacks the phosphate group, and is in fact a 5′-OH terminus (Figure 4.26a). DNA ligase is unable to form a phosphodiester bridge between 5′-OH and 3′-OH ends. The result is that, although base pairing is always occurring between the sticky ends of adaptor molecules, the association is never stabilized by ligation (Figure 4.26b).

Adaptors can therefore be ligated to a blunt-ended DNA molecule but not to themselves. After the adaptors have been attached, the abnormal 5′-OH terminus is converted

Figure 4.25

The distinction between the 5′ and 3′ termini of a polynucleotide.

(a) The structure of a polynucleotide strand showing the chemical distinction between the 5′-P and 3′-OH termini

(b) In the double helix the polynucleotide strands are antiparallel

(c) Ligation takes place between 5′-P and 3′-OH termini

to the natural 5′-P form by treatment with the enzyme polynucleotide kinase (p. 53), producing a sticky-ended fragment that can be inserted into an appropriate vector.

Homopolymer tailing

The technique of homopolymer tailing offers a radically different approach to the production of sticky ends on a blunt-ended DNA molecule. A homopolymer is simply a polymer in which all the subunits are the same. A DNA strand made up entirely of, say, deoxyguanosine is an example of a homopolymer, and is referred to as polydeoxyguanosine or poly(dG).

(a) The precise structure of an adaptor

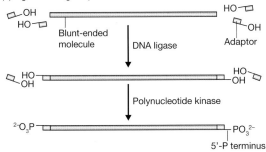

Figure 4.26

The use of adaptors. (a) The actual structure of an adaptor, showing the modified 5'-OH terminus. (b) Conversion of blunt ends to sticky ends through the attachment of adaptors.

(b) Ligation using adaptors

Tailing involves using the enzyme terminal deoxynucleotidyl transferase (p. 53) to add a series of nucleotides onto the 3'-OH termini of a double-stranded DNA molecule. If this reaction is carried out in the presence of just one deoxyribonucleotide, a homopolymer tail is produced (Figure 4.27a). Of course, in order for two tailed molecules to be able to ligate one another, the homopolymers must be complementary. Frequently, polydeoxycytosine (poly(dC)) tails are attached to the vector and poly(dG) to the DNA to be cloned. Base pairing between the two occurs when the DNA molecules are mixed (Figure 4.27b).

In practice, the poly(dG) and poly(dC) tails are not usually exactly the same length, and the base-paired recombinant molecules that result have nicks as well as discontinuities (Figure 4.27c). Repair is therefore a two-step process, using Klenow polymerase to fill in the nicks, followed by DNA ligase to synthesize the final phosphodiester bonds. This repair reaction does not always have to be performed in the test tube. If the complementary homopolymer tails are longer than about 20 nucleotides, then quite stable base-paired associations are formed. A recombinant DNA molecule, held together by base pairing although not completely ligated, is often stable enough to be introduced into the host cell in the next stage of the cloning experiment (see Figure 1.1). Once inside the host, the cell's own DNA polymerase and DNA ligase repair the recombinant DNA molecule, completing the construction begun in the test tube.

4.3.4 Blunt end ligation with a DNA topoisomerase

A more sophisticated, but easier and generally more efficient way of carrying out blunt end ligation, is to use a special type of enzyme called a DNA topoisomerase. In the cell, DNA topoisomerases are involved in processes that require turns of the double helix to be removed or added to a double-stranded DNA molecule. Turns are removed during DNA replication in order to unwind the helix and enable each polynucleotide to be replicated, and are added to newly synthesized circular molecules to introduce supercoiling. DNA topoisomerases are able to separate the two strands of a DNA molecule without actually rotating the double helix. They achieve this feat by causing transient single- or

Figure 4.27

Homopolymer tailing. (a) Synthesis of a homopolymer tail. (b) Construction of a recombinant DNA molecule from a tailed vector plus tailed insert DNA. (c) Repair of the recombinant DNA molecule.

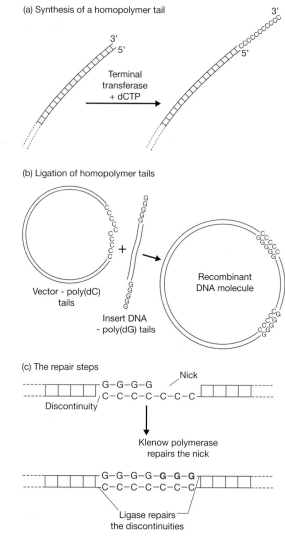

(a) Synthesis of a homopolymer tail

Terminal transferase + dCTP

(b) Ligation of homopolymer tails

Vector - poly(dC) tails

Insert DNA - poly(dG) tails

Recombinant DNA molecule

(c) The repair steps

Nick

G–G–G–G
C–C–C–C–C–C–C–C

Discontinuity

Klenow polymerase repairs the nick

G–G–G–G–**G–G–G**
C–C–C–C–C–C–C

Ligase repairs the discontinuities

double-stranded breakages in the DNA backbone (Figure 4.28). DNA topoisomerases therefore have both nuclease and ligase activities.

To carry out blunt end ligation with a topoisomerase, a special type of cloning vector is needed. This is a plasmid that has been linearized by the nuclease activity of the DNA topoisomerase enzyme from vaccinia virus. The vaccinia topoisomerase cuts DNA at the sequence CCCTT, which is present just once in the plasmid. After cutting the plasmid, topoisomerase enzymes remain covalently bound to the resulting blunt ends. The reaction can be stopped at this point, enabling the vector to be stored until it is needed.

Cleavage by the topoisomerase results in 5′-OH and 3′-P termini (Figure 4.29a). If the blunt-ended molecules to be cloned have been produced from a larger molecule by cutting with a restriction enzyme, then they will have 5′-P and 3′-OH ends. Before mixing these molecules with the vector, their terminal phosphates must be removed to

Figure 4.28

The mode of action of a Type 1 DNA topoisomerase, which removes or adds turns to a double helix by making a transient break in one of the strands.

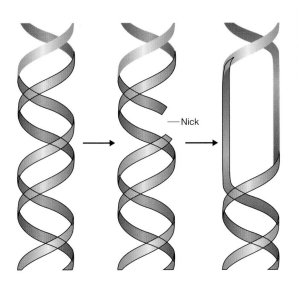

Nick

give 5′-OH ends that can ligate to the 3′-P termini of the vector. The molecules are therefore treated with alkaline phosphatase (Figure 4.29b).

Addition of the phosphatase-treated molecules to the vector reactivates the bound topoisomerases, which proceed to the ligation phase of their reaction. Ligation occurs between the 3′-P ends of the vector and the 5′-OH ends of a phosphatase-treated molecule. The blunt-ended molecule therefore becomes inserted into the vector. Only one strand is ligated at each junction point (Figure 4.29c), but this is not a problem because the discontinuities will be repaired by cellular enzymes after the recombinant molecules have been introduced into the host bacteria.

(a) The ends of the vector resulting from topoisomerase cleavage

$$3' \ PO_3^{2-} \quad HO \ 5'$$
$$\text{C-C-C-T-T} \diagup \quad \diagdown \text{A-A-G-G-G}$$
$$\text{G-G-G-A-A} \diagdown \quad \diagup \text{T-T-C-C-C}$$
$$5' \ OH \quad ^{2-}O_3P \ 3'$$

(b) Removal of terminal phosphates from the molecule to be cloned

$$^{2-}O_3P \diagdown 5' \qquad\qquad 3' \diagup OH$$

Alkaline phosphatase

$$HO \diagdown \qquad\qquad \diagup OH$$
$$HO \diagup 3' \qquad\qquad 5' \diagdown PO_3^{2-} \qquad\qquad HO \diagup \qquad\qquad \diagdown OH$$

(c) Structure of the ligation product

$$\begin{array}{l} HO \\ OH \diagdown \end{array}$$
$$\text{C-C-C-T-T} \qquad\qquad\qquad \text{A-A-G-G-G}$$
$$\text{G-G-G-A-A} \qquad\qquad\qquad \text{T-T-C-C-C}$$
$$\diagdown \begin{array}{l} HO \\ OH \end{array}$$

Figure 4.29

Blunt end ligation with a DNA topoisomerase. (a) Cleavage of the vector with the topoisomerase leaves blunt ends with 5′-OH and 3′-P termini. (b) The molecule to be cloned must therefore be treated with alkaline phosphatase to convert its 5′-P ends into 5′-OH termini. (c) The topoisomerase ligates the 3′-P and 5′-OH ends, creating a double-stranded molecule with two discontinuities, which are repaired by cellular enzymes after introduction into the host bacteria.

Further reading

Deng, G. and Wu, R. (1981) An improved procedure for utilizing terminal transferase to add homopolymers to the 3′ termini of DNA. *Nucleic Acids Research*, **9**, 4173–4188.

Helling, R.B., Goodman, H.M., and Boyer, H.W. (1974) Analysis of endonuclease R·*Eco*RI fragments of DNA from lambdoid bacteriophages and other viruses by agarose-gel electrophoresis. *Journal of Virology*, **14**, 1235–1244.

Heyman, J.A., Cornthwaite, J., Foncerrada, L. *et al.* (1999) Genome-scale cloning and expression of individual open reading frames using topoisomerase I-mediated ligation. *Genome Research*, **9**, 383–392. [A description of ligation using topoisomerase.]

Jacobsen, H., Klenow, H., and Overgaard-Hansen, K. (1974) The N-terminal amino acid sequences of DNA polymerase I from *Escherichia coli* and of the large and small fragments obtained by a limited proteolysis. *European Journal of Biochemistry*, **45**, 623–627. [Production of the Klenow fragment of DNA polymerase I.]

Lehnman, I.R. (1974) DNA ligase: structure, mechanism, and function. *Science*, **186**, 790–797.

Pingoud, A., Fuxreiter, M., Pingoud, V., and Wende, W. (2005) Type II restriction endonucleases: structure and mechanism. *Cellular and Molecular Life Sciences*, **62**, 685–707.

REBASE: http://rebase.neb.com/rebase/ [A comprehensive list of all the known restriction endonucleases and their recognition sequences.]

Rothstein, R.J., Lau, L.F., Bahl, C.P., Narang, N.A., and Wu, R. (1979) Synthetic adaptors for cloning DNA. *Methods in Enzymology*, **68**, 98–109.

Schwartz, D.C. and Cantor, C.R. (1984) Separation of yeast chromosome-sized DNAs by pulsed field gradient gel electrophoresis. *Cell*, **37**, 67–75.

Smith, H.O. and Wilcox, K.W. (1970) A restriction enzyme from *Haemophilus influenzae*. *Journal of Molecular Biology*, **51**, 379–391. [One of the first full descriptions of a restriction endonuclease.]

Zipper, H., Brunner, H., Bernhagen, J., and Vitzthum, F. (2004) Investigations on DNA intercalation and surface binding by SYBR Green I, its structure determination and methodological implications. *Nucleic Acids Research*, **32**, e103. [Details of one the DNA dyes now used as an alternative to ethidium bromide for staining agarose gels.]

Chapter 5

Introduction of DNA into Living Cells

Chapter contents

The manipulations described in Chapter 4 allow the molecular biologist to create novel recombinant DNA molecules. The next step in a gene cloning experiment is to introduce these molecules into living cells, usually bacteria, which then grow and divide to produce clones (see Figure 1.1). Strictly speaking, the word 'cloning' refers only to the later stages of the procedure, and not to the construction of the recombinant DNA molecule itself.

Cloning serves two main purposes. First, it allows a large number of recombinant DNA molecules to be produced from a limited amount of starting material. At the outset only a few nanograms of recombinant DNA may be available, but each bacterium that takes up a plasmid subsequently divides numerous times to produce a colony, each cell of which contains multiple copies of the molecule. Several micrograms of recombinant DNA can usually be prepared from a single bacterial colony, representing a thousandfold increase over the starting amount (Figure 5.1). If the colony is used not as a source of DNA but as an inoculum for a liquid culture, the resulting cells may provide milligrams of DNA, a millionfold increase in yield. In this way, cloning can supply the large amounts of DNA needed for molecular biological studies of gene structure and expression (see Chapters 10 and 11).

The second important function of cloning can be described as purification. The manipulations that result in a recombinant DNA molecule can only rarely be controlled to the extent that no other DNA molecules are present at the end of the procedure. The

Gene Cloning and DNA Analysis: An Introduction, Seventh Edition. T.A. Brown.
© 2016 John Wiley & Sons, Ltd. Published 2016 by John Wiley & Sons, Ltd.
Companion Website: www.wiley.com/go/brown/cloning

Single cell containing
multiple copies of a
recombinant DNA molecule

Provides several µg
of recombinant DNA

Provides several mg
of recombinant DNA

Figure 5.1

Cloning can supply large amounts of recombinant DNA.

ligation mixture may contain, in addition to the desired recombinant molecule, any number of the following (Figure 5.2a):

- Unligated vector molecules.
- Unligated DNA fragments.
- Vector molecules that have recircularized without new DNA being inserted ('self-ligated' vector).
- Recombinant DNA molecules that carry the wrong inserted DNA fragment.

Unligated molecules rarely cause a problem because, even though they may be taken up by bacterial cells, only under exceptional circumstances will they be replicated. It is much more likely that enzymes within the host bacteria will degrade these pieces of DNA. Self-ligated vector molecules and incorrect recombinant plasmids are more important because they are replicated just as efficiently as the desired molecule (Figure 5.2b). Purification of the desired molecule can still be achieved through cloning because it is extremely unusual for any one cell to take up more than one DNA molecule. As each cell gives rise to a single colony, each of the resulting clones will be made up of cells that all contain the same molecule. Of course, different colonies contain different molecules: some contain the desired recombinant DNA molecule, some have different recombinant molecules, and some contain self-ligated vector. The problem therefore becomes a question of identifying those colonies that contain the correct recombinant plasmids.

This chapter is concerned with the way in which plasmid and phage vectors, and recombinant molecules derived from them, are introduced into bacterial cells. During the course of the chapter it will become apparent that selection for colonies containing recombinant molecules, as opposed to colonies containing self-ligated vector, is relatively easy. The more difficult proposition of how to distinguish clones containing the correct recombinant DNA molecule from all the other recombinant clones will be tackled in Chapter 8.

5.1 Transformation: The uptake of DNA by bacterial cells

Most species of bacteria are able to take up DNA molecules from the medium in which they grow. Often, a DNA molecule taken up in this way will be degraded, but

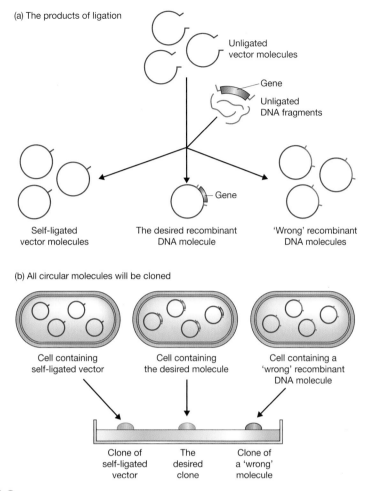

(a) The products of ligation

Unligated vector molecules

Gene

Unligated DNA fragments

Gene

Self-ligated vector molecules

The desired recombinant DNA molecule

'Wrong' recombinant DNA molecules

(b) All circular molecules will be cloned

Cell containing self-ligated vector

Cell containing the desired molecule

Cell containing a 'wrong' recombinant DNA molecule

Clone of self-ligated vector

The desired clone

Clone of a 'wrong' molecule

Figure 5.2

Cloning is analogous to purification. From a mixture of different molecules, clones containing copies of just one molecule can be obtained.

occasionally it is able to survive and replicate in the host cell. In particular, this will happen if the DNA molecule is a plasmid with an origin of replication that is recognized by the host.

5.1.1 Not all species of bacteria are equally efficient at DNA uptake

In nature, transformation is probably not a major process by which bacteria obtain genetic information. This is reflected by the fact that, in the laboratory, only a few species (notably members of the genera *Bacillus* and *Streptococcus*) can be transformed with ease. Close studies of these organisms have revealed that they possess sophisticated mechanisms for DNA binding and uptake.

Figure 5.3

The binding and uptake of DNA by a competent bacterial cell.

Most species of bacteria, including *E. coli*, take up only limited amounts of DNA under normal circumstances. In order to transform these species efficiently, the bacteria have to undergo some form of physical and/or chemical treatment that enhances their ability to take up DNA. Cells that have undergone this treatment are said to be competent.

5.1.2 Preparation of competent E. coli cells

As with many breakthroughs in recombinant DNA technology, the key development as far as transformation is concerned occurred during the early 1970s, when it was observed that *E. coli* cells that had been soaked in an ice-cold salt solution were more efficient at DNA uptake than unsoaked cells. A solution of 50 mM calcium chloride ($CaCl_2$) is traditionally used for this purpose, although other salts (notably rubidium chloride) are also effective.

Exactly why this treatment works is not understood. At one time it was thought that the $CaCl_2$ causes the DNA to precipitate onto the outside of the cells, or is responsible for some type of physical change in the cell wall that improves DNA binding. More recent studies have suggested that the salt treatment induces an over-production of certain outer membrane proteins, including one or more that bind DNA. In any case, soaking in $CaCl_2$ affects only DNA binding, and not the actual uptake into the cell. When DNA is added to treated cells it remains attached to the cell exterior, and is not at this stage transported into the cytoplasm (Figure 5.3). The actual movement of DNA into competent cells is stimulated by briefly raising the temperature to 42 °C. It is possible that this heat shock changes the permeability of the membrane to DNA, or, as with $CaCl_2$ treatment, the heat shock might induce the activity of a membrane protein that transports DNA into the cell.

5.1.3 Selection for transformed cells

The transformation of competent cells is an inefficient procedure, however carefully the cells have been prepared. Although 1 ng of the plasmid vector called pUC8 (p. 97) can yield between 1000 and 10 000 transformants, this represents the uptake of only 0.01% of all the available molecules. Furthermore, 10 000 transformants is only a very small proportion of the total number of cells that are present in a competent culture. This latter fact means that some way must be found to distinguish a cell that has taken up a plasmid from the many thousands that have not been transformed.

The uptake and stable retention of a plasmid is usually detected by looking for expression of the genes carried by the plasmid. For example, *E. coli* cells are normally sensitive

Figure 5.4

Selecting cells that contain pBR322 plasmids by plating onto agar medium containing ampicillin and/or tetracycline.

to the growth-inhibitory effects of the antibiotics ampicillin and tetracycline. Cells that contain the plasmid pBR322 (p. 94), which was one of the first cloning vectors to be developed back in the 1970s, are resistant to these antibiotics. This is because pBR322 carries two sets of genes: one gene set that codes for a β-lactamase enzyme which modifies ampicillin into a form that is non-toxic to the bacterium, and a second set of genes that code for enzymes that detoxify tetracycline. After a transformation experiment with pBR322, only those *E. coli* cells that have taken up a plasmid are $amp^R tet^R$ and able to form colonies on an agar medium that contains ampicillin or tetracycline (Figure 5.4). Non-transformants, which are still $amp^S tet^S$, do not produce colonies on the selective medium. Transformants and non-transformants are therefore easily distinguished from one another.

Most plasmid cloning vectors carry at least one gene that confers antibiotic resistance on the host cells, with the selection of transformants being achieved by plating onto an agar medium that contains the relevant antibiotic. Bear in mind, however, that resistance to the antibiotic is not due merely to the presence of the plasmid in the transformed cells. The resistance gene on the plasmid must also be expressed, so that the enzyme which detoxifies the antibiotic is synthesized. Although expression of the resistance gene begins immediately after transformation, it takes a few minutes before the cell contains enough of the enzyme to be able to withstand the toxic effects of the antibiotic. For this reason, the transformed bacteria should not be plated onto the selective medium immediately after a heat-shock treatment, but should first be placed in a small volume of liquid medium, in the absence of the antibiotic, and incubated for a short time. Plasmid replication and expression can then start so that, when the cells are plated out and encounter the antibiotic, they will already have synthesized sufficient resistance enzymes to be able to survive (Figure 5.5).

5.2 Identification of recombinants

Plating onto a selective medium enables transformants to be distinguished from non-transformants. The next problem is to determine which of the transformed colonies comprise cells that contain recombinant DNA molecules, and which contain self-ligated

Figure 5.5

Phenotypic expression. Incubation at 37 °C for 1 h before plating out improves the survival of transformants on a selective medium, because the bacteria have had time to begin synthesis of the antibiotic-resistance enzymes.

vector molecules (see Figure 5.2). With most cloning vectors, the insertion of a DNA fragment into the plasmid destroys the integrity of one of the genes present on the molecule. Recombinants can therefore be identified because the characteristic coded by the inactivated gene is no longer displayed by the host cells (Figure 5.6). We will explore the general principles of insertional inactivation by looking at the different methods used with the two cloning vectors mentioned in Section 5.1, namely pBR322 and pUC8.

5.2.1 Recombinant selection with pBR322: Insertional inactivation of an antibiotic resistance gene

pBR322 has several unique restriction sites that can be used to open up the vector before the insertion of a new DNA fragment (Figure 5.7a). *Bam*HI, for example, cuts pBR322 at just one position, within the cluster of genes that code for resistance to tetracycline. A recombinant pBR322 molecule, one that carries an extra piece of DNA in the *Bam*HI site (Figure 5.7b), is no longer able to confer tetracycline resistance on its host, as one of the necessary genes is now disrupted by the inserted DNA. Cells containing this recombinant pBR322 molecule are still resistant to ampicillin, but are sensitive to tetracycline ($amp^R tet^S$).

Screening for pBR322 recombinants is performed in the following way. After transformation, the cells are plated onto an ampicillin medium and incubated until colonies appear (Figure 5.8a). All of these colonies are transformants (remember, untransformed cells are amp^S and so do not produce colonies on the selective medium) but only a few contain recombinant pBR322 molecules; most will contain the normal, self-ligated plasmid. To identify the recombinants the colonies are replica plated onto agar

Figure 5.6

Insertional inactivation. (a) The normal, non-recombinant vector molecule carries a gene whose product confers a selectable or identifiable characteristic on the host cell. (b) This gene is disrupted when new DNA is inserted into the vector; as a result the recombinant host does not display the relevant characteristic.

(a) Normal vector molecule   (b) Recombinant vector molecule

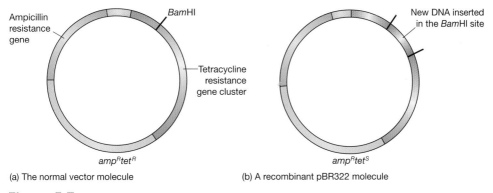

Ampicillin resistance gene

*Bam*HI

Tetracycline resistance gene cluster

New DNA inserted in the *Bam*HI site

$amp^R tet^R$

$amp^R tet^S$

(a) The normal vector molecule

(b) A recombinant pBR322 molecule

Figure 5.7

The cloning vector pBR322. (a) The normal vector molecule. (b) A recombinant molecule containing an extra piece of DNA inserted into the *Bam*HI site. For a more detailed map of pBR322, see Figure 6.1.

medium that contains tetracycline (Figure 5.8b). After incubation, some of the original colonies regrow, but others do not (Figure 5.8c). Those that do grow consist of cells that carry the normal pBR322 with no inserted DNA and therefore a functional tetracycline resistance gene cluster ($amp^R tet^R$). The colonies that do not grow on tetracycline agar are recombinants ($amp^R tet^S$). Reference back to the original ampicillin agar plate reveals the positions of these colonies, enabling samples to be recovered for further study.

5.2.2 Insertional inactivation does not always involve antibiotic resistance

Although the insertional inactivation of an antibiotic resistance gene provides an effective means of recombinant identification, the method is made inconvenient by the need to carry out two screenings: one with the antibiotic that selects for transformants; and a second screen, after replica plating, with the antibiotic that distinguishes recombinants. Most modern plasmid vectors therefore make use of a different system. An example is pUC8 (Figure 5.9a), which carries the ampicillin resistance gene and a gene called *lacZ'*, which codes for part of the enzyme β-galactosidase. Cloning with pUC8 involves insertional inactivation of the *lacZ'* gene, with recombinants identified because of their inability to synthesize β-galactosidase (Figure 5.9b).

β-Galactosidase is one of a series of enzymes involved in the breakdown of lactose to glucose plus galactose. It is normally coded by the gene *lacZ*, which resides on the *E. coli* chromosome. Some strains of *E. coli* have a modified *lacZ* gene, one that lacks the segment referred to as *lacZ'* and coding for the α-peptide portion of β-galactosidase (Figure 5.10a). These mutants can synthesize the enzyme only when they harbour a plasmid, such as pUC8, that carries the missing *lacZ'* segment of the gene.

A cloning experiment with pUC8 involves the selection of transformants on ampicillin agar, followed by screening for β-galactosidase activity to identify recombinants. Cells that harbour a normal pUC8 plasmid are *amp*^{R} and able to synthesize β-galactosidase. Recombinants are also *amp*^{R} but unable to make β-galactosidase.

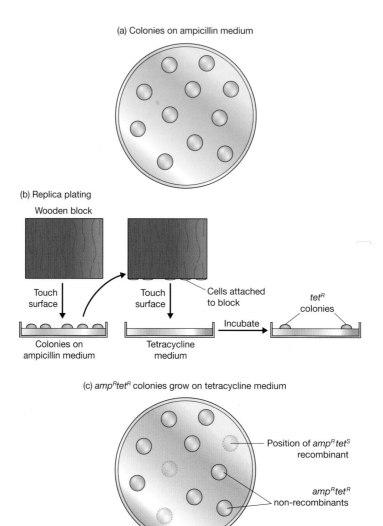

(a) Colonies on ampicillin medium

(b) Replica plating

Wooden block

Touch surface — Colonies on ampicillin medium

Touch surface — Cells attached to block — Tetracycline medium

Incubate — tet^R colonies

(c) $amp^R tet^R$ colonies grow on tetracycline medium

Position of $amp^R tet^S$ recombinant

$amp^R tet^R$ non-recombinants

Figure 5.8

Screening for pBR322 recombinants by insertional inactivation of the tetracycline resistance gene. (a) Cells are plated onto ampicillin agar: all the transformants produce colonies. (b) The colonies are replica plated onto tetracycline medium. (c) The colonies that grow on tetracycline medium are $amp^R tet^R$ and therefore non-recombinants. Recombinants ($amp^R tet^S$) do not grow, but their position on the ampicillin plate is now known.

Screening for the presence or absence of β-galactosidase is in fact quite easy. Rather than assay for lactose being split to glucose and galactose, the test is for a slightly different reaction that is also catalysed by β-galactosidase. This involves a lactose analogue termed X-gal (5-bromo-4-chloro-3-indolyl-β-D-galactopyranoside) which is broken down by β-galactosidase to a product that is coloured deep blue. If X-gal (plus an inducer of the enzyme such as isopropylthiogalactoside, IPTG) is added to the agar, along with ampicillin, then non-recombinant colonies, the cells of which synthesize β-galactosidase, will be coloured blue, whereas recombinants with a disrupted *lacZ'* gene and unable to make β-galactosidase, will be white. This system, which is known as Lac selection, is summarized in Figure 5.10b. Note that both ampicillin resistance and

(a) pUC8 $amp^R\beta\text{-}gal^+$

Figure 5.9

The cloning vector pUC8. (a) The normal vector molecule. (b) A recombinant molecule containing an extra piece of DNA inserted into the *Bam*HI site. For a more detailed map of pUC8, see Figure 6.3.

Ampicillin resistance gene

*Bam*HI

lacZ' gene

New DNA inserted in the *Bam*HI site

$amp^R\beta\text{-}gal^-$

(b) A recombinant pUC8 molecule

the presence or absence of β-galactosidase are tested for on a single agar plate. The two screenings are therefore carried out together and there is no need for the time-consuming replica plating step that is necessary with plasmids such as pBR322.

5.3 Introduction of phage DNA into bacterial cells

There are two different methods by which a recombinant DNA molecule constructed with a phage vector can be introduced into a bacterial cell, namely transfection and *in vitro* packaging.

5.3.1 Transfection

Transfection is equivalent to transformation, the only difference being that phage DNA rather than a plasmid is involved. Just as with a plasmid, the purified phage DNA or recombinant phage molecule is mixed with competent *E. coli* cells and DNA uptake is induced by heat shock. Transfection is the standard method for introducing the double-stranded RF form of an M13 cloning vector into *E. coli*.

5.3.2 In vitro *packaging of λ cloning vectors*

Transfection with λ DNA molecules is not a very efficient process when compared with the infection of a culture of cells with mature λ phage particles. It would therefore be

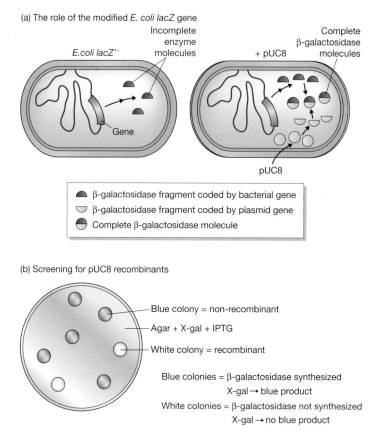

(a) The role of the modified *E. coli lacZ* gene

Incomplete enzyme molecules

E.coli lacZ'⁻

Gene

+ pUC8

Complete β-galactosidase molecules

pUC8

🔴 β-galactosidase fragment coded by bacterial gene

⬜ β-galactosidase fragment coded by plasmid gene

🔵 Complete β-galactosidase molecule

(b) Screening for pUC8 recombinants

Blue colony = non-recombinant

Agar + X-gal + IPTG

White colony = recombinant

Blue colonies = β-galactosidase synthesized
X-gal → blue product

White colonies = β-galactosidase not synthesized
X-gal → no blue product

Figure 5.10

The rationale behind insertional inactivation of the *lacZ'* gene carried by pUC8. (a) The bacterial and plasmid genes complement each other to produce a functional β-galactosidase molecule. (b) Recombinants are screened by plating onto agar containing X-gal and IPTG.

useful if recombinant λ molecules could be packaged into their λ head-and-tail structures in the test tube.

This may sound difficult but it is actually relatively easy to achieve. Packaging requires a number of different proteins coded by the λ genome, but these can be prepared at a high concentration from cells infected with defective λ phage strains. Currently, two different systems are in use.

With the single strain system, the defective λ phage carries a mutation in the *cos* sites, so that these are not recognized by the endonuclease that normally cleaves the λ catenanes during phage replication (p. 21). This means that the defective phage cannot replicate, though it does direct the synthesis of all the proteins needed for packaging. These proteins accumulate in the bacterium and can be purified from cultures of *E. coli* infected with the mutated λ. The protein preparation is then used for *in vitro* packaging of recombinant λ molecules (Figure 5.11a).

With the second system, two defective λ strains are needed, both of which carry a mutation in a gene for one of the components of the phage protein coat. With one strain the mutation is in gene *D*, and with the second strain it is in gene *E* (see Figure 2.9).

(a) A single-strain packaging system

λ proteins accumulate in the cell

λ DNA

E. coli SMR10 – λ DNA has defective *cos* sites

(b) A two-strain packaging system

λ DNA   λ proteins   λ DNA   λ proteins

E. coli BHB2688 - λ defective for synthesis of protein E (○)

E. coli BHB2690 - λ defective for synthesis of protein D (○)

(c) *In vitro* packaging

cos   *cos*   *cos*   *cos*

λ DNA catenanes

λ proteins from SMR10, or a mixture from BHB2688 and BHB2690

λ phage particles carrying packaged DNA molecules

Figure 5.11

In vitro packaging. (a) Synthesis of λ capsid proteins by *E. coli* strain SMR10, which carries a λ phage that has defective *cos* sites. (b) Synthesis of incomplete sets of λ capsid proteins by *E. coli* strains BHB2688 and BHB2690. (c) The cell lysates provide the complete set of capsid proteins and can package λ DNA molecules in the test tube.

Neither strain is able to complete an infection cycle in *E. coli* because, in the absence of the product of the mutated gene, the complete capsid structure cannot be made. Instead, the products of all the other coat protein genes accumulate (Figure 5.11b). An *in vitro* packaging mix can therefore be prepared by combining lysates of two cultures of cells, one infected with the λ D^- strain, the other infected with the E^- strain. The mixture now contains all the necessary components for *in vitro* packaging.

With both systems, the formation of phage particles is achieved simply by mixing the packaging proteins with λ DNA, because assembly of the particles occurs automatically in the test tube (Figure 5.11c). The packaged λ DNA is then introduced into *E. coli* cells simply by adding the assembled phages to the bacterial culture and allowing the normal λ infective process to take place.

Figure 5.12

Bacteriophage plaques. (a) The appearance of plaques on a lawn of bacteria. (b) Plaques produced by a phage that lyses the host cell (e.g., λ in the lytic infection cycle); the plaques contain lysed cells plus many phage particles. (c) Plaques produced by M13; these plaques contain slow-growing bacteria plus many M13 phage particles.

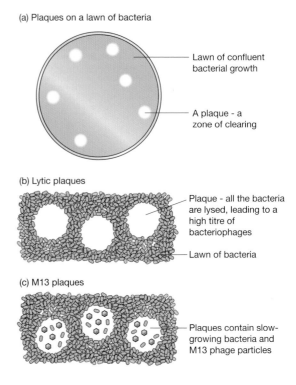

(a) Plaques on a lawn of bacteria

— Lawn of confluent bacterial growth

— A plaque - a zone of clearing

(b) Lytic plaques

— Plaque - all the bacteria are lysed, leading to a high titre of bacteriophages

— Lawn of bacteria

(c) M13 plaques

— Plaques contain slow-growing bacteria and M13 phage particles

5.3.3 Phage infection is visualized as plaques on an agar medium

The final stage of the phage infection cycle is cell lysis (p. 18). If infected cells are spread onto a solid agar medium immediately after addition of the phage particles, or immediately after transfection with phage DNA, cell lysis can be visualized as plaques on a lawn of bacteria (Figure 5.12a). Each plaque is a zone of clearing produced as the phages lyse the cells and move on to infect and eventually lyse the neighbouring bacteria (Figure 5.12b).

Both, λ and M13 form plaques. In the case of λ these are true plaques, produced by cell lysis, but M13 plaques are slightly different as M13 does not lyse the host cells (p. 19). Instead, M13 causes a decrease in the growth rate of infected cells, sufficient to produce a zone of relative clearing on a bacterial lawn. Although not true plaques, these zones of clearing are visually identical to normal phage plaques (Figure 5.12c).

The end result of a gene cloning experiment using a λ or M13 vector is therefore an agar plate covered in phage plaques, each of which is derived from a single transfected or infected cell and therefore contains identical phage particles. These may contain self-ligated vector molecules, or they may be recombinants.

5.4 Identification of recombinant phages

A variety of methods for distinguishing recombinant plaques have been devised, the following being the most important.

(a) Insertional activation of the *lacZ'* gene

— Agar + X-gal + IPTG
— Clear plaque = recombinant
— Blue plaque = non-recombinant

Figure 5.13
Strategies for the selection of recombinant phage.

(b) Insertional activation of the λ *cI* gene

— Clear plaque = recombinant
— Turbid plaque = non-recombinant

(c) Selection using the Spi phenotype

— Only recombinant λ phage can infect

— Non-recombinant λ – cannot infect

P2 prophage

(d) Selection on the basis of λ genome size

cos sites

λ catenane

Correct size for packaging

Too small to package

5.4.1 Insertional inactivation of a lacZ' gene carried by the phage vector

All M13 cloning vectors, as well as several λ vectors, carry a copy of the *lacZ'* gene. The insertion of new DNA into this gene inactivates β-galactosidase synthesis, just as with the plasmid vector pUC8. Recombinants are distinguished by plating cells onto X-gal agar. Plaques comprising normal phages are blue; recombinant plaques are clear (Figure 5.13a).

5.4.2 Insertional inactivation of the λ cI gene

Several types of λ cloning vector have unique restriction sites in the *cI* gene, the insertional inactivation of which causes a change in plaque morphology. Normal plaques appear 'turbid', whereas recombinants with a disrupted *c*I gene are 'clear' (Figure 5.13b). The difference is readily apparent to the experienced eye.

5.4.3 Selection using the Spi phenotype

λ phages cannot normally infect *E. coli* cells that already possess an integrated form of a related phage called P2; λ is therefore said to be Spi+ (sensitive to P2 prophage inhibition). Some λ cloning vectors are designed so that the insertion of new DNA causes a change from Spi+ to Spi−, enabling the recombinants to infect cells that carry P2 prophages. Such cells are used as the host for cloning experiments with these vectors. As only the recombinants are Spi−, only recombinants will form plaques (Figure 5.13c).

5.4.4 Selection on the basis of λ genome size

The λ packaging system, which assembles the mature phage particles, can only insert DNA molecules of between 37 and 52 kb into the head structure; anything less than 37 kb is not packaged. Many λ vectors have been constructed by deleting large segments of the λ DNA molecule and so are less than 37 kb in length. These can only be packaged into mature phage particles after extra DNA has been inserted, bringing the total genome size up to 37 kb or more (Figure 5.13d). Therefore, with these vectors only recombinant phages are able to replicate.

5.5 Introduction of DNA into non-bacterial cells

Various ways of introducing DNA into yeast, fungi, animals and plants are also needed if these organisms are to be used as the hosts for gene cloning. Strictly speaking, these processes are not 'transformation', as that term has a specific meaning that applies only to the uptake of DNA by bacteria, but molecular biologists have forgotten this over the years and 'transformation' is now used to describe uptake of DNA by any organism.

In general terms, soaking cells in salt is effective only with a few species of bacteria, although treatment with lithium chloride or lithium acetate does enhance DNA uptake by yeast cells, and is frequently used in the transformation of *Saccharomyces cerevisiae*. However, for most higher organisms, more sophisticated methods are needed.

5.5.1 Transformation of individual cells

With most organisms the main barrier to DNA uptake is the cell wall. Cultured animal cells, which usually lack cell walls, are easily transformed, especially if the DNA is precipitated onto the cell surface with calcium phosphate (Figure 5.14a) or enclosed in liposomes that fuse with the cell membrane (Figure 5.14b). For other types of cell, the answer is often to remove the cell wall, and enzymes that degrade yeast, fungal and plant cell walls are currently available. Under the correct conditions, intact protoplasts can be obtained (Figure 5.14c) that generally take up DNA quite readily. Transformation can

(a) Precipitation of DNA on to animal cells

Calcium phosphate solution

DNA precipitated onto the cell surface

Monolayer of animal cells

(b) Fusion with DNA-containing liposomes

Animal cell

Fusion

Fused liposome

DNA

Liposomes

DNA transfer to the nucleus

(c) Transformation of plant protoplasts

Entry of DNA

Degrade cell wall

Reform cell wall

Regenerate plant

Nucleus

Vacuole

Plant cell

Protoplast

Transformed plant cell

Transformed plant

Figure 5.14

Strategies for introducing new DNA into animal and plant cells. (a) Precipitation of DNA onto animal cells. (b) Introduction of DNA into animal cells by liposome fusion. (c) Transformation of plant protoplasts.

be stimulated by applying specialized techniques such as electroporation, whereby the cells are subjected to a short electrical pulse that is thought to induce the transient formation of pores in the cell membrane, through which DNA molecules are able to enter the cell. After transformation, the protoplasts are washed to remove the degradative enzymes, after which their cell walls are re-formed spontaneously.

In contrast to the transformation systems described above, two physical methods are available for introducing DNA into cells. The first of these methods is microinjection, in which a very fine pipette is used to inject DNA molecules directly into the nucleus of the cells to be transformed (Figure 5.15a). This technique was initially applied to animal cells but has subsequently been used successfully with plant cells. The second method involves bombarding the cells with high-velocity microprojectiles, usually particles of gold or tungsten that have been coated with DNA. The microprojectiles are fired at the cells using a particle gun (Figure 5.15b). This unusual technique is called biolistics and has been used with several different types of cell.

Figure 5.15

Two physical methods for introducing DNA into cells.

5.5.2 Transformation of whole organisms

With animals and plants, the desired end product might not be transformed cells but a transformed organism. Plants are relatively easy to regenerate from cultured cells, although problems have been experienced in developing regeneration procedures for monocotyledonous species such as cereals and grasses. A single transformed plant cell can therefore give rise to a transformed plant, which carries the cloned DNA in every cell, and passes the cloned DNA on to its progeny following flowering and seed formation. Animals, of course, cannot be regenerated from cultured cells, so the creation of transformed animals requires a rather more subtle approach. One possible technique that could be used in mammals such as mice would be to remove fertilized eggs from the oviduct, to microinject DNA, and then to reimplant the transformed cells into the mother's reproductive tract. We will look more closely at these methods for obtaining transformed animals in Chapter 13.

Further reading

FURTHER READING

Aich, P., Patra, M., Chatterjee, A.K. *et al.* (2012) Calcium chloride made *E. coli* competent for uptake of extraneous DNA through overproduction of OmpC protein. *Protein Journal*, **31**, 366–373. [Describes the effect of calcium chloride on *E. coli* cells and how this treatment might stimulate DNA uptake.]

Calvin, N.M. and Hanawalt, P.C. (1988) High efficiency transformation of bacterial cells by electroporation. *Journal of Bacteriology*, **170**, 2796–2801.

Capecchi, M.R. (1980) High efficiency transformation by direct microinjection of DNA into cultured mammalian cells. *Cell*, **22**, 479–488.

Hammer, R.E., Pursel, V.G., Rexroad, C.E. *et al.* (1985) Production of transgenic rabbits, sheep and pigs by microinjection. *Nature*, **315**, 680–683.

Hohn, B. (1979) *In vitro* packaging of lambda and cosmid DNA. *Methods in Enzymology*, **68**, 299–309.

Klein, T.M., Wolf, E.D., Wu, R., and Sanford, J.C. (1987) High velocity microprojectiles for delivering nucleic acids into living cells. *Nature*, **327**, 70–73. [Biolistics.]

Lederberg, J. and Lederberg, E.M. (1952) Replica plating and indirect selection of bacterial mutants. *Journal of Bacteriology*, **63**, 399–406.

Mandel, M. and Higa, A. (1970) Calcium-dependent bacteriophage DNA infection. *Journal of Molecular Biology*, **53**, 159–162. [The first description of the use of calcium chloride to prepare competent *E. coli* cells.]

Chapter 6

Cloning Vectors for *Escherichia coli*

Chapter contents

The basic experimental techniques involved in gene cloning have now been described. In Chapters 3, 4 and 5, we have seen how DNA is purified from cell extracts, how recombinant DNA molecules are constructed in the test tube, how DNA molecules are reintroduced into living cells, and how recombinant clones are distinguished. Now, we must look more closely at the cloning vector itself, in order to consider the range of vectors available to the molecular biologist, and to understand the properties and uses of each individual type.

The greatest variety of cloning vectors exists for use with *E. coli* as the host organism. This is not surprising in view of the central role that this bacterium has played in basic research over the past 50 years. The tremendous wealth of information that exists concerning the microbiology, biochemistry and genetics of *E. coli* has meant that virtually all fundamental studies of gene structure and function were initially carried out with this bacterium as the experimental organism. Even when a eukaryote is being studied, *E. coli* is still used for the construction of recombinant genes that will subsequently be placed back in the eukaryotic host in order to study their function and expression. During recent years, gene cloning and molecular biological research have become mutually synergistic: breakthroughs in gene cloning have acted as a stimulus to research, and the needs of research have spurred on the development of new, more sophisticated cloning vectors.

Gene Cloning and DNA Analysis: An Introduction, Seventh Edition. T.A. Brown.
© 2016 John Wiley & Sons, Ltd. Published 2016 by John Wiley & Sons, Ltd.
Companion Website: www.wiley.com/go/brown/cloning

In this chapter the most important types of *E. coli* cloning vector will be described, and their specific uses outlined. Cloning vectors for yeast, fungi, plants and animals will be considered in Chapter 7.

6.1 Cloning vectors based on *E. coli* plasmids

The simplest cloning vectors, and those most widely used in gene cloning, are based on small bacterial plasmids. A large number of different plasmid vectors are available for use with *E. coli*, many of which are obtainable from commercial suppliers. These combine ease of purification with desirable properties such as high transformation efficiency, convenient selectable markers for transformants and recombinants, and the ability to clone reasonably large (up to about 8 kb) pieces of DNA. Most routine gene cloning experiments make use of one or other of these plasmid vectors.

One of the first vectors to be developed was pBR322, which was introduced in Chapter 5 to illustrate the general principles of transformant selection and recombinant identification (p. 79). Although pBR322 lacks the more sophisticated features of the newest cloning vectors, and consequently is no longer used extensively in research, it still illustrates very well the important and fundamental properties of any plasmid cloning vector. We will therefore begin our study of *E. coli* vectors by looking more closely at pBR322.

6.1.1 The nomenclature of plasmid cloning vectors

The name 'pBR322' conforms with the standard rules for vector nomenclature:

- 'p' indicates that this is indeed a plasmid.
- 'BR' identifies the laboratory in which the vector was originally constructed (BR stands for Bolivar and Rodriguez, the two investigators who developed pBR322).
- '322' distinguishes this plasmid from others developed in the same laboratory (there are also plasmids called pBR325, pBR327, pBR328, etc.).

6.1.2 The useful properties of pBR322

The genetic and physical map of pBR322 (Figure 6.1) gives an indication of why this plasmid was such a popular cloning vector.

Figure 6.1

A map of pBR322 showing the positions of the ampicillin resistance (*amp*^{R}) and tetracycline resistance (*tet*^{R}) genes, the origin of replication (ori) and some of the most important restriction sites.

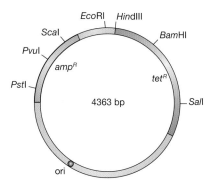

The first useful feature of pBR322 is its size. In Chapter 2, it was stated that a cloning vector should be less than 10 kb in size in order to avoid problems such as DNA breakdown during purification. pBR322 is 4363 bp, which means that not only can the vector itself be purified with ease, but so too can any recombinant DNA molecules constructed with it. Even with 6 kb of additional DNA, a recombinant pBR322 molecule is still of a manageable size.

The second feature of pBR322 is that, as described in Chapter 5, it carries two sets of antibiotic resistance genes. Either ampicillin or tetracycline resistance can be used as a selectable marker for cells containing the plasmid, and each marker gene contains unique restriction sites that can be used in cloning experiments. The insertion of new DNA into pBR322 that has been restricted with *Pst*I, *Pvu*I or *Sca*I inactivates the *amp*^{R} gene, and insertion using any one of eight restriction endonucleases (notably *Bam*HI and *Hind*III) inactivates tetracycline resistance. This great variety of restriction sites that can be used for insertional inactivation means that pBR322 can be used to clone DNA fragments with any of several types of sticky end.

A third advantage of pBR322 is that it has a reasonably high copy number. Generally, there are about 15 molecules present in a transformed *E. coli* cell, but this number can be increased up to between 1000 and 3000 by plasmid amplification in the presence of a protein synthesis inhibitor such as chloramphenicol (p. 39). An *E. coli* culture therefore provides a good yield of recombinant pBR322 molecules.

6.1.3 The pedigree of pBR322

The remarkable convenience of pBR322 as a cloning vector did not arise by chance. The plasmid was in fact designed in such a way that the final construct would possess these desirable properties. An outline of the scheme used to construct pBR322 is shown in Figure 6.2a. It is clear that the production of pBR322 was a tortuous business that required the full and skilful use of the DNA manipulative techniques described in Chapter 4. A summary of the results of these manipulations is provided in Figure 6.2b, from which it can be seen that pBR322 comprises DNA derived from three different naturally occurring plasmids:

- The *amp*^{R} gene originally resided on the plasmid R1, a typical antibiotic resistance plasmid that occurs in natural populations of *E. coli* (p. 16).
- The *tet*^{R} gene is derived from R6-5, a second antibiotic resistance plasmid.
- The replication origin of pBR322, which directs multiplication of the vector in host cells, is originally from pMB1, which is closely related to the colicin-producing plasmid ColE1 (p. 17).

6.1.4 More sophisticated E. coli plasmid cloning vectors

pBR322 was developed during the late 1970s, with the first report describing its use appearing in 1977. Since then, many other plasmid cloning vectors have been constructed, the majority of these derived from pBR322 by manipulations similar to those summarized in Figure 6.2a. One of the first of these was pBR327, which was produced by removing a 1089 bp segment from pBR322. This deletion left the *amp*^{R} and *tet*^{R} genes intact, but changed the replicative and conjugative abilities of the resulting plasmid. This was an important step in the development of more sophisticated vectors, for two reasons:

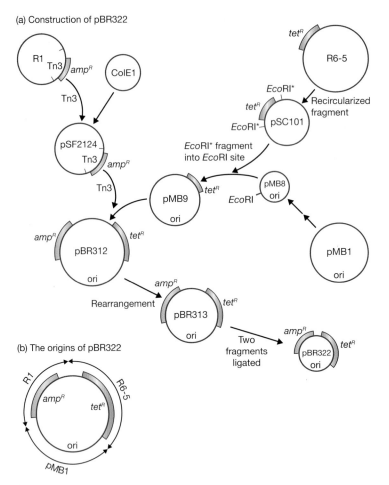

Figure 6.2

The pedigree of pBR322. (a) The manipulations involved in construction of pBR322. The *amp*^{R} gene was obtained from Tn3, a type of **transposable element** carried by the R1 plasmid. The *tet*^{R} gene was excised from pSC101 by treatment with *Eco*R1 in a low-salt solution, which decreases the specificity of the enzyme so that, as well as cutting at its standard GAATTC recognition sequence, it also cuts at related sequences such TAATTC. This is called **star activity**, and the related sequences are referred to as *Eco*RI* sites. (b) A summary of the origins of pBR322.

- The change in the replicative function means that pBR327 has a higher copy number than pBR322, being present at about 30–45 molecules per *E. coli* cell. This is not of great relevance as far as plasmid yield is concerned, as both plasmids can be amplified to copy numbers greater than 1000. The higher copy number of pBR327 in normal cells makes this vector more suitable if the aim of the experiment is to study the function of the cloned gene. In these cases, gene dosage becomes important because the more copies there are of a cloned gene the more likely it is that the effect of the cloned gene on the host cell will be detectable.
- The deletion also destroys the conjugative ability of pBR322, making pBR327 a non-conjugative plasmid that cannot direct its own transfer to other *E. coli* cells.

This is important for biological containment, averting the possibility of a recombinant pBR327 molecule escaping from the test tube and colonizing bacteria in the gut of a careless molecular biologist. In contrast, pBR322 could in theory be passed to natural populations of *E. coli* by conjugation, although in fact pBR322 also has safeguards (though less comprehensive ones) to minimize the chances of this happening.

Although pBR327, like pBR322, is no longer widely used, its properties have been inherited by most of today's modern plasmid vectors. There are a great number of these, and it would be pointless to attempt to describe them all. Two additional examples will suffice to illustrate the most important features.

pUC8: A Lac selection plasmid

This vector was mentioned in Chapter 5 when the identification of recombinants by insertional inactivation of the β-galactosidase gene was described (p. 81). pUC8 (Figure 6.3a) is descended from pBR322, although only the replication origin and the *amp*^{R} gene remain. The nucleotide sequence of the *amp*^{R} gene has been changed so that it no longer contains the unique restriction sites. All of these cloning sites are now clustered into a short segment of the *lacZ'* gene carried by pUC8.

pUC8 has two important advantages that have led to it becoming one of the most popular *E. coli* cloning vectors. The first of these was fortuitous. The manipulations

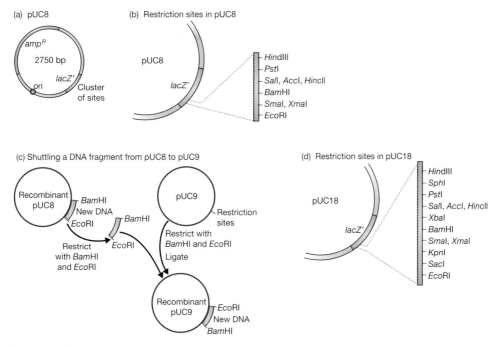

Figure 6.3

The pUC plasmids. (a) The structure of pUC8. (b) The restriction site cluster in the *lacZ'* gene of pUC8. (c) Shuttling a DNA fragment between pUC8 and pUC9, which enables the fragment to be cloned in two different directions. (d) The restriction site cluster in puC18.

involved in constructing pUC8 were accompanied by a chance mutation, within the origin of replication, which results in the plasmid having a copy number of 500–700, even before amplification. This has a significant effect on the yield of cloned DNA obtainable from *E. coli* cells transformed with recombinant pUC8 plasmids.

The second advantage is that identification of recombinant cells can be achieved by a single-step process, by plating onto agar medium containing ampicillin plus X-gal (p. 82). With both pBR322 and pBR327, selection of recombinants is a two-step procedure, requiring replica plating from one antibiotic medium to another (p. 80). A cloning experiment with pUC8 can therefore be carried out in half the time needed with pBR322 or pBR327.

The cluster of cloning sites are contained in a short artificial oligonucleotide, called a polylinker, which was inserted into the *lacZ'* gene when the first pUC8 plasmid was created. The polylinker is designed so that it does not totally disrupt the *lacZ'* gene, the reading frame being maintained throughout the polylinker, so that a functional (though altered) β-galactosidase enzyme is still produced. The polylinker in pUC8 contains nine unique restriction sites, ones that are not found elsewhere in the vector (Figure 6.3b). Because these sites are clustered, a DNA fragment with two different sticky ends (say *Eco*RI at one end and *Bam*HI at the other) can be cloned without resorting to additional manipulations such as linker attachment.

pUC8 is one of a family of vectors, differing only in the identity of the cloning sites. A second member of the family, pUC9, has the same polylinker as pUC8, but inserted into the *lacZ'* gene in the opposite orientation. This pair of vectors can therefore be used to clone a DNA fragment in both forward and reverse directions (Figure 6.3c), which enables antisense RNA to be prepared (p. 288). Other pUC vectors carry different combinations of restriction sites and provide even greater flexibility in the types of DNA fragment that can be cloned (Figure 6.3d).

pGEM3Z: In vitro transcription of cloned DNA

pGEM3Z (Figure 6.4a) is very similar to a pUC vector in that it carries the *amp*^{R} and *lacZ'* genes, the latter containing a cluster of restriction sites, and it is almost exactly the same size. The distinction is that pGEM3Z has two additional short pieces of DNA, each of which acts as the recognition site for attachment of an RNA polymerase enzyme. These two promoter sequences lie on either side of the cluster of restriction sites used for the introduction of new DNA into the pGEM3Z molecule. This means that if a recombinant pGEM3Z molecule is mixed with purified RNA polymerase in the test tube, transcription occurs and RNA copies of the cloned fragment are synthesized (Figure 6.4b). The RNA that is produced could be used as a hybridization probe (p. 143), or it might be required for experiments aimed at studying RNA processing (e.g., the removal of introns) or protein synthesis.

The promoters carried by pGEM3Z and other vectors of this type are not the standard sequences recognized by the *E. coli* RNA polymerase. Instead, one of the promoters is specific for the RNA polymerase coded by T7 bacteriophage and the other for the RNA polymerase of SP6 phage. These RNA polymerases are synthesized during the infection of *E. coli* with one or other of the phages, and are responsible for transcribing the phage genes. They are chosen for *in vitro* transcription as they are very active enzymes (remember that the entire lytic infection cycle takes only 20 min – see p. 18 – so the phage genes must be transcribed very quickly). These polymerases are

(a) pGEM3Z

Figure 6.4

pGEM3Z. (a) Map of the vector. 'R' indicates a polylinker containing restriction sites for *Eco*RI, *Sac*I, *Kpn*I, *Ava*I, *Sma*I, *Bam*HI, *Xba*I, *Sal*I, *Acc*I, *Hinc*II, *Pst*I, *Sph*I, and *Hin*dIII. (b) *In vitro* RNA synthesis from the T7 promoter.

(b) *In vitro* RNA synthesis

able to synthesize 1–2 mg of RNA per minute, which is substantially more than can be produced by the standard *E. coli* enzyme.

6.2 Cloning vectors based on λ bacteriophage

As well as plasmid cloning vectors, a variety of vectors based on λ bacteriophage have been developed for cloning DNA in *E. coli*. The primary use of these vectors is to clone large pieces of DNA, from 5 to 25 kb, most of which are too large to be handled by plasmid vectors.

6.2.1 Segments of the λ genome can be deleted without impairing viability

The first problem that had to be solved when λ cloning vectors were first developed was the capacity of the phage capsid. The λ DNA molecule can be increased in size by only about 5%, representing the addition of only 3 kb of new DNA. If the total size of the molecule is more than 52 kb, then it cannot be packaged into the λ head structure and infective phage particles are not formed. This severely limits the size of a DNA fragment that can be inserted into an unmodified λ vector (Figure 6.5).

The way forward for the development of λ cloning vectors was provided by the discovery that a large segment in the central region of the λ DNA molecule can be removed without affecting the ability of the phage to infect *E. coli* cells and produce new phage particles. A means of replicating within the host cell is, of course, the most essential

Figure 6.5

The size limitation placed on the λ genome by the need to package it into the phage head.

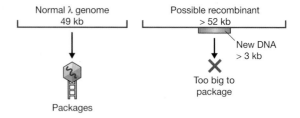

requirement for any cloning vector. For plasmid vectors this requirement is easy to satisfy, as relatively short DNA sequences are able to act as plasmid origins of replication, and most – if not all – of the enzymes needed for replication are provided by the host cell. Elaborate manipulations, such as those that resulted in pBR322 (see Figure 6.2a), are therefore possible as long as the final construction has an intact, functional replication origin. With a bacteriophage, the situation regarding replication is more complex. Phage DNA molecules generally carry several genes that are essential for replication, including genes coding for components of the phage protein coat and phage-specific DNA replicative enzymes. The alteration or deletion of any of these genes will impair or destroy the replicative ability of the resulting molecule.

The segment that can be removed from the λ DNA molecule is called the 'non-essential' region. This contains most of the genes involved in the integration and excision of the λ prophage from the *E. coli* chromosome. Removal of this segment therefore destroys the ability of the phage to follow its lysogenic infection cycle, but the lytic cycle is unaffected. This in itself is desirable for a cloning vector as it means that induction is not needed before plaques are formed (p. 41).

The non-essential region lies between positions 20 and 35 on the map shown in Figure 2.9. Its removal decreases the size of the resulting λ molecule by up to 15 kb, which means that as much as 18 kb of new DNA can now be added before the cut-off point for packaging is reached (Figure 6.6).

6.2.2 Natural selection was used to isolate modified λ that lack certain restriction sites

A second essential requirement for a cloning vector is the presence of restriction sites into which new DNA can be cloned. Unfortunately, the λ genome is so large that it has more than one recognition sequence for virtually every restriction endonuclease. Consequently, restriction cannot be used to cleave the normal λ molecule in a way that will allow the insertion of new DNA, because the molecule would be cut into several small fragments that would be very unlikely to re-form a viable λ genome on religation (Figure 6.7). Even a deleted λ genome, with the non-essential region removed, has multiple recognition sites for most restriction endonucleases. This is a problem that is often

Figure 6.6

The λ genetic map, showing the position of the main non-essential region that can be deleted without affecting the ability of the phage to follow the lytic infection cycle. There are other, much shorter non-essential regions in other parts of the genome.

Figure 6.7

λ DNA has multiple recognition sites for almost all restriction endonucleases.

encountered when a new vector is being developed. If just one or two sites need to be removed, then *in vitro* mutagenesis (p. 216) can be used to change their sequences. For example, an *Eco*RI site, GAATTC, could be changed to GGATTC, which is not recognized by the enzyme. However, *in vitro* mutagenesis was in its infancy when the first λ vectors were under development, and even today it would not be an efficient means of changing more than a few sites in a single molecule.

Instead, natural selection was used to provide strains of λ that lack the unwanted restriction sites. Natural selection can be brought into play by using as a host an *E. coli* strain that produces *Eco*RI. Most λ DNA molecules that invade the cell are destroyed by this restriction endonuclease, but a few survive and produce plaques. These are mutant phages, from which one or more *Eco*RI sites have been lost spontaneously (Figure 6.8).

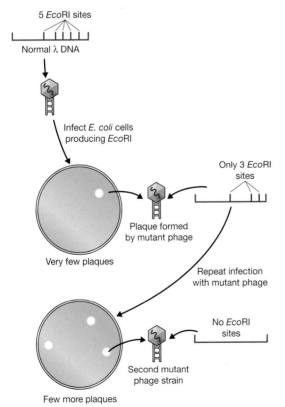

Figure 6.8

Using natural selection to isolate λ phage lacking *Eco*RI restriction sites.

Figure 6.9

λ insertion vectors. P = polylinker in the *lacZ'* gene of λZAPII, containing unique restriction sites for *Sac*I, *Not*I, *Xba*I, *Spe*I, *Eco*RI, and *Xho*I.

(a) Construction of a λ insertion vector

Normal λ DNA (49 kb)   Cleave, ligate →   λ insertion vector (35–40 kb)

Non-essential region

(b) λgt10

*Eco*RI

cI   Deletion   40 kb

(c) λZAPII

P

lacZ'   Deletion   41 kb

Several cycles of infection will eventually result in λ molecules that lack all or most of the *Eco*RI sites.

6.2.3 Insertion and replacement vectors

Once the problems posed by packaging constraints and by the multiple restriction sites had been solved, the way was open for the development of different types of λ-based cloning vectors. The first two classes of vector to be produced were λ insertion and λ replacement (or substitution) vectors.

Insertion vectors

With an insertion vector (Figure 6.9a), a large segment of the non-essential region has been deleted, and the two arms ligated together. An insertion vector possesses at least one unique restriction site into which new DNA can be inserted. The size of the DNA fragment that an individual vector can carry depends, of course, on the extent to which the non-essential region has been deleted. Two popular insertion vectors are:

- λgt10 (Figure 6.9b), which can carry up to 8 kb of new DNA, inserted into a unique *Eco*RI site located in the *c*I gene. Insertional inactivation of this gene means that recombinants are distinguished as clear rather than turbid plaques (p. 87).
- λZAPII (Figure 6.9c), with which insertion of up to 10 kb DNA into any of six restriction sites within a polylinker inactivates the *lacZ'* gene carried by the vector. Recombinants give clear rather than blue plaques on X-gal agar.

Replacement vectors

A λ replacement vector has two recognition sites for the restriction endonuclease used for cloning. These sites flank a segment of DNA that is replaced by the DNA to be cloned (Figure 6.10a). Often, the replaceable fragment (called the stuffer fragment) carries additional restriction sites that can be used to cut it up into small pieces, so that its own re-insertion during a cloning experiment is very unlikely. Replacement vectors are generally designed to carry larger pieces of DNA than insertion vectors can handle. Recombinant selection is often on the basis of size, with non-recombinant vectors being too small to be packaged into λ phage heads (p. 88).

An example of a replacement vector is:

- λDASHII (Figure 6.10b), which can carry inserted DNA of between 9 and 23 kb by replacing a segment flanked by various restriction sites, any of which can be used to remove the stuffer fragment, so that DNA fragments with a variety of sticky ends

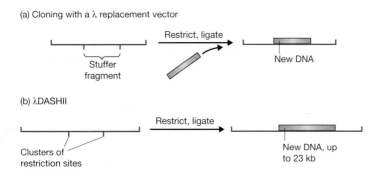

Figure 6.10

λ replacement vectors. (a) Cloning with a λ replacement vector. (b) Cloning with λDASHII.

can be cloned. Recombinant selection with λDASHII can be on the basis of size, or can utilize the Spi phenotype (p. 88). As the inserted DNA is flanked by the promoter sequences of T3 and T7 phage, RNA copies can be obtained in the same way as for pGEM3Z.

6.2.4 Cloning experiments with λ insertion or replacement vectors

A cloning experiment with a λ vector can proceed along the same lines as with a plasmid vector: the λ molecules are restricted, new DNA is added, the mixture is ligated, and the resulting molecules are used to transfect a competent *E. coli* host (Figure 6.11a). This type of experiment requires that the vector be in its circular form, with the *cos* sites hydrogen-bonded to each other.

Although satisfactory for many purposes, a procedure based on transfection is not particularly efficient, and a greater number of recombinants will be obtained if one or two refinements are introduced. The first improvement would be to use the linear form of the vector. When the linear form of the vector is digested with the relevant restriction endonuclease, the left and right arms are released as separate fragments. A recombinant molecule can be constructed by mixing together the DNA to be cloned with the vector arms. Ligation results in several molecular arrangements, including catenanes comprising left arm–DNA–right arm repeated many times (Figure 6.11b). If the inserted DNA is the correct size, the *cos* sites that separate these structures will be the correct distance apart for *in vitro* packaging (p. 83). In this way, recombinant phage are produced in the test tube and can be used to infect an *E. coli* culture. This strategy – and in particular the use of *in vitro* packaging – results in a large number of recombinant plaques.

6.2.5 Long DNA fragments can be cloned using a cosmid

The final and most sophisticated type of λ-based vector is the cosmid. Cosmids are hybrids between a phage DNA molecule and a bacterial plasmid, and their design centres on the fact that the enzymes which package the λ DNA molecule into the phage protein coat need only the *cos* sites in order to function (p. 20). The *in vitro* packaging reaction works not only with λ genomes, but also with any molecule that carries *cos* sites separated by 37–52 kb of DNA.

Figure 6.11

Different strategies for cloning with a λ vector. (a) Using the circular form of λ as a plasmid. (b) Using the left and right arms of the λ vector, plus *in vitro* packaging, to achieve a greater number of recombinant plaques.

A cosmid is basically a plasmid that carries a *cos* site (Figure 6.12a). It also needs a selectable marker (such as the ampicillin resistance gene) and a plasmid origin of replication, since cosmids lack all the λ genes and so do not produce plaques. Instead, colonies are formed on selective media, just as with a plasmid vector.

A cloning experiment with a cosmid is carried out as follows (Figure 6.12b). The cosmid is opened at its unique restriction site and new DNA fragments are inserted. These fragments are usually produced by partial digestion with a restriction endonuclease, as total digestion will almost invariably result in fragments that are too small to be cloned with a cosmid. Ligation is carried out so that catenanes are formed. Provided that the inserted DNA is the correct size, *in vitro* packaging cleaves the *cos* sites and places the recombinant cosmids into mature phage particles. These λ phage are then used to infect an *E. coli* culture, although of course plaques are not formed. Instead, infected cells are plated onto a selective medium and antibiotic-resistant colonies are grown. All colonies are recombinants, as non-recombinant linear cosmids are too small to be packaged into λ heads.

6.2.6 λ and other high-capacity vectors enable genomic libraries to be constructed

The main use of λ-based vectors is to clone DNA fragments that are too long to be handled by plasmid vectors. A replacement vector, such as λDASHII, can carry up to 23 kb

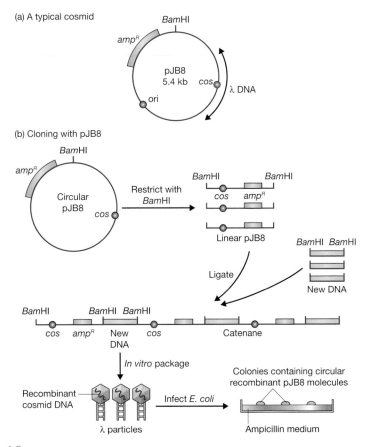

Figure 6.12

A typical cosmid and the way it is used to clone long fragments of DNA.

of new DNA, and some cosmids can manage fragments up to 40 kb. This compares with a maximum insert size of about 8 kb for most plasmids.

The ability to clone such long DNA fragments means that genomic libraries can be generated. A genomic library is a set of recombinant clones that contains all of the DNA present in an individual organism. An *E. coli* genomic library, for example, contains all the *E. coli* genes, so any desired gene can be withdrawn from the library and studied. Genomic libraries can be retained for many years, and propagated so that copies can be sent from one research group to another.

The big question is how many clones are needed for a genomic library? The answer can be calculated with the formula:

$$N = \frac{\ln(1-p)}{\ln\left(1-\frac{a}{b}\right)}$$

where N is the number of clones that are required, p is probability that any given gene will be present, a is the average size of the DNA fragments inserted into the vector, and b is the total size of the genome.

Table 6.1
Number of clones needed for genomic libraries of a variety of organisms.

SPECIES	GENOME SIZE (bp)	NUMBER OF CLONES[a]	
		17 kb FRAGMENTS[b]	35 kb FRAGMENTS[c]
E. coli	4.6×10^6	820	410
Saccharomyces cerevisiae	1.8×10^7	3225	1500
Drosophila melanogaster	1.2×10^8	21 500	10 000
Rice	5.7×10^8	100 000	49 000
Human	3.2×10^9	564 000	274 000
Frog	2.3×10^{10}	4 053 000	1 969 000

[a] Calculated for a probability (p) of 95% that any particular gene will be present in the library.
[b] Fragments suitable for a replacement vector such as λDASHII.
[c] Fragments suitable for a cosmid.

Table 6.1 shows the number of clones needed for genomic libraries of a variety of organisms, constructed using a λ replacement vector or a cosmid. For humans and other mammals, several hundred thousand clones are required. It is by no means impossible to obtain several hundred thousand clones, and the methods used to identify a clone carrying a desired gene (see Chapter 8) can be adapted to handle such large numbers, so genomic libraries of these sizes are by no means unreasonable. Nonetheless, ways of reducing the numbers of clones needed for mammalian genomic libraries are continually being sought.

One solution is to develop new cloning vectors capable of handling longer DNA inserts. The most popular of these vectors are bacterial artificial chromosomes (BACs), which are based on the F plasmid (p. 16). The F plasmid is relatively large, and vectors derived from it have a higher capacity than normal plasmid vectors. BACs can handle DNA inserts up to 300 kb in size, reducing the size of the human genomic library to just 30 000 clones. Other high-capacity vectors have been constructed from bacteriophage P1, which has the advantage over λ of being able to squeeze 110 kb of DNA into its capsid structure. Cosmid-type vectors based on P1 have been designed and used to clone DNA fragments ranging in size from 75 to 100 kb. Vectors that combine the features of P1 vectors and BACs, called P1-derived artificial chromosomes (PACs), also have a capacity of up to 300 kb.

6.3 Cloning vectors for the synthesis of single-stranded DNA

Cloning vectors capable of providing single-stranded versions of cloned DNA were extremely important for many years, as the DNA sequencing methods used during most of the 1980s and 1990s required single-stranded DNA as the starting material. Although those sequencing methods have now been superseded by techniques that use double-stranded DNA, the special vectors designed during the 1980s and 1990s for the synthesis of single-stranded DNA are still important because this type of DNA is also needed for specialist applications such as *in vitro* mutagenesis (p. 216) and phage display (p. 241).

Figure 6.13

The M13 genome, showing the positions of genes I to X.

6.3.1 Vectors based on M13 bacteriophage

The first single-stranded DNA vectors were based on M13 bacteriophage. The normal M13 genome is 6.4 kb in length, with most of the DNA taken up by ten closely packed genes (Figure 6.13), each essential for the replication of the phage. There is only a single 507-nucleotide intergenic sequence into which new DNA could be inserted without disrupting one of these genes, and this region includes the replication origin which must itself remain intact. This means that there is only limited scope for modifying the M13 genome.

The first step in the construction of an M13 cloning vector was to introduce the *lacZ'* gene into the intergenic sequence. This gave rise to M13mp1, which forms blue plaques on X-gal agar (Figure 6.14a). Polylinkers were then inserted into the *lacZ'* gene to create a series of M13 vectors with different sets of cloning sites (Figure 6.14b). The polylinkers are the same as those used in the pUC plasmids, which means that the cloned DNA can be shuttled between the equivalent M13 and pUC vectors, so that single- and double-stranded versions can be obtained.

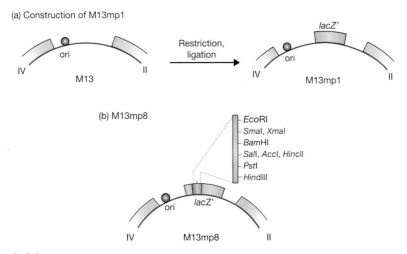

Figure 6.14

M13 vectors. (a) Construction of M13mp1. (b) M13mp8, which contains the same set of restriction sites as pUC8.

A cloning experiment with an M13 vector simply involves inserting the new DNA into one of the restriction sites in the polylinker, and then transfecting competent *E. coli* cells. After plating out on X-gal agar, recombinant plaques are identified by the blue-white colour test.

6.3.2 Hybrid plasmid–M13 vectors

Although M13 vectors are very useful for the production of single-stranded versions of cloned genes, they do suffer one major disadvantage. There is a limit to the size of DNA fragment which can be cloned with an M13 vector. Typically, 1500 bp is considered to be the maximum capacity, although fragments of up to 3 kb have occasionally been cloned. To overcome this problem, a number of hybrid vectors (phagemids) have been developed by combining a part of the M13 genome with plasmid DNA.

An example is provided by pEMBL8 (Figure 6.15a), which was made by transferring into pUC8 a 1300 bp fragment of the M13 genome. This piece of M13 DNA

Figure 6.15

pEMBL8: a hybrid plasmid–M13 vector that can be converted into single-stranded DNA.

(a) pEMBL8

M13 DNA fragment

3997 bp

amp^R

lacZ'

Cluster of sites (see fig. 6.3b)

(b) Conversion of pEMBL8 into single-stranded DNA

M13 region

M13 replication protein

The M13 protein replicates pEMBL8 into single-stranded DNA

Double-stranded pEMBL8

Single-stranded pEMBL8 molecules

pEMBL8 'phage' particles

contains the signal sequence recognized by the enzymes that convert the normal double-stranded M13 molecule into single-stranded DNA before the secretion of new phage particles. This signal sequence is still functional even though detached from the rest of the M13 genome, so pEMBL8 molecules are also converted into single-stranded DNA and secreted as defective phage particles (Figure 6.15b). All that is necessary is that the *E. coli* cells used as hosts for a pEMBL8 cloning experiment are subsequently infected with normal M13 to act as a helper phage, providing the necessary replicative enzymes and phage coat proteins. pEMBL8, being derived from pUC8, has the polylinker cloning sites within the *lacZ'* gene, so recombinant plaques can be identified in the standard way on agar containing X-gal. With pEMBL8, single-stranded versions of cloned DNA fragments up to 10 kb in length can be obtained, greatly extending the range of the M13 cloning system.

6.4 Vectors for other bacteria

Cloning vectors have also been developed for several other species of bacteria, including *Streptomyces*, *Bacillus*, and *Pseudomonas*. These vectors are used for basic studies of gene function and activity in these species, and also for the production of recombinant protein. A recombinant protein is a protein synthesized from a cloned gene, with cloning being used in this way to obtain large quantities of important proteins (e.g., insulin) that are used as pharmaceuticals to treat human disorders (see Chapter 13). As some recombinant proteins cannot be synthesized efficiently in *E. coli*, other species – both bacterial and eukaryotic – are also used for recombinant protein synthesis (p. 259).

Some cloning vectors for these other types of bacteria are based on plasmids specific to the host species, and some on broad host range plasmids that are able to replicate in a variety of bacteria. The origin of replication of the RK2 plasmid, for example, functions in most Gram-negative bacteria, and is used in cloning vectors that replicate in *Pseudomonas* species as well as *E. coli*. A cloning vector that carries the RK2 origin is therefore a type of shuttle vector, one that can be used in two different species. Shuttle vectors are useful because the initial stages of a gene cloning experiment, up to the point when the correct recombinant DNA molecule has been identified, are usually easier to carry out with *E. coli* as the host. Once the desired clone has been obtained, the broad host range properties of the shuttle vector can then be exploited to transfer the molecule to the second species.

Bacteriophages have also been used as cloning vectors for bacteria other than *E. coli*. An example is the SV1 phage of *Streptomyces venezuelae*, which has been developed into a set of cloning vectors for *Streptomyces* species. SV1 is a lysogenic phage, but rather than deleting the genes responsible for the lysogenic infection cycle, vectors based on SV1 retain this function so that the cloned gene is inserted into the *Streptomyces* genome. This results in a very stable transformant, with little possibility of the cloned gene being lost from the host even after many cycles of bacterial replication. Stability is an important consideration if the aim of the gene cloning project is to obtain a recombinant bacterial strain that creates a valuable pharmaceutical protein (p. 269).

Further reading

Bolivar, F., Rodriguez, R.L., Green, P.J., *et al.* (1977) Construction and characterization of new cloning vectors. II. A multipurpose cloning system. *Gene*, **2**, 95–113. [pBR322.]

Choi, K.-H., Trunck, L.A., Kumar, A., *et al.* (2008) Genetic tools for *Pseudomonas*, in *Pseudomonas Genomics and Molecular Biology* (ed. P. Cornells), Caister Academic Press, Norfolk, pp. 65–86. [Describes cloning vectors for *Pseudomonas*.]

Fayed, B., Younger, E., Taylor, G., and Smith, M.C.M. (2014) A novel *Streptomyces* spp. integration vector, derived from the *S. venezuelae* phage, SV1. *BMC Biotechnology*, **14**, 51.

Iouannou, P.A., Amemiya, C.T., Garnes, J., *et al.* (1994) P1-derived vector for the propagation of large human DNA fragments. *Nature Genetics*, **6**, 84–89.

Melton, D.A., Krieg, P.A., Rebagliati, M.R., Maniatis, T., Zinn, K., and Green, M.R. (1984) Efficient in vitro synthesis of biologically active RNA and RNA hybridization probes from plasmids containing a bacteriophage SP6 promoter. *Nucleic Acids Research*, **12**, 7035–7056. [RNA synthesis from DNA cloned in a plasmid such as pGEM3Z.]

Sanger, F., Coulson, A.R., Barrell, B.G., *et al.* (1980) Cloning in single-stranded bacteriophage as an aid to rapid DNA sequencing. *Journal of Molecular Biology*, **143**, 161–178. [M13 vectors.]

Shiyuza, H., Birren, B., Kim, U.J., *et al.* (1992) Cloning and stable maintenance of 300 kilobase-pair fragments of human DNA in *Escherichia coli* using an F-factor-based vector. *Proceedings of the National Academy of Sciences of the United States of America*, **89**, 8794–8797. [The first description of a BAC.]

Sternberg, N. (1992) Cloning high molecular weight DNA fragments by the bacteriophage P1 system. *Trends in Genetics*, **8**, 11–16.

Yanisch-Perron, C., Vieira, J., and Messing, J. (1985) Improved M13 phage cloning vectors and host strains: nucleotide sequences of the M13mp18 and pUC19 vectors. *Gene*, **33**, 103–119.

Chapter 7

Cloning Vectors for Eukaryotes

Chapter contents

Most cloning experiments are carried out with *E. coli* as the host, and the widest variety of cloning vectors are available for this organism. *E. coli* is particularly popular when the aim of the cloning experiment is to study the basic features of molecular biology such as gene structure and function. However, under some circumstances it may be desirable to use a eukaryotic host for a gene cloning experiment. This is especially true in biotechnology (Chapter 13), where the aim may not be to study a gene, but to use cloning to obtain recombinant protein, or to change the properties of an organism (e.g., to introduce herbicide resistance into a crop plant). We must therefore consider cloning vectors for eukaryotes.

7.1 Vectors for yeast and other fungi

The yeast *Saccharomyces cerevisiae* is one of the most important organisms in biotechnology. As well as its role in brewing and breadmaking, yeast is used as a host organism for the production of important pharmaceuticals from cloned genes (p. 260). The development of cloning vectors for yeast was initially stimulated by the discovery of a plasmid that is present in most strains of *S. cerevisiae* (Figure 7.1). The 2 µm plasmid, as it is called, is one of only a very limited number of plasmids found in eukaryotic cells.

Gene Cloning and DNA Analysis: An Introduction, Seventh Edition. T.A. Brown.
© 2016 John Wiley & Sons, Ltd. Published 2016 by John Wiley & Sons, Ltd.
Companion Website: www.wiley.com/go/brown/cloning

Figure 7.1

The yeast 2 μm plasmid. *REP1* and *REP2* are involved in replication of the plasmid, and *FLP* codes for a protein that can convert the A form of the plasmid (shown here) to the B form, in which the gene order has been rearranged by intramolecular recombination. The function of gene *D* is not exactly known.

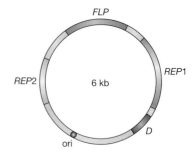

7.1.1 Selectable markers for the 2 μm plasmid

The 2 μm plasmid is an excellent basis for a cloning vector. It is 6 kb in size, which is ideal for a vector, and exists in the yeast cell at a copy number of between 70 and 200. Replication makes use of a plasmid origin, several enzymes provided by the host cell, and the proteins coded by the *REP1* and *REP2* genes carried by the plasmid.

However, all is not perfectly straightforward in using the 2 μm plasmid as a cloning vector. First, there is the question of a selectable marker. Some yeast cloning vectors carry genes conferring resistance to inhibitors such as methotrexate and copper, but most of the popular yeast vectors make use of a radically different type of selection system. In practice, a normal yeast gene is used, generally one that codes for an enzyme involved in amino acid biosynthesis. An example is the gene *LEU2*, which codes for β-isopropyl-malate dehydrogenase, one of the enzymes involved in the conversion of pyruvic acid to leucine.

In order to use *LEU2* as a selectable marker, a special kind of host organism is needed. The host must be an auxotrophic mutant that has a non-functional *LEU2* gene. Such a *leu2⁻* yeast is unable to synthesize leucine and can survive only if this amino acid is supplied as a nutrient in the growth medium (Figure 7.2a). Selection is possible because transformants contain a plasmid-borne copy of the *LEU2* gene, and so are able to grow in the absence of the amino acid. In a cloning experiment, cells are plated out onto minimal medium, which contains no added amino acids. Only transformed cells are able to survive and form colonies (Figure 7.2b).

7.1.2 Vectors based on the 2 μm plasmid: Yeast episomal plasmids

Vectors derived from the 2 μm plasmid are called yeast episomal plasmids (YEps). Some YEps contain the entire 2 μm plasmid while others include just the 2 μm origin of replication. An example of the latter type is YEp13 (Figure 7.3).

As well as the 2 μm origin of replication and the selectable *LEU2* gene, YEp13 also includes the entire pBR322 sequence. Yep13 is therefore a shuttle vector, and can replicate and be selected for in both yeast and *E. coli*. Most yeast cloning vectors are shuttle vectors because it is sometimes difficult to recover the recombinant DNA molecule from a transformed yeast colony. This is not a major problem with YEps as they are present in yeast cells primarily as plasmids, but with other yeast vectors that may integrate into one of the yeast chromosomes, purification might be impossible. This is a disadvantage because in many cloning experiments the purification of recombinant DNA

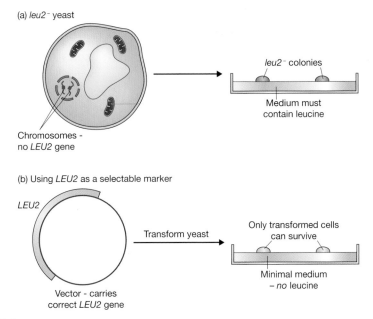

Figure 7.2

Using the *LEU2* gene as a selectable marker in a yeast cloning experiment.

is essential in order for the correct construct to be identified by, for example, DNA sequencing.

The standard procedure when cloning in yeast is therefore to perform the initial cloning experiment with *E. coli*, and to select recombinants in this organism. Recombinant plasmids can then be purified, characterized, and the correct molecule introduced into yeast (Figure 7.4).

7.1.3 A YEp may insert into yeast chromosomal DNA

The word 'episomal' indicates that a YEp can replicate as an independent plasmid, but also implies that the integration into one of the yeast chromosomes can occur (see the definition of 'episome' on p. 14). Integration occurs because the gene carried on the

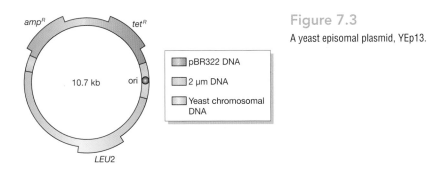

Figure 7.3

A yeast episomal plasmid, YEp13.

Figure 7.4

Cloning with an *E. coli*–yeast shuttle vector such as YEp13.

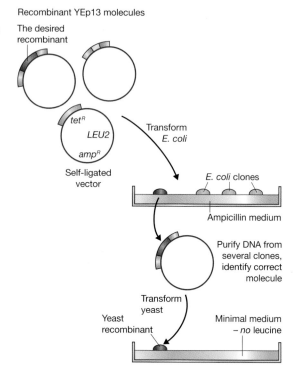

Figure 7.5

Recombination between plasmid and chromosomal *LEU2* genes can integrate YEp13 into yeast chromosomal DNA. After integration there are two copies of the *LEU2* gene. Usually one of these copies is functional, and the other mutated.

vector as a selectable marker is very similar to the mutant version of the gene present in the yeast chromosomal DNA. With YEp13, for example, homologous recombination can occur between the plasmid *LEU2* gene and the yeast mutant *LEU2* gene, resulting in insertion of the entire plasmid into one of the yeast chromosomes (Figure 7.5). The plasmid may remain integrated, or a later recombination event may result in it being excised again.

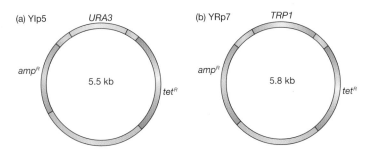

Figure 7.6
A YIp and a YRp.

7.1.4 Other types of yeast cloning vector

In addition to YEps, there are several other types of cloning vector for use with *S. cerevisiae*. Two important ones are as follows:

- Yeast integrative plasmids (YIps) are basically bacterial plasmids carrying a yeast gene. An example is YIp5, which is pBR322 with an inserted *URA3* gene (Figure 7.6a). This gene codes for orotidine-5′-phosphate decarboxylase (an enzyme that catalyses one of the steps in the biosynthesis of pyrimidine nucleotides) and is used as a selectable marker in exactly the same way as *LEU2*. A YIp cannot replicate independently as it does not contain any parts of the 2 µm plasmid, and instead depends for its survival on integration into yeast chromosomal DNA. Integration occurs just as described for a YEp (Figure 7.5).

- Yeast replicative plasmids (YRps) are able to multiply as independent plasmids because they carry a chromosomal DNA sequence that includes an origin of replication. Replication origins are known to be located very close to several yeast genes, including one or two which can be used as selectable markers. YRp7 (Figure 7.6b) is an example of a replicative plasmid. It is made up of pBR322 plus the yeast gene *TRP1*. This gene, which is involved in tryptophan biosynthesis, is located adjacent to a chromosomal origin of replication. The yeast DNA fragment present in YRp7 contains both *TRP1* and the origin.

Three factors come into play when deciding which type of yeast vector is most suitable for a particular cloning experiment. The first of these, transformation frequency, is a measure of the number of transformants that can be obtained per microgram of plasmid DNA. A high transformation frequency is necessary if a large number of recombinants are needed, or if the starting DNA is in short supply. YEps have the highest transformation frequency, providing between 10 000 and 100 000 transformed cells per microgram. YRps are also quite productive, giving between 1000 and 10 000 transformants per microgram, but a YIp yields less than 1000 transformants per microgram, and only 1–10 unless special procedures are used. The low transformation frequency of a YIp reflects the fact that the rather rare chromosomal integration event is necessary before the vector can be retained in a yeast cell.

The second important factor is copy number. YEps and YRps have the highest copy numbers of 20–50 and 5–100, respectively, whereas a YIp is usually present at just one copy per cell. These figures are important if the objective is to obtain protein from the

Figure 7.7

Chromosome structure.

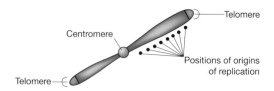

cloned gene, as the more copies there are of the gene, the greater the expected yield of the protein product.

So why would one ever wish to use a YIp? The answer is because YIps produce very stable recombinants, as the loss of a YIp that has become integrated into a chromosome occurs at only a very low frequency. In contrast, YRp recombinants are extremely unstable, the plasmids tending to congregate in the mother cell when a daughter cell buds off, so the daughter cell is non-recombinant. YEp recombinants suffer from similar problems, though an improved understanding of the biology of the 2 μm plasmid has enabled more stable YEps to be developed in recent years. Nevertheless, a YIp is the vector of choice if the needs of the experiment dictate that the recombinant yeast cells must retain the cloned gene for long periods in culture.

7.1.5 Artificial chromosomes can be used to clone long pieces of DNA in yeast

The final type of yeast cloning vector to consider is the yeast artificial chromosome (YAC), which presents a totally different approach to gene cloning. The development of YACs was a spin-off from fundamental research into the structure of eukaryotic chromosomes, work that has identified the key components of a chromosome as being (Figure 7.7):

- The centromere, which is required for the chromosome to be distributed correctly to daughter cells during cell division.
- Two telomeres, the structures at the ends of a chromosome, which are needed in order for the ends to be replicated correctly and which also prevent the chromosome from being nibbled away by exonucleases.
- The origins of replication, which are the positions along the chromosome at which DNA replication initiates, similar to the origin of replication of a plasmid.

Once chromosome structure had been defined in this way, the possibility arose that the individual components might be isolated by recombinant DNA techniques and then joined together again in the test tube, creating an artificial chromosome. As the DNA molecules present in natural yeast chromosomes are several hundred kilobases in length, it might be possible with an artificial chromosome to clone long pieces of DNA.

The structure and use of a YAC vector

Although several YAC vectors have been developed, each one is constructed along the same lines, with pYAC3 being a typical example (Figure 7.8a). At first glance, pYAC3 does not resemble an artificial chromosome but on closer examination its unique features become apparent. pYAC3 is essentially a pBR322 plasmid into which a number of

(a) pYAC3

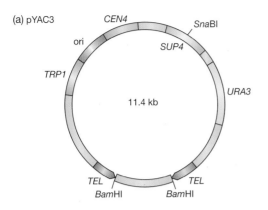

Figure 7.8

A YAC vector and the way it is used to clone
large pieces of DNA.

(b) The cloning strategy with pYAC3

yeast genes have been inserted. Two of these genes, *URA3* and *TRP1*, have been encountered already as the selectable markers for YIp5 and YRp7, respectively. As in YRp7, the
DNA fragment that carries *TRP1* also contains an origin of replication, but in pYAC3
this fragment is extended even further to include the sequence called *CEN4*, which is the
DNA from the centromere region of chromosome 4. The *TRP1*–origin–*CEN4* fragment
therefore contains two of the three components of the artificial chromosome.

The third component, the telomeres, is provided by the two sequences called *TEL*.
These are not themselves complete telomere sequences, but once inside the yeast nucleus
they act as seeding sequences onto which telomeres will be built. This just leaves one
other part of pYAC3 that has not been mentioned: *SUP4*, which is the selectable marker
into which new DNA is inserted during the cloning experiment.

The cloning strategy with pYAC3 is as follows (Figure 7.8b). The vector is first
restricted with a combination of *Bam*HI and *Sna*BI, cutting the molecule into three
fragments. The fragment flanked by *Bam*HI sites is discarded, leaving two arms, each
bounded by one *TEL* sequence and one *Sna*BI site. The DNA to be cloned, which must
have blunt ends (*Sna*BI is a blunt end cutter, recognizing the sequence TACGTA), is
ligated between the two arms, producing the artificial chromosome. Protoplast transformation (p. 88) is then used to introduce the artificial chromosome into *S. cerevisiae*.
The yeast strain that is used is a double auxotrophic mutant, $trp1^-$ $ura3^-$, which is converted to $trp1^+$ $ura3^+$ by the two markers on the artificial chromosome. Transformants

are therefore selected by plating onto minimal medium, on which only cells containing a correctly constructed artificial chromosome are able to grow. Any cell transformed with an incorrect artificial chromosome, containing two left or two right arms rather than one of each, is not able to grow on minimal medium as one of the markers is absent. The presence of the insert DNA in the vector can be checked by testing for insertional inactivation of *SUP4*, which is carried out using a simple colour test: white colonies are recombinants, red colonies are not.

Applications for YAC vectors

The initial stimulus in designing artificial chromosomes originated from yeast geneticists who wished to use them for studying various aspects of chromosome structure and behaviour, for instance to examine the segregation of chromosomes during meiosis. These experiments established that artificial chromosomes are stable during propagation in yeast cells, and raised the possibility that they might be used as vectors for genes that are too long to be cloned as a single fragment in an *E. coli* vector. Several important mammalian genes are greater than 100 kb in length (e.g., the human cystic fibrosis gene is 250 kb), which is beyond the capacity of all but the most sophisticated *E. coli* cloning systems (p. 106) but well within the range of a YAC vector. Yeast artificial chromosomes therefore opened the way to studies of the functions and modes of expression of genes that previously had been intractable to analysis by recombinant DNA techniques. A new dimension to these experiments was provided by the discovery that, under some circumstances, YACs can be propagated in mammalian cells, enabling the functional analysis to be carried out in the organism in which the gene normally resides.

Yeast artificial chromosomes are equally important in the production of gene libraries. Recall that with fragments of 300 kb, the maximum insert size for the highest capacity *E. coli* vector, some 30 000 clones are needed for a human gene library (p. 106). YAC vectors are routinely used to clone 600 kb fragments, and special types are able to handle DNA up to 1400 kb in length, the latter bringing the size of a human gene library down to just 6500 clones. Unfortunately, these 'mega-YACs' have run into problems of insert stability, the cloned DNA sometimes becoming rearranged by intramolecular recombination. Nevertheless, YACs have been of immense value in providing long pieces of cloned DNA for use in large-scale DNA-sequencing projects.

7.1.6 Vectors for other yeasts and fungi

Cloning vectors for other species of yeast and fungi are needed for basic studies of the molecular biology of these organisms, and to extend the possible uses of yeasts and fungi in biotechnology. Episomal plasmids based on the *S. cerevisiae* 2 μm plasmid are able to replicate in a few other types of yeast, but the range of species is not broad enough for 2 μm vectors to be of general value. In any case, the requirements of biotechnology are better served by integrative plasmids, equivalent to YIps, as these provide stable recombinants that can be grown for long periods. Efficient integrative vectors are now available for a number of species, including yeasts such as *Pichia pastoris* and *Kluveromyces lactis*, and the filamentous fungi such as *Aspergillus nidulans* and *Neurospora crassa*. We will look at these vectors in more detail when we study the use of yeasts and fungi in recombinant protein production (p. 260).

7.2 Cloning vectors for higher plants

Cloning vectors for higher plants were developed during the 1980s, and their use has led to the genetically modified (GM) crops that are in the headlines today. We will examine the genetic modification of crops and other plants in Chapter 15. Here we look at the cloning vectors and how they are used.

Three types of vector system have been used with varying degrees of success with higher plants:

- Vectors based on naturally occurring plasmids of *Agrobacterium*.
- Direct gene transfer using various types of plasmid DNA.
- Vectors based on plant viruses.

7.2.1 Agrobacterium tumefaciens: *nature's smallest genetic engineer*

Although no naturally occurring plasmids are known in higher plants, one bacterial plasmid – the Ti plasmid of *Agrobacterium tumefaciens* – is of great importance.

A. tumefaciens is a soil microorganism that causes crown gall disease in many species of dicotyledonous plants. Crown gall occurs when a wound on the stem allows *A. tumefaciens* bacteria to invade the plant. After infection, the bacteria cause a cancerous proliferation of the stem tissue in the region of the crown (Figure 7.9).

The ability to cause crown gall disease is associated with the presence of the Ti (tumour-inducing) plasmid within the bacterial cell. This is a large (>200 kb) plasmid that carries numerous genes involved in the infective process (Figure 7.10a). A remarkable feature of the Ti plasmid is that, after infection, part of the molecule is integrated into the plant chromosomal DNA (Figure 7.10b). This segment, called the T-DNA, is between 15 and 30 kb in size, depending on the strain. The T-DNA is maintained in a stable form in the plant cell, and is passed on to daughter cells as an integral part of the chromosomes. The most remarkable feature of the Ti plasmid is that the T-DNA contains eight or so genes that are expressed in the plant cell and are responsible for the cancerous properties of the transformed cells. These genes also direct the synthesis of unusual compounds, called opines, that the bacteria use as nutrients (Figure 7.10c). In short, *A. tumefaciens* genetically engineers the plant cell for its own purposes.

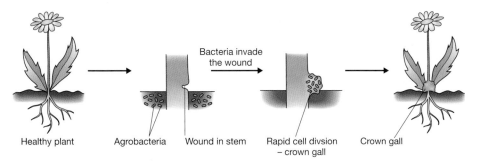

Healthy plant   Agrobacteria   Wound in stem   Rapid cell divsion   Crown gall
– crown gall

Bacteria invade the wound

Figure 7.9
Crown gall disease.

Figure 7.10

The Ti plasmid and its integration into the plant chromosomal DNA after *A. tumefaciens* infection.

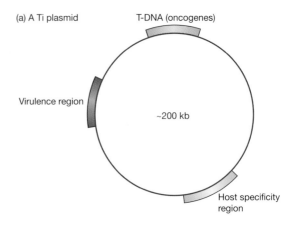

(a) A Ti plasmid

T-DNA (oncogenes)

Virulence region

~200 kb

Host specificity region

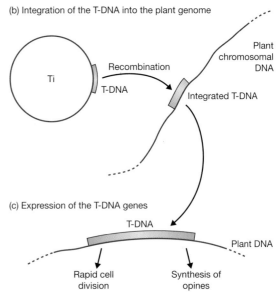

(b) Integration of the T-DNA into the plant genome

Plant chromosomal DNA

Recombination

Ti

T-DNA

Integrated T-DNA

(c) Expression of the T-DNA genes

T-DNA

Plant DNA

Rapid cell division

Synthesis of opines

Using the Ti plasmid to introduce new genes into a plant cell

It was realized very quickly that the Ti plasmid could be used to transport new genes into plant cells. All that would be necessary would be to insert the new genes into the T-DNA, after which the bacterium could do the 'hard work' of integrating them into the plant chromosomal DNA. In practice this has proved a tricky proposition, mainly because the large size of the Ti plasmid makes manipulation of the molecule very difficult.

The main problem is, of course, that it is very unlikely that there would be any unique restriction sites in a plasmid 200 kb in size, and modifying the plasmid to change all but one of the, for example, *Bam*HI sites would be impracticable. Novel strategies therefore had to be developed for inserting new DNA into the plasmid. Two are in general use:

- The binary vector strategy (Figure 7.11) is based on the observation that the T-DNA does not need to be physically attached to the rest of the Ti plasmid.

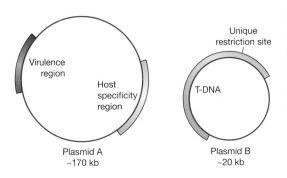

Figure 7.11

The binary vector strategy. Plasmids A and B complement each other when present together in the same *A. tumefaciens* cell. The T-DNA carried by plasmid B is transferred to the plant chromosomal DNA by proteins coded by genes carried by plasmid A.

A two-plasmid system, with the T-DNA on a relatively small molecule, and the rest of the plasmid in normal form, is just as effective at transforming plant cells. In fact, some strains of *A. tumefaciens*, and related agrobacteria, have natural binary plasmid systems. The T-DNA plasmid is small enough to have a unique restriction site and to be manipulated using standard techniques.

- The co-integration strategy (Figure 7.12) uses an entirely new plasmid, based on an *E. coli* vector, but carrying a small portion of the T-DNA. The homology between the new molecule and the Ti plasmid means that if both are present in the same *A. tumefaciens* cell, recombination can integrate the *E. coli* plasmid into the T-DNA

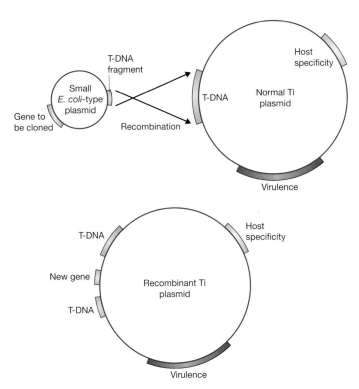

Figure 7.12

The co-integration strategy.

region. The gene to be cloned is therefore inserted into a unique restriction site on the small *E. coli* plasmid, introduced into *A. tumefaciens* cells carrying a Ti plasmid, and the natural recombination process left to integrate the new gene into the T-DNA. Infection of the plant leads to insertion of the new gene, along with the rest of the T-DNA, into the plant chromosomes.

Production of transformed plants with the Ti plasmid

If *A. tumefaciens* bacteria that contain an engineered Ti plasmid are introduced into a plant in the natural fashion, by infection of a wound in the stem, then only the cells in the resulting crown gall will possess the cloned gene (Figure 7.13a). This is obviously of little value to the biotechnologist, and an alternative method of introducing the new gene into every cell in the plant is clearly needed.

There are several solutions to this problem, the simplest being to infect not the mature plant but rather a culture of plant cells or protoplasts (p. 88) in liquid medium (Figure 7.13b). Plant cells and protoplasts whose cell walls have re-formed can be treated in the same way as microorganisms. In particular, they can be plated onto a selective medium

Figure 7.13

Transformation of plant cells by recombinant *A. tumefaciens*. (a) Infection of a wound: transformed plant cells are present only in the crown gall. (b) Transformation of a cell suspension: all the cells in the resulting plant are transformed.

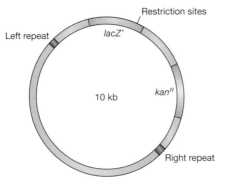

Figure 7.14

The binary Ti vector pBIN19. kan^R = kanamycin resistance gene.

in order to isolate transformants. A mature plant regenerated from transformed cells will contain the cloned gene in every cell and will pass the cloned gene to its offspring. However, the regeneration of a transformed plant can occur only if the Ti vector has been 'disarmed' so that the transformed cells do not display cancerous properties. Disarming is possible because the cancer genes, all of which lie in the T-DNA, are not needed for the infection process, infectivity being controlled mainly by the virulence region of the Ti plasmid. In fact, the only parts of the T-DNA that are involved in infection are two 25 bp repeat sequences found at the left and right borders of the region integrated into the plant DNA. Any DNA placed between these two repeat sequences will be treated as 'T-DNA' and transferred to the plant. It is therefore possible to remove all the cancer genes from the normal T-DNA, and to replace them with an entirely new set of genes, without disturbing the infection process.

Today, a number of disarmed Ti cloning vectors are available, a typical example being the binary vector pBIN19 (Figure 7.14). The left and right T-DNA borders present in this vector flank a copy of the *lacZ'* gene, containing a number of cloning sites, and a kanamycin resistance gene that functions after integration of the vector sequences into the plant chromosome. As with a yeast shuttle vector (p. 112), the initial manipulations that result in insertion of the gene to be cloned into pBIN19 are carried out in *E. coli*, the correct recombinant pBIN19 molecule then being transferred to *A. tumefaciens* and thence into the plant. Transformed plant cells are selected by plating onto agar medium containing kanamycin.

The Ri plasmid

Over the years there has also been much interest in developing plant cloning vectors based on the Ri plasmid of *Agrobacterium rhizogenes*. The Ri and Ti plasmids are very similar, the main difference being that transfer of the T-DNA from an Ri plasmid to a plant results not in a crown gall but in hairy root disease, which is typified by a massive proliferation of a highly branched root system. The possibility of growing transformed roots at high density in liquid culture has been explored by biotechnologists as a potential means of obtaining large amounts of protein from genes cloned in plants.

Limitations of cloning with Agrobacterium plasmids

Higher plants are divided into two broad categories, the monocotyledons (monocots) and the dicotyledons (dicots). Several factors have combined to make it much easier to

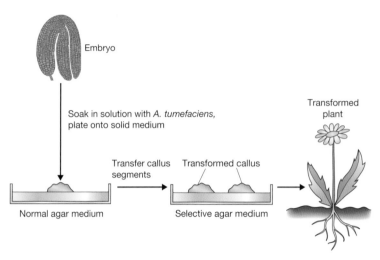

Figure 7.15

Embryo transformation.

clone genes in dicots such as tomato, tobacco, potato, peas and beans, but much more difficult to obtain the same results with monocots. This has been frustrating because monocots include wheat, barley, rice and maize, which are the most important crop plants and hence the most desirable targets for genetic engineering projects.

One difficulty stems from the fact that, in nature, *A. tumefaciens* and *A. rhizogenes* infect only dicotyledonous plants, and monocots are therefore outside of the normal host range. For some time this natural barrier was thought to be insurmountable, and that monocots were totally resistant to transformation with Ti and Ri vectors, but eventually techniques for achieving T-DNA transfer were eventually devised. But this was not the end of the story, as transformation with an *Agrobacterium* vector normally involves the regeneration of an intact plant from a transformed protoplast, cell, or callus culture. The ease with which a plant can be regenerated depends very much on the particular species involved and, once again, the most difficult plants are the monocots. With monocots, the focus has been on callus derived from embryos, as it is easier to regenerate plants from embryogenic callus than it is from callus derived from somatic tissues. An embryo is first soaked in a solution containing recombinant *A. tumefaciens*, and then placed on an agar medium that induces callus formation (Figure 7.15). As not all of the cells in the embryo become transformed, pieces of callus are cultured on a selective medium on which only the transformants can grow. Mature plants are then regenerated from the transformed callus. This approach has been used successfully with maize, barley, and several other important monocots.

7.2.2 Cloning genes in plants by direct gene transfer

Although the Ti plasmid has always been looked on as the most promising system for cloning genes in plants, limitations with the natural host range of *A. tumefaciens* have stimulated the search for alternative methods for introducing new DNA into plant cells; one such method is direct gene transfer.

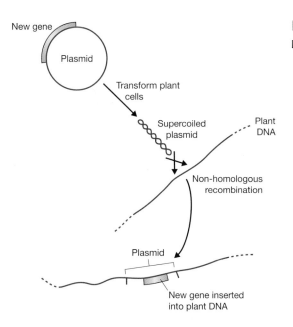

Figure 7.16

Direct gene transfer.

Direct gene transfer into the nucleus

Direct gene transfer is based on the observation, first made in 1984, that a supercoiled bacterial plasmid, although unable to replicate in a plant cell on its own, can become integrated by recombination into one of the plant chromosomes. The recombination event is poorly understood but is almost certainly distinct from the processes responsible for T-DNA integration. It is also distinct from the chromosomal integration of a yeast vector (p. 113), as there is no requirement for a region of similarity between the bacterial plasmid and the plant DNA. Integration appears to occur randomly at any position in any of the plant chromosomes (Figure 7.16).

Direct gene transfer therefore makes use of supercoiled plasmid DNA, possibly a simple bacterial plasmid, into which an appropriate selectable marker (e.g., a kanamycin resistance gene) and the gene to be cloned have been inserted. Biolistics is frequently used to introduce the plasmid DNA into plant embryos, which are then regenerated as described above for *Agrobacterium* transformation. Alternatively, if the species being engineered can be regenerated from protoplasts or single cells, then other strategies, possibly more efficient than biolistics, are possible.

One method involves resuspending protoplasts in a viscous solution of polyethylene glycol (PEG), a polymeric, negatively charged compound that is thought to precipitate DNA onto the surfaces of the protoplasts and to induce uptake by endocytosis (Figure 7.17). Electroporation (p. 89) is also sometimes used to increase transformation frequency. After treatment, the protoplasts are left for a few days in a solution that encourages regeneration of the cell walls. The cells are then spread onto selective medium to identify transformants and to provide callus cultures from which intact plants can be grown.

Transfer of genes into the chloroplast genome

If biolistics is used to introduce DNA into a plant embryo, then some particles may penetrate one or more of the chloroplasts present in the cells. Chloroplasts contain their

Figure 7.17

Direct gene transfer by precipitation of DNA onto the surfaces of protoplasts.

Plant protoplasts

own genomes, distinct from (and much shorter) than the DNA molecules in the nucleus, and under some circumstances plasmid DNA can become integrated into this chloroplast genome. Unlike the integration of DNA into nuclear chromosomes, integration into the chloroplast genome will not occur randomly. Instead, the DNA to be cloned must be flanked by sequences similar to the region of the chloroplast genome into which the DNA is to be inserted, so that insertion can take place by homologous recombination (see p. 114). Each of these flanking sequences must be 500 bp or so in length. A low level of chloroplast transformation can also be achieved after PEG-induced DNA delivery into protoplasts if the plasmid that is taken up carries these flanking sequences.

As a plant cell contains tens of chloroplasts, and probably only one per cell becomes transformed, the inserted DNA must carry a selectable marker such as the kanamycin resistance gene, and the embryos must be treated with the antibiotic for a considerable period to ensure that the transformed genomes propagate within the cell. Although this means that chloroplast transformation is a difficult method to carry out successfully, it is becoming an important adjunct to the more traditional methods for obtaining GM crops. As each cell has many chloroplasts, but only one nucleus, a gene inserted into the chloroplast genome is likely to be expressed at a higher level than one placed in the nucleus. This is particularly important when the engineered plants are to be used for the production of pharmaceutical proteins (see Chapter 14). To date, this approach has been most successful with tobacco, though chloroplast transformation has also been achieved with more useful crops such as soybean and cotton.

7.2.3 Attempts to use plant viruses as cloning vectors

Modified versions of λ and M13 bacteriophages are important cloning vectors for *E. coli* (Chapter 6). Most plants are subject to viral infection, so could viruses be used to clone genes in plants? If they could, then they would be much more convenient to use than other types of vector, because with many viruses transformation can be achieved simply by rubbing the virus nucleic acid onto the surface of a leaf. The natural infection process then spreads the virus throughout the plant.

The potential of plant viruses as cloning vectors has been explored for several years, but without great success. One problem here is that the vast majority of plant viruses have genomes not of DNA but of RNA. RNA viruses are less useful as potential cloning vectors because manipulations with RNA are more difficult to carry out. Only two classes of DNA virus are known to infect higher plants – the caulimoviruses and geminiviruses – and neither is ideally suited to gene cloning.

Caulimovirus vectors

Although one of the first successful plant genetic engineering experiments, performed back in 1984, used a caulimovirus vector to clone a new gene into turnip plants, two general difficulties with these viruses have limited their usefulness.

The first problem is that the total size of a caulimovirus genome is, like that of λ, constrained by the need to package it into its protein coat. Even after the deletion of non-essential sections of the virus genome the capacity for carrying inserted DNA is still very limited. Recent studies have shown that it might be possible to circumvent this problem by adopting a helper virus strategy, similar to that used with phagemids (p. 108). In this strategy, the cloning vector is a cauliflower mosaic virus (CaMV) genome that lacks several of the essential genes, which means that it can carry a large DNA insert but cannot, by itself, direct infection. Plants are inoculated with the vector DNA along with a normal CaMV genome. The normal viral genome then provides the genes needed for the cloning vector to be packaged into virus proteins and spread through the plant.

Although this approach has considerable potential it does not solve the second problem, which is the extremely narrow host range of caulimoviruses. This restricts cloning experiments to just a few plants, mainly brassicas such as turnips, cabbages and cauliflowers. Caulimoviruses have, however, been important in genetic engineering as the source of highly active promoters that function in all plants and that are used to obtain expression of genes introduced by Ti plasmid cloning or direct gene transfer.

Geminivirus vectors

The geminiviruses are particularly interesting because their natural hosts include plants such as maize and wheat, and they could therefore serve as potential vectors for these and other monocots. But geminiviruses have presented their own set of difficulties, one problem being that during the infection cycle the genomes of some geminiviruses undergo rearrangements and deletions, which would scramble up any additional DNA that has been inserted – an obvious disadvantage for a cloning vector. Investigations performed over the years have addressed these problems and geminiviruses are now beginning to find some specialist applications in plant gene cloning. One such application is virus-induced gene silencing (VIGS), a technique used to investigate the functions of individual plant genes. This method exploits one of the natural defence mechanisms that plants use to protect themselves against viral attack. This method, termed RNA silencing, results in the degradation of viral mRNAs. If one of the viral RNAs is transcribed from a cloned gene contained within a geminivirus genome, then not only are the viral transcripts degraded but also the cellular mRNAs derived from the plant's copy of the gene (Figure 7.18). Consequently, the plant gene becomes silenced and the effect of its inactivation on the phenotype of the plant can be studied.

7.3 Cloning vectors for animals

During recent years, considerable effort has been expended into the development of vector systems for cloning genes in animal cells. These vectors are required in biotechnology for the synthesis of recombinant protein from genes that are not expressed correctly when cloned in *E. coli* or yeast (see Chapter 13), and methods for cloning in humans are being sought by clinical molecular biologists attempting to devise techniques

Figure 7.18

The use of a geminivirus vector to silence a plant gene via virus-induced gene silencing.

for gene therapy (p. 286), in which a disease is treated by the introduction of a cloned gene into a patient.

The clinical aspect of this has led to most attention being directed at cloning systems for mammals, although important progress has also been made with insects. Cloning in insects is interesting because it makes use of a novel type of vector that we have not met so far. We will therefore examine insect vectors before concluding the chapter with an overview of the cloning methods used with mammals.

7.3.1 Cloning vectors for insects

The fruit fly, *Drosophila melanogaster*, has been – and still is – one of the most important model organisms used by biologists. Its potential was first recognized by the famous geneticist Thomas Hunt Morgan, who in 1910 started to carry out genetic crosses between fruit flies with different eye colours, body shapes, and other inherited characteristics. These experiments led to the techniques still used today for gene mapping in insects and other animals. More recently, the discovery that the homeotic selector genes of *Drosophila* – the genes that control the overall body plan of the fly – are closely related to equivalent genes in mammals, has led to *D. melanogaster* being used as a model for the study of human developmental processes. The importance of the fruit fly in modern biology makes it imperative that vectors for cloning genes in this organism are available.

P elements as cloning vectors for Drosophila

The development of cloning vectors for *Drosophila* has taken a different route to that followed with bacteria, yeast, plants and mammals. No plasmids are known in *Drosophila*, and although fruit flies are, like all organisms, susceptible to infection with viruses, these have not been used as the basis for cloning vectors. Instead, cloning in *Drosophila* makes use of a transposon called the P element.

Transposons, which are common in all types of organism, are short pieces of DNA (usually less than 10 kb in length) that can move from one position to another in the chromosomes of a cell. P elements, which are one of several types of transposon in *Drosophila*, are 2.9 kb in length and contain three genes flanked by short, inverted repeat sequences at either end of the element (Figure 7.19a). The genes code for

(a) The structure of a P element

(b) P element transposition

(c) The structure of a P element cloning vector

Figure 7.19

Cloning in *Drosophila* with a P element vector. (a) The structure of a P element. (b) Transposition of a P element from a plasmid to a fly chromosome. (c) The structure of a P element cloning vector. The left-hand P element contains a cloning site (R) that disrupts its transposase gene. The right-hand P element has an intact transposase gene but cannot itself transpose because it is 'wings-clipped' – it lacks terminal inverted repeats.

transposase, the enzyme that carries out the transposition process, and the inverted repeats form the recognition sequences that enable the enzyme to identify the two ends of the inserted transposon.

As well as moving from one site to another within a single chromosome, P elements can also jump between chromosomes, or between a plasmid carrying a P element and one of the fly's chromosomes (Figure 7.19b). The latter is the key to the use of P elements as cloning vectors. The vector is a plasmid that carries two P elements, one of which contains the insertion site for the DNA that will be cloned. Insertion of the new DNA into this P element results in disruption of its transposase gene, so this element is inactive. The second P element carried by the plasmid is therefore one that has an intact version of the transposase gene. Ideally, this second element should not itself be transferred to the *Drosophila* chromosomes, so it has its 'wings clipped', which means that its inverted repeats are removed so that the transposase does not recognize it as being a real P element (Figure 7.19c). Once the gene to be cloned has been inserted into the vector, the plasmid DNA is microinjected into fruit fly embryos. The transposase from the wings-clipped P element directs transfer of the engineered P element into one of the fruit fly chromosomes. If this happens within a germline nucleus, then the adult fly that develops from the embryo will carry copies of the cloned gene in all its cells. P element cloning was first developed during the 1980s and has made a number of important contributions to *Drosophila* genetics.

Cloning vectors based on insect viruses

Although virus vectors have not been developed for cloning genes in *Drosophila*, one type of virus – the baculovirus – has played an important role in gene cloning with other insects. The main use of baculovirus vectors is in the production of recombinant protein, and we will return to them when we consider this topic in Chapter 13.

7.3.2 Cloning in mammals

At present, gene cloning in mammals is carried out for one of three reasons:

- To achieve a gene knockout, which is an important technique used to help determine the function of an unidentified gene (p. 233). These experiments are usually carried out with rodents, such as mice.
- For the production of recombinant protein in a mammalian cell culture, and in the related technique of pharming. The latter method involves the genetic engineering of a farm animal so that it synthesizes an important protein such as a pharmaceutical, often in its milk (p. 263).
- In gene therapy, in which human cells are engineered in order to treat a disease (p. 286).

Viruses as cloning vectors for mammals

For many years it was thought that viruses would prove to be the key to cloning in mammals, but this expectation has only partially been realized. The first cloning experiment involving mammalian cells was carried out in 1979 with a vector based on simian virus 40 (SV40). This virus is capable of infecting several mammalian species, following a lytic cycle in some hosts and a lysogenic cycle in others. The genome is 5.2 kb in size (Figure 7.20a) and contains two sets of genes, the 'early' genes, which are expressed early in the infection cycle and code for proteins involved in viral DNA replication, and the 'late' genes, which code for the viral capsid proteins. SV40 suffers from the same problem as λ and the plant caulimoviruses, in that packaging constraints limit the amount of new DNA that can be inserted into the genome. Cloning with SV40 therefore involves replacing one or more of the existing genes with the DNA to be cloned. In the original experiment, a segment of the late gene region was replaced (Figure 7.20b), but early gene replacement is also an option.

Figure 7.20

SV40 and an example of its use as a cloning vector. To clone the rabbit β-globin gene, the *Hind*III to *Bam*HI restriction fragment was deleted (resulting in SVGT-5) and replaced with the rabbit gene.

During the years since 1979, a number of other types of virus have been used to clone genes in mammals, including:

- **Adenoviruses**, which enable DNA fragments of up to 8 kb to be cloned, longer than is possible with an SV40 vector, though adenoviruses are more difficult to handle because their genomes are bigger.

- **Papillomaviruses**, which also have a relatively high capacity for inserted DNA. Bovine papillomavirus (BPV), which causes warts on cattle, is particularly attractive because it has an unusual infection cycle in mouse cells, taking the form of a multicopy plasmid with about 100 molecules present per cell. It does not cause death of the mouse cell, and BPV molecules are passed to daughter cells on cell division, giving rise to a permanently transformed cell line. Shuttle vectors consisting of BPV and *E. coli* sequences, and capable of replication in both mouse and bacterial cells, have been used for the production of recombinant proteins in mouse cell lines.

- **Adeno-associated virus (AAV)**, which is unrelated to adenovirus but often found in the same infected tissues, mainly because AAV utilizes some of the proteins synthesized by adenovirus in order to complete its replication cycle. In the absence of this helper virus, the AAV genome inserts into its host's DNA. With most integrative viruses this is a random event, but AAV has the unusual property of always inserting at the same position, within human chromosome 19. Knowing exactly where the cloned gene will be in the host genome is important if the outcome of the cloning experiment must be checked rigorously, as is the case for applications such as gene therapy. AAV vectors are therefore considered to have major potential in this area.

- **Retroviruses**, which are the most commonly used vectors for gene therapy. Although they insert at random positions, the resulting integrants are very stable, which means that the therapeutic effects of the cloned gene will persist for some time. We will return to gene therapy in Chapter 14.

Gene cloning without a vector

One of the reasons why virus vectors have not become widespread in mammalian gene cloning is because it was discovered in the early 1990s that the most effective way of transferring new genes into mammalian cells is by microinjection. Although a difficult procedure to carry out, microinjection of bacterial plasmids or linear DNA copies of genes into mammalian nuclei results in the DNA being inserted into the chromosomes, possibly as multiple copies in a tandem, head-to-tail arrangement (Figure 7.21). This procedure is generally looked on as being more satisfactory than the use of a viral vector, because it avoids the possibility that viral DNA will infect the cells and cause defects of one type or another.

Microinjection of DNA forms the basis to creation of a transgenic animal, one that contains a cloned gene in all of its cells. A transgenic mouse can be generated by microinjection of a fertilized egg cell which is subsequently cultured *in vitro* for several cell

Multiple copies of the cloned DNA

Chromosomal DNA

Figure 7.21

Multiple copies of cloned DNA molecules inserted as a tandem array in a chromosomal DNA molecule.

divisions and then implanted into a foster mother. Alternatively, an embryonic stem (ES) cell can be used. These are obtained from within an early embryo and, unlike most mammalian cells, are totipotent, which means that their developmental pattern is not preset and cells descended from them can form many different structures in the adult mouse. After microinjection, the ES cell is placed back in an embryo which is then implanted into the foster mother. The resulting mouse is a chimera, comprising a mixture of engineered and non-engineered cells, because the embryo that receives the ES cell also contains a number of ordinary cells that contribute, along with the ES cell, to the make-up of the adult mouse. Non-chimeric mice, which contain the cloned gene in all their cells, are obtained by allowing the chimera to reproduce, as some of the offspring will be derived from egg cells that contain the cloned gene.

Further reading

Brisson, N., Paszkowski, J., Penswick, J.R., *et al.* (1984) Expression of a bacterial gene in plants by using a viral vector. *Nature*, **310**, 511–114. [The first cloning experiment with a caulimovirus.]

Broach, J.R. (1982) The yeast 2 μm circle. *Cell*, **28**, 203–204.

Burke, D.T., Carle, G.F., and Olson, M.V. (1987) Cloning of large segments of exogenous DNA into yeast by means of artificial chromosome vectors. *Science*, **236**, 806–812.

Carrillo-Tripp, J., Shimada-Beltrán, H., and Rivera-Bustamante, R. (2006) Use of geminiviral vectors for functional genomics. *Current Opinion in Plant Biology*, **9**, 209–215. [Virus-induced gene silencing.]

Chilton, M.D. (1983) A vector for introducing new genes into plants. *Scientific American*, **248** (June), 50–59. [The Ti plasmid.]

Clarke, J.L., Daniell, H., and Nugent, J.M. (2011) Chloroplast biotechnology, genomics and evolution: current status, challenges and future directions. *Plant Molecular Biology*, **76**, 207–209. [Information on chloroplast transformation.]

Colosimo, A., Goncz, K.K., Holmes, A.R., *et al.* (2000) Transfer and expression of foreign genes in mammalian cells. *Biotechniques*, **29**, 314–321.

Crystal, R.G. (2014) Adenovirus: the first effective *in vivo* gene delivery vector. *Human Gene Therapy*, **25**, 3–11.

Evans, M. (2011) Discovering pluripotency: 30 years of mouse embryonic stem cells. *Nature Reviews Molecular Cell Biology*, **12**, 680–686.

Guillon, S., Trémouillaux-Guiller, J., Pati, P.K., Rideau, M., and Gantet, P. (2006) Hairy root research: recent scenario and exciting prospects. *Current Opinion in Plant Biology*, **9**, 341–346. [Applications of the Ri plasmid.]

Hamer, D.H. and Leder, P. (1979) Expression of the chromosomal mouse β-maj-globin gene cloned in SV40. *Nature*, **281**, 35–40.

Lee, L.-Y. and Gelvin, S.B. (2008) T-DNA binary vectors and systems. *Plant Physiology*, **146**, 325–332.

Päcurar, D.I., Thordal-Christensen, H., Päcurer, M.L., *et al.* (2011) *Agrobacterium tumefaciens*: from crown gall tumors to genetic transformation. *Physiological and Molecular Plant Pathology*, **76**, 76–81.

Parent, S.A., Fenimore, C.M., and Bostian, K.A. (1985) Vector systems for the expression, analysis and cloning of DNA sequences in *S. cerevisiae*. *Yeast*, **1**, 83–138. [Details of different yeast cloning vectors.]

Paszkowski, J., Shillito, R.D., Saul, M., *et al*. (1984) Direct gene transfer to plants. *EMBO Journal*, **3**, 2717–2722.

Rubin, G.M. and Spradling, A.C. (1982) Genetic transformation of *Drosophila* with transposable element vectors. *Science*, **218**, 348–353. [Cloning with P elements.]

Viaplana, R., Turner, D.S., and Covey, S.N. (2001) Transient expression of a GUS reporter gene from cauliflower mosaic virus replacement vectors in the presence and absence of helper virus. *Journal of General Virology*, **82**, 59–65. [The helper virus approach to cloning with CaMV.]

Chapter 8

How to Obtain a Clone of a Specific Gene

Chapter contents

In the preceding chapters we have examined the basic methodology used to clone genes, and surveyed the range of vector types that are used with bacteria, yeast, plants and animals. Now, we must look at the methods available for obtaining a clone of an individual, specified gene. This is the critical test of a gene cloning experiment, where success or failure often depends on whether or not a strategy can be devised by which clones of the desired gene can be selected directly, or alternatively, distinguished from other recombinants. Once this problem has been resolved, and a clone has been obtained, the molecular biologist is able to make use of a wide variety of different techniques that will extract information about the gene. The most important of these will be described in Chapters 10 and 11.

8.1 The problem of selection

The problem faced by the molecular biologist wishing to obtain a clone of a single, specified gene was illustrated in Figure 1.4. Even the simplest organisms, such as *E. coli*, contain several thousand genes, and a restriction digest of total cell DNA produces not only the fragment carrying the desired gene but also many other fragments carrying all the other genes (Figure 8.1a). During the ligation reaction there is no selection for an individual fragment, with numerous different recombinant DNA molecules being produced, all of which contain different pieces of DNA (Figure 8.1b). Consequently, a

Gene Cloning and DNA Analysis: An Introduction, Seventh Edition. T.A. Brown.
© 2016 John Wiley & Sons, Ltd. Published 2016 by John Wiley & Sons, Ltd.
Companion Website: www.wiley.com/go/brown/cloning

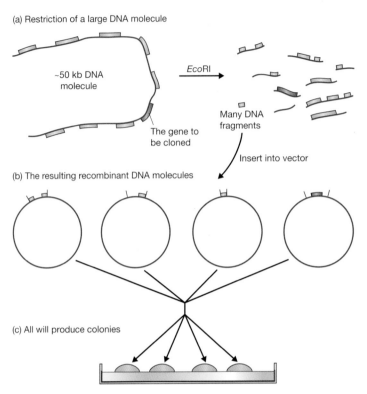

(a) Restriction of a large DNA molecule

~50 kb DNA molecule

EcoRI

Many DNA fragments

The gene to be cloned

Insert into vector

(b) The resulting recombinant DNA molecules

(c) All will produce colonies

Figure 8.1
The problem of selection.

variety of recombinant clones are obtained after transformation and plating out (Figure 8.1c). Somehow, the correct one must be identified.

8.1.1 There are two basic strategies for obtaining the clone you want

Although there are many different procedures by which the desired clone can be obtained, all are variations on two basic themes:

- Direct selection for the desired gene (Figure 8.2a), which means that the cloning experiment is designed in such a way that the only clones that are obtained are those of the required gene. Almost invariably, selection occurs at the plating-out stage.
- Identification of the clone from a gene library (Figure 8.2b). This entails an initial shotgun cloning experiment to produce a clone library representing all or most of the genes present in the cell, followed by analysis of the individual clones to identify the correct one.

In general terms, direct selection is the preferred method, as it is quick and usually unambiguous. Unfortunately, as we shall see, it is not applicable to all genes and techniques for clone identification are therefore very important.

(a) Direct selection

Only the correct
recombinant can survive

Figure 8.2

The basic strategies that can be used to obtain a particular clone. (a) Direct selection. (b) Identification of the desired recombinant from a clone library.

(b) Clone identification

A clone
library

Correct clone

8.2 Direct selection

To be able to select for a cloned gene it is necessary to plate the transformants onto an agar medium on which only the desired recombinants, and no others, can grow. The only colonies that are obtained will therefore be ones that comprise cells containing the desired recombinant DNA molecule.

The simplest example of direct selection occurs when the desired gene specifies resistance to an antibiotic. As an example, we will consider an experiment to clone the gene for kanamycin resistance from plasmid R6-5. This plasmid carries genes for resistances to four antibiotics: kanamycin, chloramphenicol, streptomycin, and sulphonamide. The kanamycin resistance gene lies within one of the 13 fragments produced when R6-5 is digested with *Eco*RI (Figure 8.3a).

To clone this gene, the *Eco*RI fragments of R6-5 could be inserted into the *Eco*RI site of a vector such as pBR322. The ligated mix will comprise many copies of 13 different recombinant DNA molecules, one set of which carries the gene for kanamycin resistance (Figure 8.3b).

Insertional inactivation cannot be used to select recombinants when the *Eco*RI site of pBR322 is used. This is because this site does not lie in either the ampicillin or the tetracycline resistance genes of this plasmid (see Figure 6.1). But this is immaterial for cloning the kanamycin resistance gene because in this case the cloned gene can be used as the selectable marker. Transformants are plated onto kanamycin agar, on which the only cells able to survive and produce colonies are those recombinants that contain the cloned kanamycin resistance gene (Figure 8.3c).

Figure 8.3

Direct selection for the cloned R6-5 kanamycin resistance (*kan^R*) gene.

(a) Plasmid R6-5

kan^R gene

EcoRI sites

EcoRI

13 different fragments

Ligate into the *EcoRI* site of pBR322

(b) Ligation gives 13 different recombinant DNA molecules

✗ ✗ ✗

Transform *E. coli*, plate out

(c) Only one allows growth on kanamycin agar

Medium contains 50 µg ml⁻¹ kanamycin

Only the recombinant carrying the *kan^R* gene can survive

8.2.1　Marker rescue extends the scope of direct selection

Direct selection would be very limited indeed if it could be used only for cloning antibiotic resistance genes. Fortunately, the technique can be extended by making use of mutant strains of *E. coli* as the hosts for transformation.

As an example, consider an experiment to clone the gene *trpA* from *E. coli*. This gene codes for the enzyme tryptophan synthase, which is involved in biosynthesis of the essential amino acid tryptophan. A mutant strain of *E. coli* that has a non-functional *trpA* gene is called *trpA⁻*, and is able to survive only if tryptophan is added to the growth medium. *E. coli trpA⁻* is therefore another example of an auxotroph (p. 112).

The *E. coli trpA⁻* auxotroph can be used to clone the correct version of the *trpA* gene. Total DNA is first purified from a normal (wild-type) strain of the bacterium. Digestion with a restriction endonuclease, followed by ligation into a vector, produces numerous recombinant DNA molecules, one of which may (with luck) carry an intact copy of the *trpA* gene (Figure 8.4a). This is, of course, the functional gene, as it has been obtained from the wild-type strain.

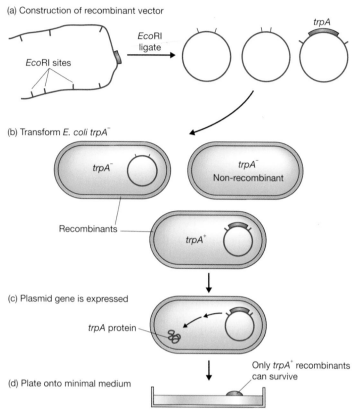

(a) Construction of recombinant vector

EcoRI sites

EcoRI
ligate

trpA

(b) Transform E. coli trpA⁻

trpA⁻

trpA⁻
Non-recombinant

Recombinants

trpA⁺

(c) Plasmid gene is expressed

trpA protein

Only trpA⁺ recombinants
can survive

(d) Plate onto minimal medium

Figure 8.4

Direct selection for the *trpA* gene cloned in a *trpA⁻* strain of *E. coli*.

The ligation mixture is now used to transform the auxotrophic *E. coli trpA⁻* cells (Figure 8.4b). The vast majority of the resulting transformants will be auxotrophic, but a few will now have the plasmid-borne copy of the correct *trpA* gene. These recombinants are non-auxotrophic; they no longer require tryptophan as the cloned gene is able to direct the production of tryptophan synthase (Figure 8.4c). Direct selection is therefore performed by plating transformants onto minimal medium, which lacks any added supplements, and in particular has no tryptophan (Figure 8.4d). As auxotrophs cannot grow on minimal medium, the only colonies to appear are recombinants that contain the cloned *trpA* gene.

8.2.2 The scope and limitations of marker rescue

Although marker rescue can be used to obtain clones of many genes, the technique is subject to two limitations:

- A mutant strain must be available for the gene in question.
- A medium on which only the wild-type can survive is needed.

Marker rescue is applicable for most genes that code for biosynthetic enzymes, as clones of these genes can be selected on minimal medium in the manner described for *trpA*. The technique is not limited to *E. coli* or bacteria – auxotrophic strains of yeast and filamentous fungi are also available, and marker rescue has been used to select genes cloned into these organisms.

In addition, *E. coli* auxotrophs can be used as hosts for the selection of some genes from other organisms. Often, there is sufficient similarity between equivalent enzymes from different bacteria, or even from yeast, for the foreign enzyme to function in *E. coli*, so that the cloned gene is able to transform the host to wild-type.

8.3 Identification of a clone from a gene library

Although marker rescue is a powerful technique, it is not all-embracing and there are many important genes that cannot be selected using this method. Many bacterial mutants are not auxotrophs, so the mutant and wild-type strains cannot be distinguished by plating onto minimal or any other special medium. In addition, neither marker rescue nor any other direct selection method is of much use in providing bacterial clones of genes from animals or plants, as in these cases the differences are usually so great that the foreign enzymes do not function in the bacterial cell.

An alternative strategy must therefore be considered. This is where a large number of different clones are obtained and the desired one identified in some way.

8.3.1 Gene libraries

Before examining the methods used to identify individual clones, the library itself must be considered. A genomic library (p. 105) is a collection of clones sufficient in number to be likely to contain every single gene present in a particular organism. Genomic libraries are prepared by purifying total cell DNA, and then making a partial restriction digest, resulting in fragments that can be cloned into a suitable vector (Figure 8.5), usually a λ replacement vector, a cosmid, or possibly a yeast artificial chromosome (YAC), bacterial artificial chromosome (BAC) or P1 vector.

Not all genes are expressed at the same time

For bacteria, yeast and fungi, the number of clones needed for a complete genomic library is not so large as to be unmanageable (see Table 6.1). For plants and animals, a complete library contains so many different clones that identification of the desired one may prove a mammoth task. With these multicellular organisms a second type of library, specific not to the whole organism but rather to a particular cell type, might be more useful. Each cell contains the same complement of genes, but in different cell types different sets of genes are switched on, while others are silent (Figure 8.6). Only those genes that are being expressed are transcribed into messenger RNA (mRNA), so if mRNA is used as the starting material then the resulting clones would comprise only a selection of the total number of genes in the cell.

A cloning method that uses mRNA would be particularly useful if the desired gene is expressed at a high rate in an individual cell type. For example, the gene for gliadin (one of the nutritionally important proteins present in wheat) is expressed at a very high level in the cells of developing wheat seeds. In these cells, over 30% of the total mRNA specifies gliadin. Clearly, if we could clone the mRNA from wheat seeds we would obtain a large number of clones specific for gliadin.

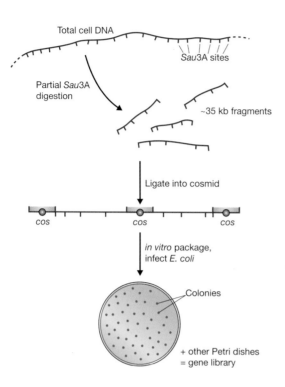

Figure 8.5

Preparation of a gene library in a cosmid vector.

Figure 8.6

Different genes are expressed in different types of cell.

mRNA can be cloned as complementary DNA

Messenger RNA cannot itself be ligated into a cloning vector, but it can be converted into DNA by complementary DNA (cDNA) synthesis. The key to this method is the enzyme reverse transcriptase (p. 52), which synthesizes a DNA polynucleotide complementary to an existing RNA strand (Figure 8.7a). Once the cDNA strand has been synthesized, the RNA member of the hybrid molecule can be partially degraded by treating with ribonuclease HI (Figure 8.7b). The remaining RNA fragments then serve as primers

Figure 8.7

One possible scheme for cDNA cloning. First strand synthesis is primed by oligo(dT), a short oligonucleotide made up entirely of deoxythymidine nucleotides, which base pairs to the poly(A) tail present at the 3′ end of a eukaryotic mRNA.

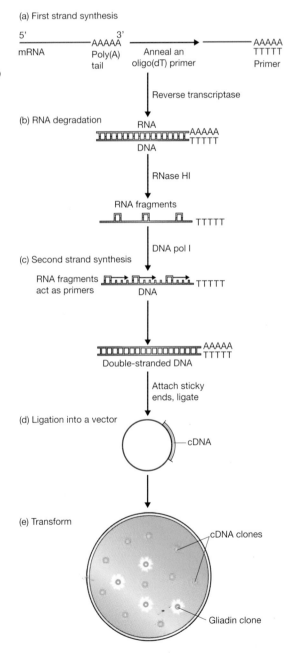

(p. 51) for DNA polymerase I, which synthesizes the second cDNA strand (Figure 8.7c). The result is a double-stranded DNA fragment that can be ligated into a vector and cloned (Figure 8.7d).

The resulting cDNA clones are representative of the mRNA present in the original preparation. In the case of mRNA prepared from wheat seeds, the cDNA library would contain a large proportion of clones representing gliadin mRNA (Figure 8.7e). Other clones will also be present, but locating the cloned gliadin cDNA is a much easier process than identifying the equivalent gene from a complete wheat genomic library.

8.4 Methods for clone identification

Once a suitable library has been prepared, a number of procedures can be employed to attempt identification of the desired clone. Although a few of these procedures are based on detecting the translation product of the cloned gene, it is usually easier to identify directly the correct recombinant DNA molecule. This can be achieved by the important technique of hybridization probing.

8.4.1 Complementary nucleic acid strands hybridize to each other

Any two single-stranded nucleic acid molecules have the potential to form base pairs with one another. With most pairs of molecules, the resulting hybrid structures are unstable as only a small number of individual interstrand bonds are formed (Figure 8.8a). If the polynucleotides are complementary, extensive base pairing can occur to form a stable double-stranded molecule (Figure 8.8b). Not only can this occur between single-stranded DNA molecules to form the DNA double helix, but also between a pair of single-stranded RNA molecules or between combinations of one DNA strand and one RNA strand (Figure 8.8c).

Nucleic acid hybridization can be used to identify a particular recombinant clone if a DNA or RNA probe, complementary to the desired gene, is available. The exact nature of the probe will be discussed later in this chapter. First we must consider the technique itself.

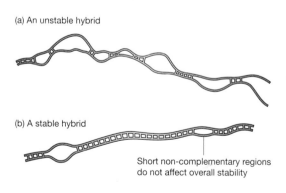

(a) An unstable hybrid

(b) A stable hybrid

Short non-complementary regions do not affect overall stability

(c) A DNA–RNA hybrid

DNA
RNA

Figure 8.8

Nucleic acid hybridization. (a) An unstable hybrid molecule formed between two non-homologous DNA strands. (b) A stable hybrid formed between two complementary strands. (c) A DNA–RNA hybrid, such as may be formed between a gene and its transcript.

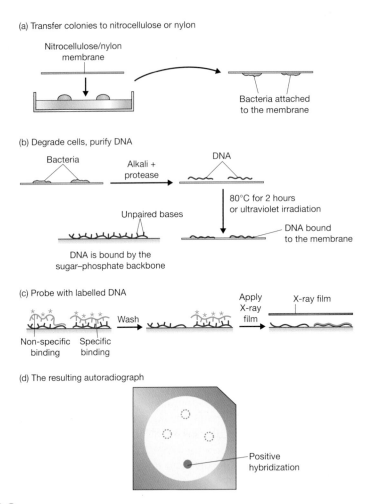

(a) Transfer colonies to nitrocellulose or nylon

Nitrocellulose/nylon membrane

Bacteria attached to the membrane

(b) Degrade cells, purify DNA

Bacteria

Alkali + protease

DNA

80°C for 2 hours or ultraviolet irradiation

DNA bound to the membrane

Unpaired bases

DNA is bound by the sugar–phosphate backbone

(c) Probe with labelled DNA

Non-specific binding   Specific binding

Wash

Apply X-ray film

X-ray film

(d) The resulting autoradiograph

Positive hybridization

Figure 8.9

Colony hybridization probing. In this example, the probe is labelled with a radioactive marker and hybridization detected by autoradiography, but other types of label and detection system can also be used.

8.4.2 Colony and plaque hybridization probing

Hybridization probing can be used to identify recombinant DNA molecules contained in either bacterial colonies or bacteriophage plaques. First, the colonies or plaques are transferred to a nitrocellulose or nylon membrane (Figure 8.9a), and then treated to remove all contaminating material, leaving just DNA (Figure 8.9b). Usually, this treatment also results in denaturation of the DNA molecules, so that the hydrogen bonds between individual strands in the double helix are broken. These single-stranded molecules can then be bound tightly to the membrane by heating for a short period at 80 °C if a nitrocellulose membrane is being used, or with a nylon membrane by ultraviolet (UV) irradiation. The molecules become attached to the membrane through their sugar–phosphate backbones, which means that the bases are free to pair with complementary nucleic acid molecules.

NH_2

Radioactive ^{32}P

Figure 8.10

The structure of α-^{32}P-deoxyadenosine triphosphate ([α-^{32}P]dATP).

The probe must now be labelled with a radioactive or other type of marker, denatured by heating, and applied to the membrane in a solution of chemicals that promote nucleic acid hybridization (Figure 8.9c). After a period to allow hybridization to take place, the filter is washed to remove any unbound probe, dried, and the label detected in order to identify the colonies or plaques to which the probe has become bound (Figure 8.9d).

Labelling with a radioactive marker

A DNA molecule is often labelled by incorporating nucleotides that carry a radioactive isotope of phosphorus, ^{32}P (Figure 8.10). Several methods are available:

- **Nick translation.** Most purified samples of DNA contain some nicked molecules, however carefully the preparation has been carried out, which means that DNA polymerase I is able to attach to the DNA and catalyse a strand replacement reaction (Figure 8.11a). This reaction requires a supply of nucleotides; if one of these nucleotides is radioactively labelled the DNA molecule will itself become labelled. Nick translation can be used to label any DNA molecule, but might under some circumstances also cause DNA cleavage.

- **End filling** is a gentler method than nick translation and rarely causes breakage of the DNA, but unfortunately it can only be used to label DNA molecules that have sticky ends. The enzyme used is the Klenow fragment (p. 51), which 'fills in' a sticky end by synthesizing the complementary strand (Figure 8.11b)). As with nick

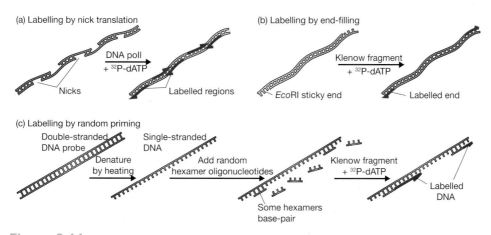

(a) Labelling by nick translation

DNA poll
+ ^{32}P-dATP

Nicks   Labelled regions

(b) Labelling by end-filling

Klenow fragment
+ ^{32}P-dATP

*Eco*RI sticky end   Labelled end

(c) Labelling by random priming

Double-stranded DNA probe   Single-stranded DNA

Denature by heating   Add random hexamer oligonucleotides

Klenow fragment
+ ^{32}P-dATP

Some hexamers base-pair

Labelled DNA

Figure 8.11

Methods for labelling DNA.

translation, if the end-filling reaction is carried out in the presence of labelled nucleotides, the DNA becomes labelled.

- Random priming results in a probe with higher activity and is therefore able to detect smaller amounts of membrane-bound DNA. The denatured DNA is mixed with a set of hexameric oligonucleotides of random sequence. By chance, these random hexamers will contain a few molecules that will base pair with the probe and prime new DNA synthesis. The Klenow fragment is used as this enzyme lacks the nuclease activity of DNA polymerase I (p. 51) and so only fills in the gaps between adjacent primers (Figure 8.11c). Labelled nucleotides are incorporated into the new DNA that is synthesized.

After hybridization, the location of the bound probe is detected by autoradiography. For this, a sheet of X-ray-sensitive photographic film is placed over the membrane. The radioactive DNA then exposes the film, which is developed to reveal the positions of the colonies or plaques to which the probe has hybridized (see Figure 8.9d).

Non-radioactive labelling

Today, radioactive labelling methods are beginning to fall out of favour, partly because of health hazards to the research workers and partly because of problems associated with the disposal of radioactive waste. Alternatively, the hybridization probe can be labelled in a non-radioactive manner. A number of methods have been developed for this purpose, two of which are illustrated in Figure 8.12. The first method utilizes deoxyuridine triphosphate (dUTP) nucleotides which have been modified by reaction with biotin, an organic molecule that has a high affinity for a protein called avidin. After hybridization, the positions of the bound biotinylated probe can be determined by washing with avidin coupled to a fluorescent marker (Figure 8.12a). This method is as sensitive as radioactive probing and is becoming increasingly popular.

The same is true for a second method for non-radioactive hybridization probing, in which the probe DNA is first complexed with the enzyme horseradish peroxidase. The probe DNA is then detected through the enzyme's ability to degrade luminol, with the emission of chemiluminescence (Figure 8.12b). The signal can be recorded on normal photographic film in a manner analogous to autoradiography.

8.4.3 Examples of the practical use of hybridization probing

Clearly, the success of colony or plaque hybridization as a means of identifying a particular recombinant clone depends on the availability of a DNA molecule that can be used as a probe. This probe must share at least a part of the sequence of the cloned gene. If the gene itself is not available (which presumably is the case if the aim of the experiment is to provide a clone of it), then what can be used as the probe?

In practice, the nature of the probe is determined by the information available about the desired gene. We will consider three possibilities:

- Where the desired gene is expressed at a high level in a cell type from which a cDNA clone library has been prepared.
- Where the amino acid sequence of the protein coded by the gene is completely or partially known.
- Where the gene is a member of a family of related genes.

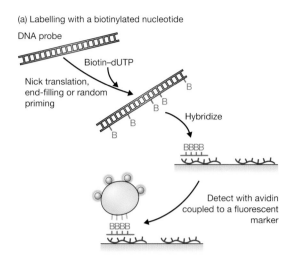

(a) Labelling with a biotinylated nucleotide

Figure 8.12

Two methods for the non-radioactive labelling of DNA probes.

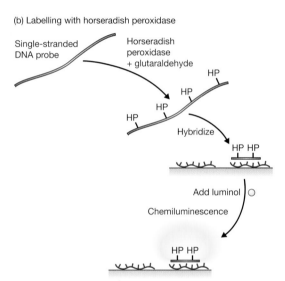

(b) Labelling with horseradish peroxidase

Abundancy probing to analyse a cDNA library

As described earlier in this chapter, a cDNA library is often prepared in order to obtain a clone of a gene expressed at a relatively high level in a particular cell type. In the example of a cDNA library from developing wheat seeds, a large proportion of the clones are copies of the mRNA transcripts of the gliadin gene (see Figure 8.7e).

Identification of the gliadin clones is simply a case of using individual cDNA clones from the library to probe all the other members of the library (Figure 8.13). A clone is selected at random and the recombinant DNA molecule purified, labelled, and used to probe the remaining clones. This is repeated with different clones as probes until one that hybridizes to a large proportion of the library is obtained. This abundant cDNA is considered a possible gliadin clone and analysed in greater detail (e.g., by DNA sequencing and isolation of the translation product) to confirm the identification.

Figure 8.13

Probing within a library to identify an abundant clone.

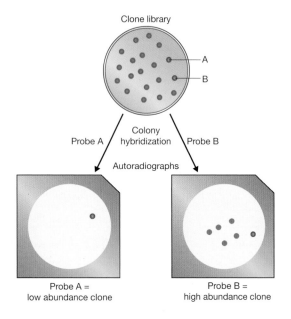

Probe A =
low abundance clone

Probe B =
high abundance clone

Oligonucleotide probes for genes whose translation products have been characterized

Often, the gene to be cloned codes for a protein that has already been studied in some detail. In particular, the amino acid sequence of the protein might have been determined. If the amino acid sequence is known, then it is possible to use the genetic code to predict the nucleotide sequence of the relevant gene. This prediction is always an approximation, as only methionine and tryptophan can be assigned unambiguously to triplet codons, all other amino acids being coded by at least two codons each. Nevertheless, in most cases, the different codons for an individual amino acid are related. Alanine, for example, is coded by GCA, GCC, GCG, and GCT, so two out of the three nucleotides of the triplet coding for alanine can be predicted with certainty.

As an example to clarify how these predictions are made, consider cytochrome *c*, a protein that plays an important role in the respiratory chain of all aerobic organisms. The cytochrome *c* protein from yeast was sequenced in 1963, with the result shown in Figure 8.14. This sequence contains a segment, starting at amino acid 59, that runs Trp–Asp–Glu–Asn–Asn–Met. The genetic code states that this hexapeptide is coded

```
                                              15
GLY–SER–ALA–LYS–LYS–GLY–ALA–THR–LEU–PHE–LYS–THR–ARG–CYS–GLU–
                                              30
LEU–CYS–HIS–THR–VAL–GLU–LYS–GLY–GLY–PRO–HIS–LYS–VAL–GLY–PRO–
                                              45
ASN–LEU–HIS–GLY–ILE–PHE–GLY–ARG–HIS–SER–GLY–GLN–ALA–GLN–GLY–
                                              60
TYR–SER–TYR–THR–ASP–ALA–ASN–ILE–LYS–LYS–ASN–VAL–LEU–TRP–ASP–
                                              75
GLU–ASN–ASN–MET–SER–GLU–TYR–LEU–THR–ASN–PRO–LYS–LYS–TYR–ILE–
                                              90
PRO–GLY–THR–LYS–MET–ALA–PHE–GLY–GLY–LEU–LYS–LYS–GLU–LYS–ASP–
                                              103
ARG–ASN–ASP–LEU–ILE–THR–TYR–LEU–LYS–LYS–ALA–CYS–GLU
```

Figure 8.14

The amino acid sequence of yeast cytochrome *c*. The hexapeptide that is highlighted red is the one used to illustrate how a nucleotide sequence can be predicted from an amino acid sequence.

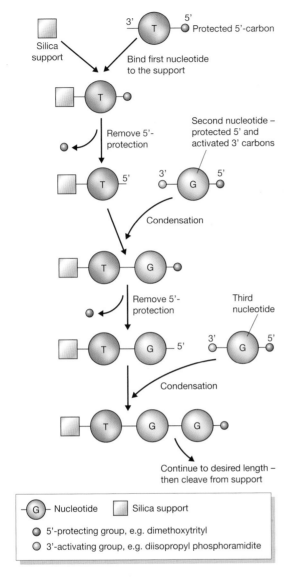

Figure 8.15

A simplified scheme for oligonucleotide synthesis. Each nucleotide is modified by attachment of an activating group to the 3′ carbon and a protecting group to the 5′ carbon. The activating group enables the normally inefficient process of nucleotide joining to proceed much more rapidly. The protecting group ensures that individual nucleotides cannot attach to one another, and instead react only with the terminal 5′ group of the growing oligonucleotide, this 5′ group being deprotected by chemical treatment at the appropriate point in each synthesis cycle.

by TGG–GA^T/$_C$–GA^A/$_G$–AA^T/$_C$–AA^T/$_C$–ATG. Although this represents a total of 16 different possible sequences, 14 of the 18 nucleotides can be predicted with certainty.

Oligonucleotides of up to about 150 nucleotides in length can easily be synthesized in the laboratory (Figure 8.15). An oligonucleotide probe could therefore be constructed according to the predicted nucleotide sequence, and this probe might be able to identify the gene coding for the protein in question. In the example of yeast cytochrome *c*, the 16 possible oligonucleotides that can code for Trp–Asp–Glu–Asn–Asn–Met would be synthesized, either separately or as a pool, and then used to probe a yeast genomic or cDNA library (Figure 8.16). One of the oligonucleotides in the probe will have the correct sequence for this region of the cytochrome *c* gene, and its hybridization signal will indicate which clones carry this gene.

Figure 8.16

The use of synthetic, end-labelled oligonucleotides to identify a clone of the yeast cytochrome *c* gene.

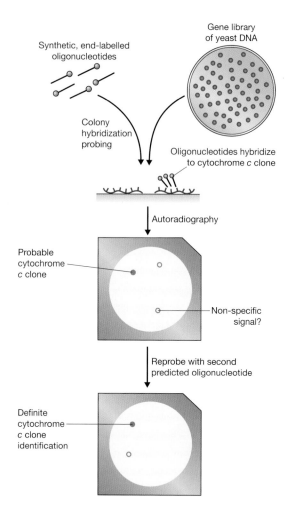

The result can be checked by carrying out a second probing with a mixture of oligonucleotides whose sequences are predicted from a different segment of the cytochrome *c* protein (Figure 8.16). However, the segment of the protein used for nucleotide sequence prediction must be chosen with care. The hexapeptide Ser–Glu–Tyr–Leu–Thr–Asn, which immediately follows our first choice, could be coded by several thousand different 18-nucleotide sequences, clearly is an unsuitable choice for a synthetic probe.

Heterologous probing allows related genes to be identified

Often, a substantial amount of nucleotide similarity is seen when two genes for the same protein, but from different organisms, are compared, this being a reflection of the conservation of gene structure during evolution. Frequently, two genes from related organisms are sufficiently similar for a single-stranded probe prepared from one gene to form a stable hybrid with the second gene. Although the two molecules are not entirely complementary, enough base pairs are formed to produce a stable structure (Figure 8.17a).

Heterologous probing makes use of hybridization between related sequences for clone identification. For example, the yeast cytochrome *c* gene, as identified in the

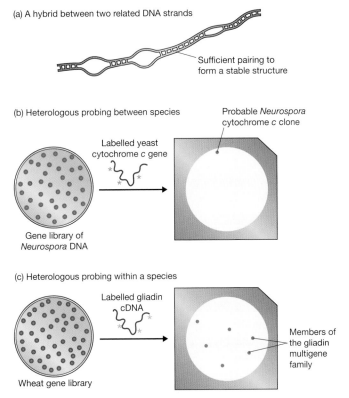

(a) A hybrid between two related DNA strands

Sufficient pairing to
form a stable structure

(b) Heterologous probing between species

Probable *Neurospora*
cytochrome *c* clone

Labelled yeast
cytochrome *c* gene

Gene library of
Neurospora DNA

(c) Heterologous probing within a species

Labelled gliadin
cDNA

Members of
the gliadin
multigene
family

Wheat gene library

Figure 8.17

Heterologous probing.

previous section by oligonucleotide probing, could itself be used as a hybridization probe to identify cytochrome *c* genes in clone libraries of other organisms. A probe prepared from the yeast gene would not be entirely complementary to the gene from, say, the fungus *Neurospora crassa*, but sufficient base pairing should occur for a hybrid to be formed (Figure 8.17b).

Heterologous probing can also identify related genes in the same organism. If the wheat gliadin cDNA clone, as identified earlier in the chapter by abundancy probing, is used to probe a genomic library, it will hybridize not only to its own gene but also to a variety of other genes (Figure 8.17c). These genes are all related to the gliadin cDNA, but have slightly different nucleotide sequences. This is because the wheat gliadins form a complex group of related proteins that are coded by the members of a multigene family. Once one gene in the family has been cloned, all the other members can be isolated by heterologous probing.

Southern hybridization enables a specific restriction fragment containing a gene to be identified

As well as colony and plaque hybridization analysis, there are also occasions when it is necessary to use hybridization probing to identify which of a series of restriction fragments contains a gene of interest. As an example, we will return to the genomic clone of

Figure 8.18

A long cloned DNA fragment may contain several genes in addition to the one in which we are interested. B = *Bam*HI restriction site.

the yeast cytochrome *c* gene, which we identified by oligonucleotide hybridization prob-ing. Let us imagine that this particular genomic library was prepared by partial restric-tion of yeast DNA with *Bam*HI followed by cloning in the cosmid vector pJB8 (p. 103). The cloned fragment containing the cytochrome *c* gene will therefore be approximately 40 kb in length, and will probably contain about ten *Bam*HI fragments, remembering that the hexanucleotide recognition site for this enzyme will be present, on average, once every $4^6 = 4096$ bp.

The cytochrome *c* gene, on the other hand, is predicted to be just 309 bp in length (we know that the protein has 103 amino acids; see Figure 8.14). The gene therefore com-prises less than 1% of the cloned DNA fragment, and it is quite possible that other genes that we are not interested in are also present in this insert (Figure 8.18). The method called Southern hybridization enables the individual restriction fragment containing the cytochrome *c* gene to be identified.

The first step in using Southern hybridization for this purpose would be to digest the clone with *Bam*HI and to then separate the restriction fragments by electrophore-sis in an agarose gel (Figure 8.19a). The aim is to use the oligonucleotide probe for cytochrome *c* to identify the fragment that contains the gene. This can be attempted

Figure 8.19

Southern hybridization.

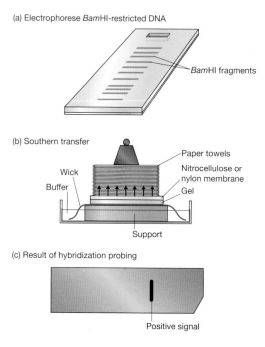

(a) Electrophorese *Bam*HI-restricted DNA

*Bam*HI fragments

(b) Southern transfer

Paper towels

Nitrocellulose or nylon membrane

Wick

Gel

Buffer

Support

(c) Result of hybridization probing

Positive signal

while the restriction fragments are still contained in the electrophoresis gel, although the results are usually not very good as the gel matrix causes a lot of spurious background hybridization that obscures the specific hybridization signal. Instead, the DNA bands in the agarose gel are transferred to a nitrocellulose or nylon membrane, which provides a much cleaner environment for the hybridization experiment.

The transfer of DNA bands from an agarose gel to a membrane employs the technique perfected in 1975 by Professor E.M. Southern, and referred to as Southern transfer. The membrane is placed on the gel, through which buffer is allowed to soak, carrying the DNA from the gel to the membrane where it becomes bound. Whilst sophisticated pieces of apparatus can be purchased to assist this process, many molecular biologists prefer a homemade set-up incorporating many paper towels and considerable balancing skills (Figure 8.19b). The same method can also be used for the transfer of RNA molecules (northern transfer) or proteins (western transfer). So far, no one has devised eastern transfers.

Southern transfer results in a membrane that carries a replica of the DNA bands from the agarose gel. If the labelled probe is now applied, hybridization occurs and autoradiography (or the equivalent detection system for a non-radioactive probe) reveals which restriction fragment contains the cloned gene (Figure 8.19c).

8.4.4 Identification methods based on detection of the translation product of the cloned gene

Hybridization probing is usually the preferred method for the identification of a particular recombinant from a clone library. The technique is easy to perform and, with modifications introduced in recent years, can be used to check up to 10 000 recombinants per experiment, allowing large genomic libraries to be screened in a reasonably short time. Nevertheless, the requirement for a probe that is at least partly complementary to the desired gene sometimes makes it impossible to use hybridization in clone identification. On these occasions, a different strategy is needed.

The main alternative to hybridization probing is immunological screening. The distinction here is that, whereas with hybridization probing the cloned DNA fragment is itself directly identified, an immunological method detects the protein coded by the cloned gene. Immunological techniques therefore presuppose that the cloned gene is being expressed, so that the protein is being made, and that this protein is not normally present in the host cells.

Antibodies are required for immunological detection methods

If a purified sample of a protein is injected into the bloodstream of a rabbit, the immune system of the animal responds by synthesizing antibodies that bind to and help degrade the foreign molecule (Figure 8.20a). This is a version of the natural defence mechanism that the animal uses to deal with invasion by bacteria, viruses and other infective agents.

Once a rabbit is challenged with a protein, the levels of antibody present in its bloodstream remain high enough over the next few days for substantial quantities to be purified. It is not necessary to kill the rabbit, because as little as 10 ml of blood will provide a considerable amount of antibody (Figure 8.20b). This purified antibody binds only to the protein with which the animal was originally challenged.

Figure 8.20

Antibodies. (a) Antibodies in the bloodstream bind to foreign molecules and help degrade them. (b) Purified antibodies can be obtained from a small volume of blood taken from a rabbit injected with the foreign protein.

(a) Antibodies bind to foreign molecules

Foreign molecule, e.g. protein

Antibodies

(b) Antibody purification

Remove 10 ml of blood

Rabbit injected with foreign protein

Blood

Purified antibody

Using a purified antibody to detect protein in recombinant colonies

There are several versions of immunological screening, but the most useful method is a direct counterpart of colony hybridization probing. Recombinant colonies are transferred to a polyvinyl or nitrocellulose membrane, the cells are lysed, and a solution containing the specific antibody is added (Figure 8.21a). In the original methods, either the antibody itself was labelled, or the membrane was subsequently washed with a solution

Figure 8.21

Using a purified antibody to detect protein in recombinant colonies. Instead of using a secondary antibody, the primary antibody itself can be labelled, or the antibody can be detected with labelled protein A.

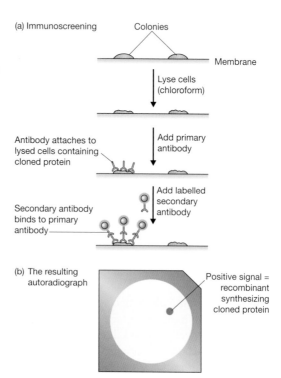

(a) Immunoscreening   Colonies

Membrane

Lyse cells (chloroform)

Antibody attaches to lysed cells containing cloned protein

Add primary antibody

Secondary antibody binds to primary antibody

Add labelled secondary antibody

(b) The resulting autoradiograph

Positive signal = recombinant synthesizing cloned protein

of labelled protein A, a bacterial protein that binds specifically to the immunoglobulins of which antibodies are made. In more modern methods, the bound antibody – the primary antibody – is detected by washing the membrane with a labelled secondary antibody, which binds specifically to the primary antibody. The secondary antibody is prepared by injecting the primary antibody into an animal, of a different species to the one from which the primary antibody was prepared. For example, if the primary antibody was prepared in a rabbit, the secondary antibody could be obtained by injecting a sample of the primary antibody into a goat. The goat's immune system will look on the primary antibody as a foreign protein antigen, and synthesize the secondary antibody to bind to it.

Several secondary antibody molecules can bind to a single primary antibody molecule, increasing the amount of signal that is produced and enabling a clearer detection of each positive colony. In all three methods, the label can be radioactive, in which case the colonies that bind the label can be detected using autoradiography (Figure 8.21b), or non-radioactive labels that result in a fluorescent or chemiluminescent signal can be used as an alternative.

The problem of gene expression

Immunological screening depends on the cloned gene being expressed so that the protein translation product is present in the recombinant cells. As will be discussed in greater detail in Chapter 13, a gene from one organism is often not expressed in a different organism. In particular, it is very unlikely that a cloned animal or plant gene will be expressed in *E. coli* cells. This problem can be circumvented by using a special type of vector, called an expression vector (p. 250), designed specifically to promote expression of the cloned gene in a bacterial host. Immunological screening of recombinant *E. coli* colonies carrying animal genes cloned into expression vectors has been very useful in identifying genes for several important hormones.

Further reading

FURTHER READING

Benton, W.D. and Davis, R.W. (1977) Screening λgt recombinant clones by hybridization to single plaques *in situ. Science*, **196**, 180–182.

Feinberg, A.P. and Vogelstein, B. (1983) A technique for labelling DNA restriction fragments to high specific activity. *Analytical Biochemistry*, **132**, 6–13. [Random priming labelling.]

Grunstein, M. and Hogness, D.S. (1975) Colony hybridization: a method for the isolation of cloned cDNAs that contain a specific gene. *Proceedings of the National Academy of Sciences of the United States of America*, **72**, 3961–3965.

Gubler, U. and Hoffman, B.J. (1983) A simple and very efficient method for generating cDNA libraries. *Gene*, **25**, 263–269.

Southern, E.M. (2000) Blotting at 25. *Trends in Biochemical Science*, **25**, 585–588. [The origins of Southern hybridization.]

Thorpe, G.H.G., Kricka, L.J., Moseley, S.B., and Whitehead, T.P. (1985) Phenols as enhancers of the chemiluminescent horseradish peroxidase–luminol–hydrogen peroxide reaction: application in luminescence-monitored enzyme immunoassays. *Clinical Chemistry*, **31**, 1335–1341. [Describes the basis to a non-radioactive labelling method.]

Young, R.A. and Davis, R.W. (1983) Efficient isolation of genes by using antibody probes. *Proceedings of the National Academy of Sciences of the United States of America*, **80**, 1194–1198.

Chapter 9

The Polymerase Chain Reaction

Chapter contents

As a result of the last seven chapters we have become familiar not only with the basic principles of gene cloning, but also with fundamental molecular biology techniques such as restriction analysis, gel electrophoresis, DNA labelling, and DNA–DNA hybridization. To complete our basic education in DNA analysis we must now return to the second major technique for studying genes, the polymerase chain reaction (PCR).

PCR is a very uncomplicated technique. All that happens is that a short region of a DNA molecule – a single gene for instance – is copied many times by a DNA polymerase enzyme (see Figure 1.2). This might seem a rather trivial exercise, but it has a multitude of applications in genetics research and in broader areas of biology.

We begin this chapter with an outline of PCR in order to understand exactly what it achieves. Then, we will look at the key issues that determine whether or not an individual PCR experiment is successful, before examining some of the methods that have been devised for studying the amplified DNA fragments that are obtained.

9.1 PCR in outline

PCR results in a selective amplification of a chosen region of a DNA molecule. Any region of any DNA molecule can be chosen, as long as the sequences at the borders of

Gene Cloning and DNA Analysis: An Introduction, Seventh Edition. T.A. Brown.
© 2016 John Wiley & Sons, Ltd. Published 2016 by John Wiley & Sons, Ltd.
Companion Website: www.wiley.com/go/brown/cloning

Figure 9.1

Hybridization of the oligonucleotide primers to the template DNA at the beginning of a PCR.

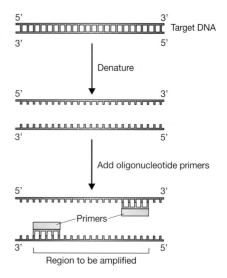

the region are known. The border sequences must be known because, in order to carry out a PCR, two short oligonucleotides must hybridize to the DNA molecule, one to each strand of the double helix (Figure 9.1). These oligonucleotides, which act as primers for the DNA synthesis reactions, delimit the region that will be amplified.

Amplification is usually carried out by the DNA polymerase I enzyme from *Thermus aquaticus*. As mentioned elsewhere (p. 52), this organism lives in hot springs and many of its enzymes, including *Taq* polymerase, are thermostable, which means they are resistant to denaturation by heat treatment. As will be apparent very soon, the thermostability of *Taq* polymerase is an essential requirement in PCR methodology.

To carry out a PCR experiment, the target DNA is mixed with *Taq* polymerase, the two oligonucleotide primers, and a supply of nucleotides. The amount of target DNA can be very small because PCR is extremely sensitive and will operate with just a single starting molecule. The reaction is started by heating the mixture to 94 °C, at which temperature the hydrogen bonds holding together the two polynucleotides of the double helix are broken, so that the target DNA becomes denatured into single-stranded molecules (Figure 9.2). The temperature is then reduced to 50–60 °C, which results in some rejoining of the single strands of the target DNA, but also allows the primers to attach to their annealing positions. DNA synthesis can now begin, so the temperature is raised to 74 °C, just below the optimum for *Taq* polymerase. In this first stage of the PCR, a set of 'long products' is synthesized from each strand of the target DNA. These polynucleotides have identical 5′ ends but random 3′ ends, the latter representing positions where DNA synthesis terminates by chance.

The cycle of denaturation–annealing–synthesis is now repeated (Figure 9.3). The long products denature and the four resulting strands are copied during the DNA synthesis stage. This gives four double-stranded molecules, two of which are identical to the long products from the first cycle, and two of which are made entirely of new DNA. During the third cycle, the latter give rise to 'short products', the 5′ and 3′ ends of which are both set by the primer annealing positions. In subsequent cycles, the number of short products accumulates in an exponential fashion (doubling during each cycle) until one of the components of the reaction becomes depleted. This means that after 30 cycles, there will be over 130 million short products derived from each starting molecule.

Figure 9.2

The first stage of a PCR, resulting in synthesis of the long products.

In real terms, this equates to several micrograms of PCR product from a few nanograms or less of target DNA.

At the end of a PCR, a sample of the reaction mixture is usually analysed using agarose gel electrophoresis, sufficient DNA having been produced for the amplified fragment to be visible as a discrete band after staining with ethidium bromide or some other DNA-binding dye. This may by itself provide useful information about the DNA region that has been amplified or, alternatively, the PCR product can be examined by techniques such as DNA sequencing.

9.2 PCR in more detail

Although PCR experiments are very easy to set up, they must be planned carefully if the results are to be of any value. The sequences of the primers are critical to the success of the experiment, as are the precise temperatures used in the heating and cooling stages of the reaction cycle.

9.2.1 Designing the oligonucleotide primers for a PCR

The primers are the key to the success or failure of a PCR experiment. If the primers are designed correctly the experiment results in amplification of a single DNA fragment, corresponding to the target region of the template molecule. If the primers are incorrectly designed the experiment will fail, possibly because no amplification occurs, or possibly because the wrong fragment, or more than one fragment, is amplified

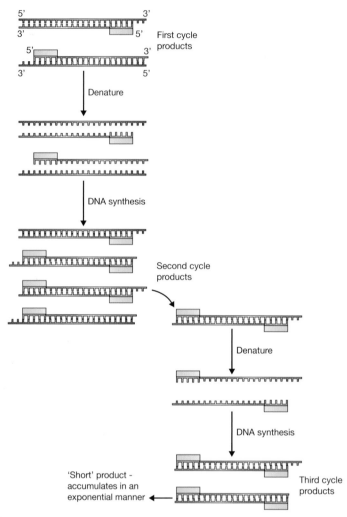

Figure 9.3

The second and third cycles of a PCR, during which the first short products are synthesized.

(Figure 9.4). Clearly, a great deal of thought must be put into the design of the primers.

Working out appropriate sequences for the primers is not a problem: they must correspond with the sequences flanking the target region on the template molecule. Each primer must, of course, be complementary (not identical) to its template strand in order for hybridization to occur, and the 3′ ends of the hybridized primers should point towards one another (Figure 9.5). The DNA fragment to be amplified should not be greater than about 3 kb in length and ideally less than 1 kb. Fragments up to 10 kb can be amplified using standard PCR techniques, but the longer the fragment the less efficient is the amplification and the more difficult it is to obtain consistent results. The amplification of very long fragments – up to 40 kb – is possible, but this requires special methods.

Figure 9.4

The results of PCRs with well-designed and poorly designed primers. Lane 1 shows a single amplified fragment of the expected size, the result of a well-designed experiment. In lane 2 there is no amplification product, suggesting that one or both of the primers were unable to hybridize to the template DNA. Lanes 3 and 4 show, respectively, an amplification product of the wrong size, and a mixture of products (the correct product plus two wrong ones). Both results are due to hybridization of one or both of the primers to non-target sites on the template DNA molecule.

The first important issue to address is the length of the primers. If the primers are too short they might hybridize to non-target sites and give undesired amplification products. To illustrate this point, imagine that total human DNA is used in a PCR experiment with a pair of primers eight nucleotides in length (in PCR jargon, these are called '8-mers'). The likely result is that a number of different fragments will be amplified. This is because attachment sites for these primers are expected to occur, on average, once every $4^8 = 65\ 536$ bp, giving approximately 49 000 possible sites in the 3 200 000 kb of nucleotide sequence that makes up the human genome. This means that it would be very unlikely that a pair of 8-mer primers would give a single, specific amplification product with human DNA (Figure 9.6a).

What if the 17-mer primers shown in Figure 9.5 are used? The expected frequency of a 17-mer sequence is once every $4^{17} = 17\ 179\ 869\ 184$ bp. This figure is more than fivefold greater than the length of the human genome, so a 17-mer primer would be expected to have just one hybridization site in total human DNA. A pair of 17-mer primers should therefore give a single, specific amplification product (Figure 9.6b).

Why not simply make the primers as long as possible? The length of the primer influences the rate at which it hybridizes to the template DNA, with long primers hybridizing at a slower rate. The efficiency of the PCR, measured by the number of amplified molecules produced during the experiment, is therefore reduced if the primers are too long, as complete hybridization to the template molecules cannot occur in the time allowed during the reaction cycle. In practice, primers longer than 30-mer are rarely used.

Figure 9.5

A pair of primers designed to amplify the human α_1-globin gene. The exons of the gene are shown as red boxes, the introns as grey boxes.

Figure 9.6

The lengths of the primers are critical for the specificity of the PCR.

(a) PCR of human DNA with 8-mer primers

(b) PCR of human DNA with 17-mer primers

9.2.2 Working out the correct temperatures to use

During each cycle of a PCR, the reaction mixture is transferred between three temperatures (Figure 9.7):

- The denaturation temperature, usually 94 °C, which breaks the base pairs and releases single-stranded DNA to act as templates in the next round of DNA synthesis.
- The hybridization or annealing temperature, at which the primers attach to the templates.
- The extension temperature, at which DNA synthesis occurs. This is usually set at 74 °C, just below the optimum for *Taq* polymerase.

Figure 9.7

A typical temperature profile for a PCR.

(a) Annealing temperature is too high

Primers and templates remain dissociated

(b) Annealing temperature is too low

Mismatched hybrid – not all the correct base pairs have formed

(c) Correct annealing temperature

Priming occurs only at the desired target sites

Figure 9.8

Temperature has an important effect on the hybridization of the primers to the template DNA.

The annealing temperature is the important one because, again, this can affect the specificity of the reaction. DNA–DNA hybridization is a temperature-dependent phenomenon. If the temperature is too high no hybridization takes place, and the primers and templates remain dissociated (Figure 9.8a). If the temperature is too low, mismatched hybrids – ones in which not all the correct base pairs have formed – are stable (Figure 9.8b). If this occurs, the earlier calculations regarding the appropriate lengths for the primers become irrelevant, as it was assumed in these calculations that only perfect primer–template hybrids are able to form. If mismatches are tolerated, the number of potential hybridization sites for each primer is greatly increased, and amplification is more likely to occur at non-target sites in the template molecule.

The ideal annealing temperature must be low enough to enable hybridization between primer and template, but high enough to prevent mismatched hybrids from forming (Figure 9.8c). This temperature can be estimated by determining the melting temperature or T_m of the primer–template hybrid. The T_m is the temperature at which the correctly base-paired hybrid dissociates ('melts'). A temperature 1–2 °C below this should be low enough to allow the correct primer–template hybrid to form, but too high for a hybrid with a single mismatch to be stable. The T_m can be determined experimentally, but is more usually calculated from the simple formula (Figure 9.9):

$$T_m = (4 \times [G + C]) + (2 \times [A + T])°C$$

Primer sequence: 5' AGACTCAGAGAGAACCC 3'

4 Gs   5 Cs   7 As   1 T

$T_m = (4 \times 9) + (2 \times 8)$
$= 36 + 16$
$= 52°C$

Figure 9.9

Calculating the T_m of a primer.

in which [G + C] is the number of G and C nucleotides in the primer sequence, and [A + T] is the number of A and T nucleotides.

The annealing temperature for a PCR experiment is therefore determined by calculating the T_m for each primer and using a temperature of 1–2 °C below this figure. Note that this means the two primers should be designed so that they have identical T_ms. If this is not the case, the appropriate annealing temperature for one primer may be too high or too low for the other member of the pair.

9.3 After the PCR: Studying PCR products

PCR is often the starting point for a longer series of experiments in which the amplification product is studied in various ways in order to obtain information about the DNA molecule that acted as the original template. We will encounter many studies of this type in Parts II and III of this book, when we examine the applications of gene cloning and PCR in research and biotechnology. Although a wide range of procedures have been devised for studying PCR products, three techniques are particularly important:

- Gel electrophoresis of PCR products.
- Cloning of PCR products.
- Sequencing of PCR products.

The first two of these techniques are dealt with in this chapter. The third technique is deferred until Chapter 10, when all aspects of DNA sequencing will be covered.

9.3.1 Gel electrophoresis of PCR products

The results of most PCR experiments are checked by running a portion of the amplified reaction mixture in an agarose gel. A band representing the amplified DNA may be visible after staining, or if the DNA yield is low the product can be detected using Southern hybridization (p. 151). If the expected band is absent, or if additional bands are present, then something has gone wrong and the experiment must be repeated.

In some cases, agarose gel electrophoresis is used not only to determine if a PCR experiment has worked, but also to obtain additional information. For example, the presence of restriction sites in the amplified region of the template DNA can be determined by treating the PCR product with a restriction endonuclease before running the sample in the agarose gel (Figure 9.10). This is a type of restriction fragment length polymorphism (RFLP) analysis.

Alternatively, the exact size of the PCR product can be used to establish if the template DNA contains an insertion or deletion mutation in the amplified region (Figure 9.10). Length mutations of this type form the basis of DNA profiling, a central technique in forensic science (Chapter 16).

In some experiments, the mere presence or absence of the PCR product is the diagnostic feature. An example of this occurs when PCR is used as the screening procedure to identify a desired gene from a genomic or cDNA library. Carrying out PCRs with every clone in a genomic library might seem a tedious task, but one of the main advantages of PCR is that individual experiments are quick to set up and many PCRs can be performed in parallel. The workload can also be reduced by combinatorial screening, an example of which is shown in Figure 9.11.

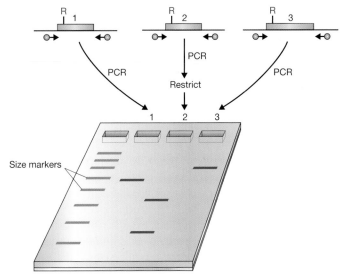

Figure 9.10

Gel electrophoresis of the PCR product can provide information on the template DNA molecule. Lanes 1 and 2 show, respectively, an unrestricted PCR product and a product restricted with the enzyme that cuts at site R. Lane 3 shows the result obtained when the template DNA contains an insertion in the amplified region.

Figure 9.11

Combinatorial screening of clones in microtitre trays. A library of 960 clones, contained in ten microtitre trays, is screened by a series of PCRs, each with a combination of clones. The clone combinations that give positive results enable the well(s) containing positive clone(s) to be identified. For example, if positive PCRs are given with row A of tray 2, row D of tray 6, column 7 of tray 2, and column 9 of tray 6, then it can be deduced that there are positive clones in well A7 of tray 2 and well D9 of tray 6. Although there are 960 clones, unambiguous identification of the positive clones is therefore achieved after just 200 PCRs.

Figure 9.12

Polynucleotides synthesized by *Taq* polymerase often have an extra adenosine at their 3′ ends.

```
5'                                          3'
AGACTCAGA...........AACTTATTTA
| | | | | | | | |        | | | | | | | | | |
ATCTGAGTCT...........TTGAATAAA
3'                                          5'
```

9.3.2 Cloning PCR products

Some applications require that after a PCR the resulting products are ligated into a vector and examined by any of the standard methods used for studying cloned DNA. This may sound easy, but there are complications.

The first problem concerns the ends of the PCR products. From an examination of Figure 9.3, it might be imagined that the short products resulting from PCR amplification are blunt-ended. If this were the case they could be inserted into a cloning vector by blunt-end ligation, or alternatively the PCR products could be provided with sticky ends by the attachment of linkers or adaptors (p. 68). Unfortunately, the situation is not so straightforward. *Taq* polymerase tends to add an additional nucleotide, usually an adenosine, to the end of each strand that it synthesizes. This means that a double-stranded PCR product is not blunt-ended, and instead most 3′ termini have a single nucleotide overhang (Figure 9.12). The overhangs could be removed by treatment with an exonuclease enzyme, resulting in PCR products with true blunt ends, but this is not a popular approach as it is difficult to prevent the exonuclease from becoming overactive and causing further damage to the ends of the molecules.

One solution is to use a special cloning vector which carries thymidine (T) overhangs and which can therefore be ligated to a PCR product (Figure 9.13). These vectors are usually prepared by restricting a standard vector at a blunt-end site, and then treating with *Taq* polymerase in the presence of just 2′-deoxythymidine 5′-triphosphate (dTTP). As no primer is present, all the polymerase can do is add a T nucleotide to the 3′ ends of the blunt-ended vector molecule, resulting in the T-tailed vector into which the PCR products can be inserted. Special vectors of this type have also been designed for use with the topoisomerase ligation method (p. 71), and this is currently the most popular way of cloning PCR products.

A second solution is to design primers that contain restriction sites. After PCR, the products are treated with the restriction endonuclease, which cuts each molecule within the primer sequence, leaving sticky-ended fragments that can be ligated efficiently into

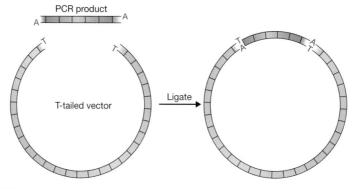

Figure 9.13

Using a special T-tailed vector to clone a PCR product.

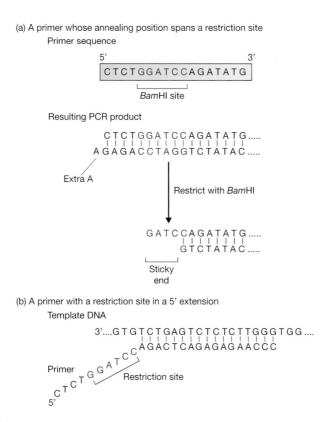

(a) A primer whose annealing position spans a restriction site

Primer sequence

5'   3'

CTCTGGATCCAGATATG

*Bam*HI site

Resulting PCR product

CTCTGGATCCAGATATG.....
AGAGACCTAGGTCTATAC.....

Extra A

Restrict with *Bam*HI

GATCCAGATATG.....
GTCTATAC.....

Sticky
end

(b) A primer with a restriction site in a 5' extension

Template DNA

3'....GTGTCTGAGTCTCTCTTGGGTGG....
AGACTCAGAGAGAACCC

Primer   GGATCC

CTCTG   Restriction site

5'

Figure 9.14

Obtaining a PCR product with a sticky end through use of a primer whose sequence includes a restriction site.

a standard cloning vector (Figure 9.14a). The approach is not limited to those instances where the primers span restriction sites that are present in the template DNA. Instead, the restriction site can be included within a short extension at the 5' end of each primer (Figure 9.14b). These extensions cannot hybridize to the template molecule, but they are copied during the PCR, resulting in PCR products that carry terminal restriction sites.

9.3.3 Problems with the error rate of *Taq* polymerase

All DNA polymerases make mistakes during DNA synthesis, occasionally inserting an incorrect nucleotide into the growing DNA strand. Most polymerases are able to rectify these errors by reversing over the mistake and resynthesizing the correct sequence. This property, which is referred to as the proofreading function, depends on the polymerase possessing a 3' to 5' exonuclease activity (p. 178).

    Taq polymerase lacks a proofreading activity and as a result is unable to correct its errors. This means that the DNA synthesized by *Taq* polymerase is not always an accurate copy of the template molecule. The error rate has been estimated at one mistake for every 9000 nucleotides of DNA that is synthesized, which might appear to be almost insignificant but translates to one error in every 300 bp for the PCR products obtained after 30 cycles. This is because PCR involves copies being made of copies of copies, so that the polymerase-induced errors gradually accumulate, and the fragments produced at the end of a PCR will contain copies of earlier errors together with any new errors introduced during the final round of synthesis.

For many applications, this high error rate does not present a problem. In particular, sequencing of a PCR product provides the correct sequence of the template, even though the PCR products contain the errors introduced by *Taq* polymerase. This is because the errors are distributed randomly, so for every molecule that has an error at a particular nucleotide position, there will be many molecules with the correct sequence. The resulting sequence will be incorrect only if an error occurred during a very early cycle in the PCR, as repeated copying will result in these errors being present in many of the molecules in the final product.

Unfortunately, the errors become much more of a problem if the PCR products are cloned. As each resulting clone contains multiple copies of a single amplified fragment, the cloned DNA will not necessarily have the same sequence as the original template molecule used in the PCR (Figure 9.15). This possibility lends an uncertainty to all experiments carried out with cloned PCR products and dictates that, whenever possible, the amplified DNA should be studied directly rather than being cloned.

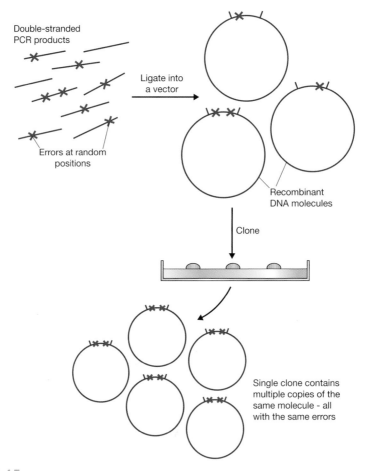

Figure 9.15

The high error rate of *Taq* polymerase becomes a factor when PCR products are cloned.

Table 9.1

Number of short products synthesized after 25 cycles of PCR with different numbers of starting molecule.

NUMBER OF STARTING MOLECULES	NUMBER OF SHORT PRODUCTS
1	4 194 304
2	8 388 608
5	20 971 520
10	41 943 040
25	104 857 600
50	209 715 200
100	419 430 400

The numbers assume that amplification is 100% efficient, and none of the reactants becoming limiting during the course of the PCR.

9.4 Real-time PCR enables the amount of starting material to be quantified

The amount of product synthesized during a set number of cycles of a PCR depends on the number of DNA molecules present in the starting mixture (Table 9.1). If only a few DNA molecules are present at the beginning of the PCR, then relatively little product will be made, but if there are many starting molecules then the product yield will be higher. This relationship enables PCR to be used to quantify the number of DNA molecules present in an extract.

9.4.1 Carrying out a quantitative PCR experiment

In quantitative PCR (qPCR) the amount of product synthesized during a test PCR is compared with the amounts synthesized during PCRs with known quantities of starting DNA. When PCR was first developed, agarose gel electrophoresis was used to make these comparisons. After staining the gel, the band intensities were examined to identify the control PCR whose product was most similar to that of the test (Figure 9.16). Although easy to perform, this type of qPCR is imprecise because large differences in the amount of starting DNA give relatively small differences in the band intensities of the resulting PCR products.

Today, quantification is carried out by real-time PCR, which is a modification of the standard PCR technique whereby synthesis of the product is measured over time, as

Figure 9.16

Using agarose gel electrophoresis to quantify the amount of DNA in a test PCR. Lanes 1 to 4 are control PCRs carried out with decreasing amounts of template DNA. The intensity of staining for the test band suggests that this PCR contained approximately the same amount of DNA as the control run in lane 2.

Figure 9.17

Hybridization of a reporter probe to its target DNA.

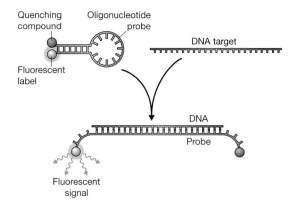

the PCR proceeds through its series of cycles. There are two ways of following product synthesis in real time:

- A dye that emits a fluorescent signal when it binds to double-stranded DNA can be included in the PCR mixture. This method is used to measure the total amount of double-stranded DNA in the PCR at any particular time, which may overestimate the actual amount of the product because the primers sometimes anneal to one another in various non-specific ways, increasing the amount of double-stranded DNA that is present.
- A short oligonucleotide called a reporter probe, which emits a fluorescent signal when it hybridizes to the PCR product, can be used. Because the probe only hybridizes to the PCR product, this method is less prone to inaccuracies caused by primer–primer annealing. Each probe molecule has a pair of labels. A fluorescent dye is attached to one end of the oligonucleotide, and a quenching compound, which inhibits the fluorescent signal, is attached to the other end (Figure 9.17). Normally, there is no fluorescence because the oligonucleotide is designed in such a way that its two ends base pair to one another, placing the quencher next to the dye. Hybridization between the oligonucleotide and the PCR product disrupts this base pairing, moving the quencher away from the dye and enabling the fluorescent signal to be generated.

Figure 9.18

Quantification by real-time PCR. The graph shows product synthesis during three PCRs, each with a different amount of starting DNA. During a PCR, product accumulates exponentially, the amount present at any particular cycle being proportional to the amount of starting DNA. The blue curve is therefore the PCR with the greatest amount of starting DNA, and the green curve is the one with the least starting DNA. If the amounts of starting DNA in these three PCRs are known, then the amount in a test PCR can be quantified by comparison with these controls. In practice, the comparison is made by identifying the cycle at which product synthesis moves above a threshold amount, indicated by the horizontal line on the graph.

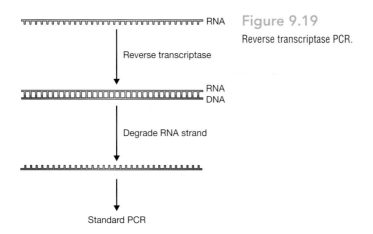

Figure 9.19

Reverse transcriptase PCR.

Both systems enable synthesis of the PCR product to be followed by measuring the fluorescent signal. Quantification again requires comparison between test and control PCRs, usually by identifying the stage in the PCR at which the amount of fluorescent signal reaches a pre-set threshold (Figure 9.18). The more rapidly the threshold is reached, the greater the amount of DNA in the starting mixture.

9.4.2 Real-time PCR can also quantify RNA

Real-time PCR is often used to quantify the amount of DNA in an extract, for example to follow the progression of a viral infection by measuring the amount of pathogen DNA that is present in a tissue. The method can also be used to measure RNA amounts, in particular to determine the extent of expression of a particular gene by quantifying its mRNA. The gene under study might be one that is switched on in cancerous cells, in which case quantifying its mRNA will enable the development of the cancer to be monitored and the effects of subsequent treatment to be assessed.

How do we carry out PCR if RNA is the starting material? The answer is to use reverse transcriptase PCR. The first step in this procedure is to convert the RNA molecules into single-stranded complementary DNA (cDNA) (Figure 9.19). Once this preliminary step has been carried out, the PCR primers and *Taq* polymerase are added and the experiment proceeds exactly as in the standard technique. Some thermostable polymerases are able to make DNA copies of both RNA and DNA molecules (i.e., they have both reverse transcriptase and DNA-dependent DNA polymerase activities) and so can carry out all the steps of this type of PCR in a single reaction.

Further reading

FURTHER READING

Higuchi, R., Dollinger, G., Walsh, P.S., and Griffith, R. (1992) Simultaneous amplification and detection of specific DNA sequences. *Biotechnology*, **10**, 413–417. [The first description of real-time PCR].

Marchuk, D., Drumm, M., Saulino, A., and Collins, F.S. (1991) Construction of T-vectors, a rapid and general system for direct cloning of unmodified PCR products. *Nucleic Acids Research*, **19**, 1154.

Rychlik, W., Spencer, W.J., and Rhoads, R.E. (1990) Optimization of the annealing temperature for DNA amplification *in vitro. Nucleic Acids Research*, **18**, 6409–6412.

Saiki, R.K., Gelfand, D.H., Stoffel, S., *et al.* (1988) Primer-directed enzymatic amplification of DNA with a thermostable DNA polymerase. *Science*, **239**, 487–491. [The first description of PCR with *Taq* polymerase.]

Tindall, K.R. and Kunkel, T.A. (1988) Fidelity of DNA synthesis by the *Thermus aquaticus* DNA polymerase. *Biochemistry*, **27**, 6008–6013. [Describes the error rate of *Taq* polymerase.]

VanGuilder, H.D., Vrana, K.E., and Freeman, W.M. (2008) Twenty-five years of quantitative PCR for gene expression analysis. *Biotechniques*, **44**, 619–624.

PART II

The Applications of Gene Cloning and DNA Analysis in Research

Chapter 10

Sequencing Genes and Genomes

Chapter contents

Part I of this book has shown how a skilfully performed cloning or PCR experiment can provide a pure sample of an individual gene, or any other DNA sequence, separated from all the other genes and DNA sequences in the cell. Now, we can turn our attention to the ways in which cloning, PCR, and other DNA analysis techniques are used to study genes and genomes. We will consider three aspects of molecular biology research:

- The techniques used to obtain the nucleotide sequence of individual genes and entire genomes (this Chapter).
- The methods used to study the expression and function of individual genes (Chapter 11).
- The techniques that are used to study entire genomes (Chapter 12).

Probably the most important technique available to the molecular biologist is DNA sequencing, by which the precise order of nucleotides in a piece of DNA can be determined. DNA sequencing methods have been around for 50 years, and since the mid-1970s rapid and efficient sequencing has been possible. Initially, these techniques were applied to individual genes, but since the early 1990s it has been possible to obtain entire genome sequences from prokaryotes and eukaryotes.

Over the years, a number of different methods for DNA sequencing have been devised, and others are currently being developed and are likely to become important in the future. The techniques in use today can be divided into two categories:

- The chain-termination method, which was devised by Fred Sanger and colleagues in the mid-1970s, and has been extensively used ever since.

Gene Cloning and DNA Analysis: An Introduction, Seventh Edition. T.A. Brown.
© 2016 John Wiley & Sons, Ltd. Published 2016 by John Wiley & Sons, Ltd.
Companion Website: www.wiley.com/go/brown/cloning

- Next-generation sequencing, which is a collection of different methods, all of which involve a massively parallel strategy that enables millions of sequences to be generated at the same time.

In this chapter we will study the methodology used in DNA sequencing and then examine how these techniques are used in genome projects.

10.1 Chain-termination DNA sequencing

Chain-termination DNA sequencing is based on the principle that single-stranded DNA molecules that differ in length by just a single nucleotide can be separated from one another by polyacrylamide gel electrophoresis. If the electrophoresis is carried out in a capillary system, then it is possible to resolve a family of molecules representing all lengths up to 1500 nucleotides, with the single-stranded molecules emerging one after another from the end of the capillary (Figure 10.1).

10.1.1 Chain-termination sequencing in outline

Chain-termination sequencing is carried out with a DNA polymerase, which makes copies of the DNA molecule being sequenced. To start the sequencing experiment, a short oligonucleotide is annealed to the template DNA, this oligonucleotide subsequently acting as the primer for synthesis of a new DNA strand that is complementary to the template (Figure 10.2a).

The strand synthesis reaction, which requires the four deoxyribonucleotide triphosphates (dNTPs – dATP, dCTP, dGTP, and dTTP) as substrates, would normally continue until several thousand nucleotides had been polymerized. This does not occur in a chain-termination sequencing experiment because, as well as the four deoxynucleotides, a small amount of each of four dideoxynucleotides (ddNTPs – ddATP, ddCTP, ddGTP,

(a) Capillary gel

(b) Electrophoresis of single-stranded DNA molecules

Load molecules to the top of the capillary

Molecules become separated according to length

Longest

Capillary, 50–80 cm in length

Gel, 0.1 mm diameter

Shortest

Figure 10.1

Polyacrylamide gel electrophoresis in a capillary system can resolve single-stranded DNA molecules that differ in length by just one nucleotide. The capillary is typically 50–80 cm in length with a gel diameter of 0.1 mm.

Figure 10.2

Chain-termination DNA sequencing.

and ddTTP) is added to the reaction. Each of these dideoxynucleotides is labelled with a different fluorescent marker.

The polymerase enzyme does not discriminate between deoxy- and dideoxynucleotides, but once incorporated a dideoxynucleotide blocks further elongation because it lacks the 3′-hydroxyl group needed to form a connection with the next nucleotide (Figure 10.2b). Because the normal deoxynucleotides are also present, but in larger amounts than the dideoxynucleotides, the strand synthesis does not always terminate close to the primer, and several hundred nucleotides may be polymerized before a dideoxynucleotide is eventually incorporated. The result is a set of new molecules, all of different lengths, and each ending in a dideoxynucleotide whose identity indicates the nucleotide – A, C, G, or T – that is present at the equivalent position in the template DNA (Figure 10.2c).

To work out the DNA sequence, all that we have to do is identify the dideoxynucleotide at the end of each chain-terminated molecule. This is where the polyacrylamide gel comes into play. The mixture is loaded into the capillary gel and electrophoresis carried out to separate the molecules according to their lengths. After separation, the molecules are run past a fluorescence detector capable of discriminating the labels attached to the dideoxynucleotides (Figure 10.3a). The detector therefore determines if each molecule ends in an A, C, G, or T. The sequence can either be printed out for examination by the operator (Figure 10.3b) or entered directly into a storage device for future analysis.

Figure 10.3

Reading the sequence generated by a chain-termination experiment. (a) Each dideoxynucleotide is labelled with a different fluorochrome, so the chain-terminated polynucleotides are distinguished as they pass by the detector. (b) An example of a sequence print out.

(a) Detection of chain-terminated polynucleotides

→ Imaging system

Detector

KEY
○ ddATP
○ ddTTP
○ ddCTP
○ ddGTP

Polynucleotides move past the detector

(b) The print out from an automated sequencer

CACCGCATCGAAATTAACTTCCAAAGTTAAGCTTGG

10   20   30

10.1.2 Not all DNA polymerases can be used for sequencing

Any DNA polymerase is capable of extending a primer that has been annealed to a single-stranded DNA molecule, but not all polymerases can be used for DNA sequencing. This is because many DNA polymerases have a mixed enzymatic activity, being able to degrade as well as synthesize DNA (p. 48). Degradation can occur in either the 5'→3' or 3'→5' direction (Figure 10.4), but it is the latter that is detrimental to chain-termination sequencing. This is because the 3'→5' activity prevents chain termination from occurring, by removing the dideoxynucleotide immediately after it has been added at the 3' end of the strand that is being synthesized.

(a) 5'→3' exonuclease activity

(b) 3'→5' exonuclease activity

Figure 10.4

The exonuclease activities of some DNA polymerases. (a) The 5'→3' activity has an important role in DNA repair in the cell, as it enables the polymerase to replace a damaged DNA strand. (b) The 3'→5' activity also has an important role in the cell, as it allows the polymerase to correct its own mistakes, by reversing and replacing a nucleotide that has been added in error (e.g., a T instead of a G). This is called 'proofreading'. During DNA sequencing, this activity can result in removal of a dideoxynucleotide that has just been added to the newly synthesized strand, so that chain termination does not occur.

In the original method for chain-termination sequencing, the Klenow polymerase was used as the sequencing enzyme. As described on p. 51, this is a modified version of the DNA polymerase I enzyme from *E. coli*, the modification removing the 5'→3' exonuclease activity of the standard enzyme. The enzyme used for sequencing was further modified by mutation to remove the 3'→5' exonuclease activity. However, the Klenow polymerase has low processivity, which means that it can only synthesize a relatively short DNA strand before dissociating from the template due to natural causes. This limits the length of sequence that can be obtained from a single experiment to about 250 bp. To avoid this problem, most sequencing today makes use of the *Taq* DNA polymerase, which has high processivity and no exonuclease activity and so is ideal for chain-termination sequencing, enabling sequences of 750 bp and longer to be obtained in a single experiment.

10.1.3 Chain-termination sequencing with Taq polymerase

This chain-termination method that uses *Taq* polymerase is called thermal cycle sequencing. It is carried out in a similar way to PCR, but just one primer is used and the reaction mixture includes the four dideoxynucleotides (Figure 10.5). Because there is only one primer, only one of the strands of the starting molecule is copied and the product accumulates in a linear fashion, not exponentially as is the case in a real PCR. The presence of the dideoxynucleotides in the reaction mixture causes chain termination, and the family of resulting strands is then separated by polyacrylamide gel electrophoresis and the sequence read as described above.

Thermal cycle sequencing is usually carried out with PCR products or with DNA that has been cloned in a plasmid or phage vector. If a PCR product is being sequenced, then one of the primers from the original PCR can be used in the sequencing reaction. If two reactions are carried out, one with each of the two PCR primers, then forward and

Figure 10.5

The basis of thermal cycle sequencing. A PCR is set up with just one primer and one of the dideoxynucleotides. One of the template strands is copied into a family of chain-terminated polynucleotides. ddA = dideoxyATP.

Figure 10.6

Strategies for thermal cycle sequencing. (a) Forward and reverse sequences are obtained by using different PCR primers. (b) Internal primers are used to obtain sequence from the middle region of a long PCR product.

reverse sequences are obtained (Figure 10.6a). This is an advantage if the PCR product is more than 750 bp and hence too long to be sequenced completely in one experiment. Alternatively, it is possible to extend the sequence in one direction by synthesizing a new primer, designed to anneal at a position within the PCR product (Figure 10.6b).

If cloned DNA is being sequenced, then a universal primer can be used, this being a primer that is complementary to the part of the vector DNA immediately adjacent to the point into which new DNA is ligated (Figure 10.7). The 3′ end of the primer points toward the inserted DNA, so the sequence that is obtained starts with a short stretch of the vector and then progresses into the cloned DNA fragment. Any fragment that is cloned in the vector can therefore be sequenced with the same primer. Once again, both forward and reverse universal primers can be used, enabling sequences to be obtained from both ends of the insert, and an internal primer can provide sequence from the middle region of a long insert.

10.1.4 Limitations of chain-termination sequencing

With the most up-to-date versions of the chain-termination method it is possible to obtain over 750 bp of sequence per experiment. Although most genes – especially

Figure 10.7

Sequencing a cloned DNA fragment with a universal primer.

Table 10.1
Sizes of representative genomes.

SPECIES	TYPE OF ORGANISM	GENOME SIZE (Mb)
Mycoplasma genitalium	Bacterium	0.58
Haemophilus influenzae	Bacterium	1.83
Escherichia coli	Bacterium	4.64
Saccharomyces cerevisiae	Yeast	12.1
Caenorhabditis elegans	Nematode worm	97
Drosophila melanogaster	Insect	180
Arabidopsis thaliana	Plant	125
Homo sapiens	Mammal	3200
Triticum aestivum	Plant (wheat)	16 000

eukaryotic ones – are longer than this it is easy to carry out two or more chain-termination experiments, directed at different parts of a gene, in order to build up the complete sequence. Chain-termination sequencing is therefore the method of choice for sequencing genes and other DNA fragments obtained by cloning or PCR.

The chain-termination method was also used to obtain the first complete genome sequences. This was a much more challenging task, because even the smallest bacterial genomes are over 1 Mb in length, and the human genome, which initially was sequenced by the chain-termination method, is 3200 Mb (Table 10.1). Furthermore, as no sequencing method is entirely accurate it is necessary to sequence each region of a genome multiple times, in order to identify errors present in individual sequence reads (Figure 10.8). With the chain-termination method, to ensure that errors are identified, at least a fivefold sequence depth or coverage is required, which means that every nucleotide is present in five different reads. So, for the human genome, a total of 5 × 3200 = 16 000 Mb of sequence would be required. This is equivalent to over 21 million chain-termination sequences averaging 750 bp in length. When the human genome was sequenced during the 1990s and early 2000s, chain-termination sequencing was the only method available, so this challenge had to be met. In fact it was exceeded, as the Human Genome Project had generated 23 147 Mb of sequence by the time that the first draft of the genome was published in 2001. This capacity was achieved by automated sequencing machines, capable of generating 384 sequences in parallel in a single run, each run taking about one hour, corresponding to an output of almost 7 Mb per day under optimal working conditions.

```
                    AGCATCGTAGCTTCAGTATGATGATGCTAG   Read 1
      ATGATCGTAGCTAGCATCGTAGCTAGC                    Read 2
        ATCGTAGCTAGCATCGTAGCTAGCATCGTAGCTT           Read 3
                         TTGTAGCTTCAGTATGATGATGCTAG   Read 4
            GCATCGTAGCTAGCATCGTAGCTTCAGT             Read 5
      ATGATCGTAGCTAGCATCGTA                          Read 6
      ATGATCGTAGCTAGCATCGTAGCTAGCATCGTAGCTTCAGTATGATGATGCTAG   Deduced sequence
```

Figure 10.8

Each region of a genome must be sequenced multiple times, in order to identify errors present in individual sequence reads. In this example, the discrepancy in Read 4 in the highlighted column can be ascribed to a sequencing error, the correct nucleotide at this position being C.

Whilst 7 Mb of sequence per day is an impressive output, it still places a limitation on the speed with which a genome sequence can be obtained, even if multiple machines are operated continually in a factory setting. Cost is also an important consideration, because the objective – especially with the human genome – was not simply to sequence the genome of one person. Everybody's genome is slightly different, with its own personal sequence variations that provide us with our individual characteristics. These characteristics include not only the pleasant things such as our eye colour and the natural shade of our hair, but also the less pleasant things such as our individual susceptibility to disease. One of the goals of personalized medicine is to use individual genome sequences to make accurate diagnoses of a person's risk of developing a disease, and to use that person's genetic characteristics to plan effective therapies and treatment regimes. In order for personalized medicine to become a reality, cost-effective and rapid ways of sequencing individual genomes are needed. It was this requirement that stimulated the development of next-generation sequencing methods, aimed at generating increasingly larger amounts of sequence data not only more quickly but also at less cost.

10.2 Next-generation sequencing

The rationale behind next-generation sequencing is quite different to that of the chain-termination method. With the latter, the sequencing experiment is preceded by a cloning project or a PCR, and each cloned insert or PCR product is sequenced individually. With a next-generation method, a large library made up of thousands or millions of DNA fragments is sequenced in a single experiment (Figure 10.9). Often, these fragments represent an entire genome, and in most projects the sequencing experiment is preceded only by DNA extraction.

10.2.1 Preparation of a next-generation sequencing library

Although several different next-generation sequencing methods are available, they all begin with the preparation of a library of DNA fragments that have been immobilized

Figure 10.9

The difference between chain-termination and next-generation sequencing.

Components of the DNA library

Solid support

Figure 10.10

A DNA library immobilized on a solid support.

on a solid support, in such a way that the individual sequencing reactions can be carried out side-by-side in an array format (Figure 10.10). There are three steps to the preparation of this library:

1 Breakage of the starting DNA into fragments of sizes that are suitable for the sequencing method being used.
2 Immobilization of the fragments on to the solid support.
3 Amplification of the immobilized fragments.

DNA fragmentation

Library preparation begins with purification of the DNA to be sequenced, which is usually total cell DNA from the organism whose genome is being studied. The DNA is then broken into fragments, usually between 100 bp and 500 bp in length, with the precise sizes depending on the lengths of the individual sequences that can be obtained by the next-generation method being employed. The standard fragmentation method is sonication, in which ultrasound is used to cause breaks in the DNA molecules (Figure 10.11). The sonicated DNA is then fractionated by agarose gel electrophoresis, and fragments of the desired size purified from the gel.

Sonication is the preferred fragmentation method because it causes breaks at random positions in DNA molecules. Random breakage is important to ensure that all parts of the starting DNA are sequenced. This is because each fragment will be sequenced from its ends. If a restriction endonuclease were used to break up the DNA, some fragments might be too long to be sequenced entirely from their ends, and the internal regions of these fragments would therefore be absent from the final set of sequence reads that is obtained. With next-generation methods it is not possible to direct the sequencing

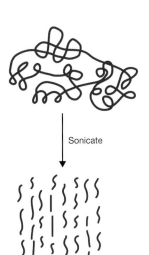

Figure 10.11

Random fragmentation of DNA by sonication.

Sonicate

Figure 10.12

Immobilization of the DNA fragments in a sequencing library by base pairing to oligonucleotides on a glass side.

towards the middle of a fragment, as can be done by designing an internal primer for the chain-termination method.

Immobilization of the library

The DNA fragments that make up the library cannot be directly immobilized on to a solid support. First, adaptors (p. 69) must be attached to the ends of the fragments. The exact role of these adaptors in library immobilization depends on the particular sequencing method being used. Two examples are as follows:

- In the first method, the solid support is a glass slide that has been coated with many copies of a short oligonucleotide (Figure 10.12). The sequence of this oligonucleotide matches the ends of the adaptors. The fragments are therefore denatured into single-stranded DNA. The single-stranded molecules then attach to the glass slide by base pairing between their adaptor sequences and the immobilized oligonucleotides.

- In the second method, the solid support is provided by small metallic beads that are coated with streptavidin. One of the adaptors has a biotin label attached to its 5′ end. DNA fragments therefore become attached to the beads by biotin–streptavidin linkages (Figure 10.13a). The ratio of DNA fragments to beads is set so that, on average, just one fragment becomes attached to each bead. The beads are then shaken in an oil–water mixture, with the conditions again carefully controlled so that there is just one bead in each aqueous droplet within the emulsion (Figure 10.13b). These aqueous droplets are transferred into wells on a plastic strip, one droplet per well, with thousands of wells in total.

Amplification of the library

With both immobilization methods, the final step in library preparation is amplification of the immobilized fragments by PCR, to produce a sufficient number of identical copies to be sequenced. The adaptors now play a second role as they provide the annealing sites for the primers for this PCR. The same pair of primers can therefore be used for all the fragments, even though the fragments themselves have many different sequences.

With the glass slide method, the PCR products become attached to oligonucleotides adjacent to the template fragment, with each template being converted into a cluster of identical fragments (Figure 10.14a). If metallic beads are used the PCR is carried out in the oil emulsion, with the products of each PCR being retained within their own water droplet, prior to deposition of the droplets into their wells (Figure 10.14b).

(a) Attachment to beads by streptavidin-biotin linkages

Streptavidin–biotin
attachment

Figure 10.13

Immobilization of the DNA fragments in a
sequencing library by base pairing to metallic
beads. (a) Each DNA fragment is attached to a
single bead via a streptavidin–biotin linkage.
(b) Individual beads, with their attached DNA
fragments, are placed within water droplets in
an oil–water emulsion.

(d) Beads in water droplets in an oil emulsion

10.2.2 Next-generation sequencing methods

Once the fragment library has been immobilized on to its solid support, the actual
sequencing phase of the next-generation project can begin. Among the various next-
generation sequencing methods that have been devised, two are particularly popular:

- Reversible terminator sequencing, which is usually performed with fragment
  clusters on a slide support.
- Pyrosequencing, which is usually carried out with fragments immobilized on
  metallic beads.

(a) Glass slide method

(b) Metallic bead method

Figure 10.14

Amplification of immobilized DNA libraries.

An important characteristic of both methods is that there is no electrophoresis or any other fragment separation step, which would be difficult – if not impossible – to carry out in a massively parallel format. Instead, each fragment in the immobilized library is copied by a DNA polymerase and, as the new strand is being made, the order in which the nucleotides are incorporated is detected so that the sequence can be read as the reaction proceeds.

Reversible terminator sequencing

This method, like chain-termination sequencing, makes use of modified nucleotides which block further strand synthesis once one has been incorporated at the end of the growing polynucleotide. There is, however, a critical difference. As the name 'reversible terminator' implies, the termination step is reversible because the chemical group that has been attached at the 3′ position of the terminator nucleotide can be removed, con-verting the position to a 3′-hydoxyl group, which allows further extension to occur (Figure 10.15a). The terminator nucleotides are labelled, each with a different fluo-rophore, so which of the four nucleotides is added at each step can be identified. Once the nucleotide has been identified, the 3′ blocking group, as well as the fluorophore, are detached, leaving a standard nucleotide (Figure 10.15b). The next step in the strand synthesis reaction can now take place, enabling the next nucleotide in the sequence to be identified.

The reversible terminator method with libraries immobilized on a slide support gen-erates relatively short sequence reads, with a maximum length of 300 bp, but is so

Figure 10.15

Reversible terminator sequencing.

(a) A reversible terminator nucleotide

(b) Reversible terminator sequencing

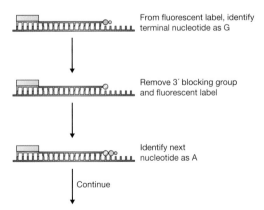

massively parallel that up to 2000 Mb of sequence can be obtained per run. This technology is usually referred to as Illumina sequencing, named after the company that markets the necessary equipment.

Pyrosequencing

Pyrosequencing takes a radically different approach to either the chain-termination or reversible terminator methods. The individual nucleotides are not labelled; instead, the sequence is read by detecting flashes of chemiluminescence. The chemiluminescence is generated by the enzyme sulphurylase from the molecule of pyrophosphate released each time DNA polymerase adds a deoxynucleotide to the 3' end of the growing strand. Hence, a flash of chemiluminescence signals the successful copying of one position in the template molecule. Of course, if all four deoxynucleotides were to be added at once, the flashes of light would be seen continually and no useful sequence information would be obtained. Consequently, each deoxynucleotide is added separately, one after the other, with a nucleotidase enzyme also present in the reaction mixture. In this way, if a deoxynucleotide is not incorporated into the polynucleotide it will be rapidly degraded before the next one is added (Figure 10.16). This procedure makes it possible to follow the order in which the deoxynucleotides are incorporated into the growing strand.

Pyrosequencing of fragment libraries immobilized on magnetic beads is the procedure used by the next-generation method termed 454 sequencing, which is named after the company that initially developed this particular technology. In its most advanced format, 454 sequencing can give sequence reads up to 1000 bp in length, with enough reads to generate a total of 700 Mb of DNA sequence per run.

Figure 10.16

Pyrosequencing.

Figure 10.17

Real-time DNA sequencing with each nucleotide addition detected with a zero-mode waveguide.

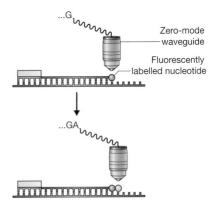

10.2.3 Third-generation sequencing

The development of new DNA sequencing methods continues apace. One unsatisfactory aspect of both the reversible terminator and pyrosequencing approaches is the time that it takes to complete a sequencing experiment. This is because the methods used for reading the sequence require a delay at each step in the strand synthesis reaction. With the reversible terminator method the delay is caused by the need to remove the 3′ blocking group and the fluorophore after each nucleotide addition, while during pyrosequencing the delay occurs because each nucleotide is presented individually to the polymerase.

Third-generation sequencing aims to improve on the next-generation methods by carrying out sequencing in real time – in other words during the normal, unimpeded progression of the polymerase along the template as it synthesizes the second strand. Real-time sequencing also enables read lengths to be longer, as the regular pauses that occur during the next-generation methods reduces the processivity of the polymerase.

One of the most promising third-generation methods is single molecule real-time sequencing. A sophisticated optical system called a zero-mode waveguide is used to observe the copying of a single DNA template, with each nucleotide addition identified from its attached fluorescent label (Figure 10.17). Because the optical system is so precise, there is no need to block the 3′ carbon of the nucleotide. Instead, the fluorescent signal is detected and the label removed immediately after nucleotide incorporation, so strand synthesis can progress without interruption. Read lengths of up to 20 000 bp have been reported with this method.

10.2.4 Directing next-generation sequencing at specific sets of genes

When used in genome sequencing, the random nature of the next-generation methods is an advantage, as it means that all parts of the genome are sequenced at the same time. Genome sequencing is hugely important, but sometimes the objective of a sequencing project is only to sequence a part of a genome. One example is in the screening of human DNA for sequence variations that might indicate that a patient has a susceptibility to particular types of cancer. This type of screen would be more efficient if the sequencing were directed at just those 500–600 genes that are associated with cancer. Those genes make up about 4 Mb of DNA, compared with the 3200 Mb of the entire human genome.

(a) DNA capture

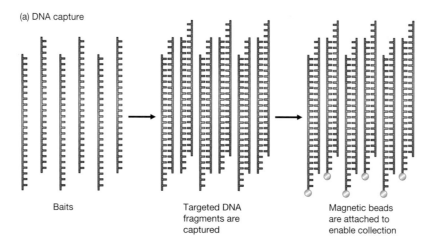

Baits

Targeted DNA
fragments are
captured

Magnetic beads
are attached to
enable collection

(b) Comparison between sequencing before and after capture

ACGTATCATGCGGATGG
TAGCATGACGTAGCGTT
GTAGCAGGTACGATGCC
GTAGACGATGCAGCATC
TAGGACCTAGCCGGACA

Many DNA fragments
are sequenced

ATCGAGACTACGTAGCC
TAGCTGATGCTACGTAA
ATCGAGACTACGTAGCC
ATCGAGACTACGTAGCC
TACTGCAGCTATAGCGG

Selected fragments
are sequenced

Figure 10.18

Target enrichment. (a) Baits are used to capture DNA fragments representing genes of interest. (b) Only the captured DNA fragments are sequenced.

Next-generation sequencing can be directed at specific sets of genes by a target enrichment method. For this, a large set of oligonucleotides, each typically 150 nucleotides in length, is synthesized, with the sequences of the oligonucleotides collectively corresponding to the sequences of the genes that are being targeted (Figure 10.18a). The oligonucleotides then act as baits that hybridize to, and hence capture, the DNA fragments representing the genes of interest during library preparation. Various strategies can be used. One approach is similar to the magnetic bead method used for the purification of total cell DNA (p. 31). One end of each bait is labelled with biotin and, after hybridization, the captured DNA is attached, via the baits, to streptavidin-coated magnetic beads. A magnet is then used to collect the beads so that the DNA from the non-targeted parts of the genome, which remains in solution, can be discarded. The captured DNA is then released from the baits and used to prepare the sequencing library (Figure 10.18b).

10.3 How to sequence a genome

The first DNA molecule to be completely sequenced was the 5386 nucleotide genome of bacteriophage φX174, which was completed in 1975. This was quickly followed by sequences for SV40 virus (5243 bp) in 1977 and pBR322 (4363 bp) in 1978. Gradually, sequencing was applied to larger molecules. Professor Sanger's group published the sequence of the human mitochondrial genome (16.6 kb) in 1981 and of bacteriophage λ (49 kb) in 1982.

During the 1990s, automated methods for chain-termination sequencing were developed, enabling much larger genomes to be sequenced. The first eukaryotic chromosome sequence, for chromosome III of the yeast *Saccharomyces cerevisiae*, was published in 1992, and the entire yeast genome was completed in 1996. This was followed by complete genome sequences for other model organisms: for the worm *Caenorhabditis elegans*, the fly *Drosophila melanogaster*, and the plant *Arabidopsis thaliana*. This phase of genome sequencing included publication of the human genome sequence in 2001. With the development of next-generation methods, genome sequencing has become more routine, and complete genome sequences are now known for over 300 eukaryotic species and several thousand bacteria and archaea, with many more sequencing projects currently in progress.

Since the 1990s, when the chain-termination method was first automated, the actual generation of sequence data has not been a limiting factor in genome sequencing projects. Rather, the main challenge lies with sequence assembly, which is the procedure used to convert the thousands or millions of short sequence reads into the contiguous genome sequence. The simplest way of doing this is the shotgun approach, in which the genome is randomly broken into short fragments which are individually sequenced. The genome sequence is then assembled by searching for overlaps between the individual short sequences (Figure 10.19).

10.3.1 Shotgun sequencing of prokaryotic genomes

The shotgun approach to sequence assembly was first used during the 1990s with bacterial and archaeal genomes, which are much shorter than eukaryotic genomes (see Table 10.1), and hence can be sequenced from a relatively small number of individual DNA fragments. The fewer the fragment number, the easier the sequence assembly will be.

Shotgun sequencing of the Haemophilus influenzae *genome*
The bacterium *Haemophilus. influenzae* was the first free-living organism whose genome was sequenced, the results being published in 1995. The first step was to

Figure 10.19

The shotgun method for sequence assembly. The DNA molecule is broken into small fragments, each of which is sequenced. The master sequence is assembled by searching for overlaps between the sequences of individual fragments. In practice, an overlap of several tens of base pairs would be needed to establish that two sequences should be linked together.

H. influenzae
genomic DNA

Sonicate

DNA fragments
of various sizes

Agarose gel electrophoresis

Lane 1: Sonicated H.
     influenzae DNA
Lane 2: DNA markers

1   2

Purify DNA
from the gel

DNA fragments –
1.6–2.0 kb

Prepare a clone library

Obtain end-sequences
of DNA inserts

End-sequences

Construct sequence
contigs

Figure 10.20

A schematic of the key steps in the *H. influenzae*
genome sequencing project.

sonicate the 1830 kb genome, breaking the DNA into short fragments that were then
cloned in plasmid and M13 vectors to provide the templates for the sequencing experiments. It was decided to concentrate on fragments of 1.6–2.0 kb because these could
yield two chain-termination DNA sequences, one from each end, reducing the amount
of cloning and DNA preparation that was required. The sonicated DNA was therefore fractionated by agarose gel electrophoresis and fragments of the desired size purified from the gel (Figure 10.20). After cloning, 28 643 chain-termination sequencing

experiments were carried out with 19 687 of the clones. A few of these sequences – 4339 in all – were rejected because they were less than 400 bp in length. The remaining 24 304 sequences were entered into a computer, which then spent 30 hours searching for overlaps among the sequences. The result was 140 contiguous sequences or contigs, each made up of a series of overlapping sequence reads.

The average contig length was 12 kb, suggesting that together they made up over 90% of the genome sequence, although of course that sequence would not be complete until the contigs had been placed in their correct order and the gaps between them closed. It might have been possible to continue sequencing more of the sonicated fragments in order eventually to close the gaps between the individual contigs. However, 11 631 485 bp of sequence had already been generated – six times the length of the genome – suggesting that a large amount of additional work would be needed before the correct fragments were, by chance, sequenced. It was also possible that some of the gaps were contained in fragments of DNA that could not be cloned because of incompatibility problems, which occur with all cloning vectors and prevent inserts with certain sequences from being propagated. If the gap sequences were absent from the clone library, then no amount of additional sequencing would close them.

A more directed strategy was therefore used to close each of the gaps individually. Several approaches were tried, the most successful involving hybridization analysis of a clone library prepared in a λ vector (Figure 10.21). Incompatibility problems tend to be specific for particular vector types, so fragments that cannot be cloned in a plasmid vector can often be cloned in λ. The λ library was probed in turn with a series of oligonucleotides whose sequences corresponded with the ends of each of the 140 contigs. In some cases, two oligonucleotides hybridized to the same λ clone, indicating that the two contig ends represented by those oligonucleotides lay adjacent to one another

Figure 10.21

Using oligonucleotide hybridization to close gaps in the *H. influenzae* genome sequence. Oligonucleotides 2 and 5 both hybridize to the same λ clone, indicating that contigs I and III are adjacent. The gap between them can be closed by sequencing the appropriate part of the λ clone.

(a) Prepare oligonucleotide probes

(b) Probe a genomic library

Figure 10.22

Using a reference genome to aid sequence assembly.

in the genome. The gap between these two ends could then be closed by sequencing the appropriate part of the λ clone.

Shotgun sequencing of other prokaryotic genomes

Shotgun sequencing has been successful with many prokaryotic genomes, the short lengths of these genomes meaning that the computational requirements for finding sequence overlaps are not too great. If a next-generation sequencing method is used, then gap closure is less of an issue, because a single 454 or Illumina experiment can easily generate sequence reads whose total length is several hundred times that of the genome.

Many prokaryotic genomes are assembled by *de novo* sequencing, as described above for *H. influenzae*, with the genome sequence worked out solely by finding overlaps between individual sequence reads. However, as the genomes of more and more species become available, a second approach to assembly is sometimes possible. This is when an existing sequence is used as a reference genome for the assembly of a related genome. Rather than just looking for overlaps among the sequence reads for the genome that is being assembled, individual reads and short contigs are also mapped on to the reference sequence by looking for regions of sequence similarity (Figure 10.22). The rationale is that if the reference genome comes from a related species, or from a different strain of the same species, then the sequences of the two genomes will be similar and the reference can therefore direct assembly of the new genome.

The reference genome approach makes sequence assembly quicker and more accurate, but care has to be taken to identify any regions in the new genome where the gene order has been changed by recombination (Figure 10.23). In these cases the only difference between the sequences will be at the boundaries of the segments that have been rearranged. If these boundaries are not identified, then reliance on the reference genome might lead to the rearrangement not being recognized in the genome being assembled.

Figure 10.23

One potential problem when using a reference genome. In this example, the positions of genes B and C are different in the reference compared with the genome being sequenced. If the sequence coverage does not include the critical regions at the boundaries for this recombination event, then the fact that the gene order is different in the genome being sequenced will not be recognized.

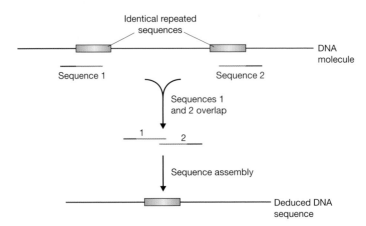

Figure 10.24

One problem with the shotgun approach. An incorrect overlap is made between two sequence reads that both terminate within a repeated element. The result is that a segment of the DNA molecule is absent from the DNA sequence.

10.3.2 Sequencing of eukaryotic genomes

As well as their small size, prokaryotic genomes have a second advantage that aids sequence assembly by the shotgun method. Prokaryotic genomes contain few or no repetitive DNA sequences. These are sequences which range from a few base pairs to several kilobases in size and are repeated at two or more places in a genome. They cause problems for the shotgun approach because a sequence read that lies partly or wholly within one repeat element might accidentally be assigned an overlap with the identical sequence present in a different repeat element (Figure 10.24). This could lead to a part of the genome sequence being placed at the incorrect position or left out entirely.

Although absent from most bacterial genomes, repetitive sequences are common in eukaryotes, and in some species make up more than 50% of the genome. For this reason, it was initially thought that a eukaryotic genome could not be assembled entirely by shotgun sequencing.

The hierarchical shotgun approach

The first eukaryotic genomes, including the human genome, were sequenced by a modification of the shotgun approach called hierarchical shotgun sequencing. This approach involves a pre-sequencing phase during which the genome is broken into large fragments, typically 300 kb in length, and these fragments then cloned into a high-capacity vector such as a BAC (p. 106). Clones that contain overlapping fragments of DNA are then identified, enabling a contiguous series to be built up (Figure 10.25). Each piece of cloned DNA is then broken into smaller fragments and these sequenced and assembled by the shotgun method. The sequence of each cloned fragment can then be placed at its appropriate position in the series of clones in order to gradually build up the overlapping genome sequence. The confusion caused by repetitive DNA now arises only if two or more copies of the same repeat are present in the same cloned fragment, and even then it might be possible to identify errors in the assembly by examining the sequences of clones that overlap with the one containing the repeats (Figure 10.26).

The problem with the hierarchical approach is that identifying clones that contain overlapping inserts is a massive task. One technique that can be used to overcome this

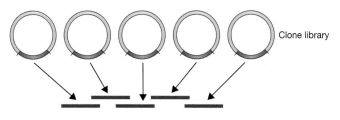

Figure 10.25

A set of clones whose fragments overlap.

difficulty is chromosome walking. To begin a chromosome walk a clone is selected at random from the library, labelled, and used as a hybridization probe against all the other clones in the library (Figure 10.27a). The clones that give hybridization signals are those that overlap with the probe. One of these overlapping clones is now labelled and a second round of probing carried out. More hybridization signals are seen, some of these indicating additional overlaps (Figure 10.27b). Gradually, the clone contig is built up in a step-by-step fashion.

The weakness of chromosome walking is that it begins at a fixed starting point and builds up the series of overlapping clones one by one, and hence slowly, from that fixed point. More rapid techniques do not use a fixed starting point but instead aim to identify pairs of overlapping clones. When enough overlapping pairs have been identified the series of clones is revealed (Figure 10.28). The various techniques that can be used to identify overlaps are collectively known as clone fingerprinting. These techniques are based on the identification of sequence features that are shared by a pair of clones. If

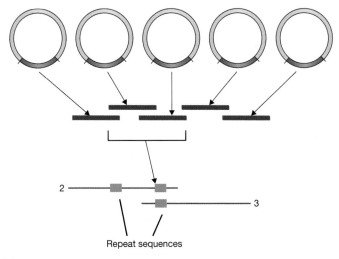

Figure 10.26

Hierarchical shotgun sequencing can avoid problems with repeat sequences. The positions of repeat sequences in clones 2 and 3 from Figure 10.25 are shown. Sequence assembly of the reads from clone 2 could result in the segment between the two repeats being lost. However, the sequence of clone 3 enables the mistake to be recognized and corrected.

Figure 10.27

Chromosome walking.

(a) Probe the library with clone A1

(b) Probe the library with clone I4

two clones contain this feature then clearly they must overlap. A sequence feature of this type is called a sequence tagged site (STS). Often, an STS is a gene that has been sequenced in an earlier project. As the sequence is known, a pair of PCR primers that are specific for that gene can be designed and used to identify which members of a clone library contain the gene. However, the STS does not have to be a gene – it can be any short piece of DNA sequence, the only requirement being that it occurs just once in the genome.

Shotgun sequencing of eukaryotic genomes

The identification of a series of overlapping clones is a time-consuming process, and there would be major advantages if this stage could be side-stepped and a eukaryotic genome sequenced entirely by the shotgun approach. Eukaryotic genomes are of course much longer than prokaryotic ones – 3200 Mb for the human genome compared with 1.83 Mb for *H. influenzae* – so a correspondingly larger number of sequence reads would have to be examined in order to build up lengthy contigs. This is a computational issue, and has been solved by the development of sophisticated algorithms for sequence assembly, concurrent with the increase in processor power for laboratory computers. But what about the problems that repetitive DNA poses for sequence assembly?

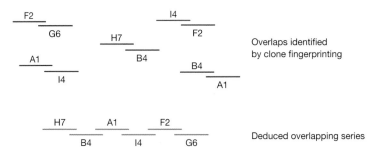

Figure 10.28

Building up a series of overlapping clones using a clone fingerprinting technique.

Repeat sequences

Both end-sequences of this 10-kb
fragment are positioned correctly on
the master sequence

Figure 10.29

The strategy used to ensure that repetitive DNA regions were
assembled correctly when the *Drosophila melanogaster* genome
was sequenced. Two identical 8-kb repeats are adjacent in the
genome. The end-sequences of the 10-kb fragment that spans one
of these repeats can be checked against the master sequence to
ensure that the segment between the two repeats has not been
lost.

Several possible ways of avoiding errors when repetitive DNA regions are assembled
have been explored. The most successful strategy is to use two or more sequencing
libraries, with one of the libraries containing fragments that are longer than the longest
repeat sequences in the genome being studied. For example, one of the clone libraries
used when this method was applied to the *D. melanogaster* genome contained inserts
with an average size of 10 kb, because most *Drosophila* repeat sequences are 8 kb or
shorter in length. Sequence jumps, from one repeat sequence to another, were avoided by
ensuring that the two end-sequences of each 10-kb fragment were at their appropriate
positions in the master sequence (Figure 10.29).

Libraries containing fragments of different sizes are still used in many eukaryotic
genome projects. The initial result of sequence assembly is a series of scaffolds, each
of which comprises a set of contigs, made up of reads from the short-fragment library,
separated by gaps which lie between paired-end reads – the mini-sequences from the
two ends of a single large fragment (Figure 10.30). As more sequences are added into
the dataset, longer and longer scaffolds are built up. This strategy was initially used
with chain-termination sequencing, with short fragments of DNA cloned into a plasmid
vector, and the longer fragments obtained as λ or BAC clones. However, with the advent
of next-generation sequencing a hybrid approach was adopted, with the next-generation
method being used to generate reads from short fragments, and the chain-termination
method retained for obtaining the paired-end reads of large fragment clones. In 2010,
the first successful genome assembly based entirely on next-generation sequencing was
reported, for the giant panda. This project used fragment libraries of 150 bp, 500 bp,
2 kb, 5 kb, and 10 kb, along with the ability of modern next-generation machines to
sequence both ends of a fragment and hence obtain paired-end reads without recourse
to chain-termination sequencing.

Contig 1   Contig 2   Contig 3

Short reads

Scaffold

Paired-end reads of long fragments

Figure 10.30

A scaffold. The paired-end reads of the long fragments are used to place the contigs in their correct order.

What do we mean by 'genome sequence'?

It is important to recognize that a 'complete' genome sequence – one in which every nucleotide is sequenced, with no errors, and every segment placed at its correct position – is currently an impossibility for a eukaryotic genome. The development of the Human Genome Project illustrates this point. The initial version of the human genome obtained by the hierarchical shotgun method, published in 2001, was a draft that covered just 90% of the genome. The missing 320 Mb was located predominantly in constitutive heterochromatin, regions of chromosomes in which the DNA is very tightly packaged and which are thought to contain few, if any, genes. Within the 90% of the genome that was covered, each part had been sequenced at least four times, providing an 'acceptable' level of accuracy, but only 25% had been sequenced the eight to ten times deemed necessary before the sequence could be considered to be 'finished'. Furthermore, this draft sequence had approximately 150 000 gaps, and some segments had probably not been ordered correctly. Even the 'finished' sequence, achieved in 2005, was not entirely complete, the standard being that at least 95% of the euchromatin – the part of the genome in which most of the genes are located – should be sequenced with a predicted error rate of less than one in 10^4 nucleotides, and all except the most refractory gaps filled.

Since 2005, a number of statistics have been devised to measure the degree of completeness of a genome sequence. One of these is the N50 size, which can be applied to either contigs or scaffolds. To calculate the contig N50 size, the total length of all the contigs added together is calculated. The contigs are then ordered by length, from the shortest to the longest. Beginning with the longest contig, the individual contig lengths are added together until their combined length is equal to half the total length of all the contigs. The N50 value is the length of the last – and hence shortest – contig whose addition makes the combined length greater than 50% of the total. A higher N50 value therefore indicates a more complete assembly. The same approach can be used with scaffolds rather than contigs. A different value, the NG50 size, is based not on the total length of contigs or scaffolds but on the actual genome size. The NG50 value therefore enables direct comparisons to be made between genome assemblies for different species.

Further reading

Fleischmann, R.D., Adams, M.D., White, O., *et al.* (1995) Whole genome random sequencing and assembly of *Haemophilus influenzae* Rd. *Science*, **269**, 496–512. [The first complete bacterial genome sequence to be published.]

Li, R., Fan, W., Tian, G., *et al.* (2010) The sequence and *de novo* assembly of the giant panda genome. *Nature*, **463**, 311–317. [Genome sequencing based entirely on next-generation methods.]

Loman, N.J., Constantinidou, C., Chan, J.Z.M., *et al.* (2012) High-throughput bacterial genome sequencing: an embarrassment of choice, a world of opportunity. *Nature Reviews Microbiology*, **10**, 599–606.

Ronaghi, M., Ehleen, M., and Nyrn, P. (1998) A sequencing method based on real-time pyrophosphate. *Science*, **281**, 363–365. [Pyrosequencing.]

Sanger, F., Nicklen, S., and Coulson, A.R. (1977) DNA sequencing with chain-terminating inhibitors. *Proceedings of the National Academy of Sciences of the United States of America*, **74**, 5463–5467. [The first description of chain-termination sequencing.]

Schatz, M.C., Delcher, A.L., and Salzberg, S.L. (2010) Assembly of large genomes using second-generation sequencing. *Genome Research*, **20**, 1165–1173.

Sears, L.E., Moran, L.S., Kisinger, C., *et al.* (1992) CircumVent thermal cycle sequencing and alternative manual and automated DNA sequencing protocols using the highly thermostable Vent (exo⁻) DNA polymerase. *Biotechniques*, **13**, 626–633. [Thermal cycle sequencing.]

van Dijk, E.L., Auger, H., Jaszczyszyn, Y., and Thermes, C. (2014) Ten years of next-generation sequencing technology. *Trends in Genetics*, **30**, 418–426.

Chapter 11

Studying Gene Expression and Function

All genes have to be expressed in order to function. The first step in expression is transcription of the gene into a complementary RNA strand (Figure 11.1a). For some genes – for example, those coding for transfer RNA (tRNA) and ribosomal RNA (rRNA) molecules – the transcript itself is the functionally important molecule. For other genes, the transcript is translated into a protein molecule.

To understand how a gene is expressed, the RNA transcript must be studied. In particular, the molecular biologist will want to know whether the transcript is a faithful copy of the gene, or whether segments of the gene are missing from the transcript (Figure 11.1b). These missing pieces are called introns, and considerable interest centres on their structure and possible function. In addition to introns, the exact locations of the start and end points of transcription are important. Most transcripts are copies not only of the gene itself but also of the nucleotide regions either side of it (Figure 11.1c). The signals that determine the start and finish of the transcription process are only partly understood, and their positions must be located if the expression of a gene is to be studied.

In this chapter we will begin by looking at the methods used for transcript analysis. These methods can be used to map the positions of the start and end points for transcription, and also to determine if a gene contains introns. We will then briefly

Gene Cloning and DNA Analysis: An Introduction, Seventh Edition. T.A. Brown.
© 2016 John Wiley & Sons, Ltd. Published 2016 by John Wiley & Sons, Ltd.
Companion Website: www.wiley.com/go/brown/cloning

Figure 11.1

Some fundamentals of gene expression.

(a) Genes are expressed by transcription and translation

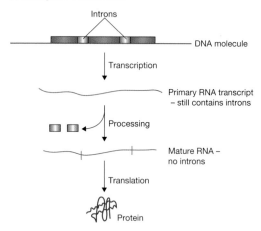
(b) Some genes contain introns

(c) RNA transcripts include regions either side of the gene

consider a few of the numerous techniques that have been developed during recent years for examining how the expression of a gene is regulated. These techniques are very important, as aberrations in gene regulation underlie many clinical disorders. Finally, we will tackle the difficult problem of how to identify the translation product of a gene.

11.1 Studying the RNA transcript of a gene

Over the years, a variety of techniques have been devised for studying RNA transcripts, including methods for detecting the presence of a transcript, for measuring its length, and obtaining its RNA sequence. More advanced techniques also enable the start and end of the transcript to be mapped and the positions of introns to be located.

11.1.1 Detecting the presence of a transcript and determining its nucleotide sequence

Before studying the more sophisticated techniques for RNA analysis, we must consider the methods used to obtain basic information about a transcript. The first of these methods is northern hybridization, the RNA equivalent of Southern hybridization (p. 151), which is used to measure the length of a transcript. An RNA extract is electrophoresed in an agarose gel, using a denaturing electrophoresis buffer (e.g., one containing formaldehyde) to ensure that the RNAs do not form intermolecular or intramolecular base pairs, as base pairing would affect the rate at which the molecules migrate through the gel. After electrophoresis, the gel is blotted onto a nylon or nitrocellulose membrane, and hybridized with a labelled probe (Figure 11.2). If the probe is a cloned gene, the band that appears in the autoradiograph is the transcript of that gene. The size of the transcript can be determined from its position within the gel, and if RNA from different tissues is run in different lanes of the gel, then the possibility that the gene is differentially expressed can be examined.

Once a transcript has been identified, cDNA synthesis (p. 142) can be used to convert it into a double-stranded DNA copy, which can then be cloned and sequenced. Comparison between the sequence of the cDNA and the sequence of its gene will reveal the positions of introns and possibly the start and end points of the transcript. For this to be possible, the cDNA must be a full-length copy of the mRNA from which it is derived. The 3′ end of the transcript will usually be represented in the cDNA, because most methods for cDNA synthesis begin with a primer that anneals to the poly(A) tail of the mRNA, which means that the 3′ end is the first part to be copied (see Figure 8.7). The cDNA synthesis might not, however, continue all the way to the 5′ end of the transcript, especially if the RNA is more than a few hundred nucleotides in length. Premature

Figure 11.2

Northern hybridization. Three RNA extracts from different tissues have been electrophoresed in an agarose gel. The extracts are made up of many RNAs of different lengths so each gives a smear of RNA, but two distinct bands are seen in the gel, one for each of the abundant ribosomal RNAs. The sizes of these rRNAs are known (e.g., 4718 and 1874 nucleotides in mammals), so they can be used as internal size markers. The gel is transferred to a membrane, probed with a cloned gene, and the results visualized, for example by autoradiography if the probe has been radioactively labelled. Only lane 1 gives a band, showing that the cloned gene is expressed only in the tissue from which this RNA extract was obtained.

Figure 11.3

The 5′ end of an mRNA cannot be identified from the sequence of an incomplete cDNA.

termination of cDNA synthesis will result in a cDNA that is not a full-length copy of its transcript, and whose 3′ end does not correspond to the true 5′ end of the mRNA (Figure 11.3). If this is the case, then the start point for transcription cannot be identified from the sequence of the cDNA. It will also be impossible to map the positions of any introns that are located near the start of the gene.

11.1.2 Transcript mapping by hybridization between gene and RNA

The problems with incomplete cDNA synthesis mean that other, more direct methods are needed to identify the start point of a transcript. One of these methods is based on examination of the hybridization product formed between a cloned gene and its RNA.

Nucleic acid hybridization occurs just as readily between complementary DNA and RNA strands as between single-stranded DNA molecules. If a hybrid is formed between a DNA strand, containing a gene, and its mRNA, then the boundaries between the double- and single-stranded regions will mark the start and end points of the mRNA

Figure 11.4

A DNA–mRNA hybrid and the effect of treating this hybrid with a single-strand-specific nuclease such as S1.

(Figure 11.4a). Introns, which are present in the DNA but not in the mRNA, will 'loop out' as additional single-stranded regions.

Now consider the result of treating the DNA–RNA hybrid with a single-strand-specific nuclease such as S1 (p. 50). The S1 nuclease degrades single-stranded DNA or RNA polynucleotides, including single-stranded regions at the ends of or within predominantly double-stranded molecules, but has no effect on double-stranded DNA or on DNA–RNA hybrids. S1 nuclease will therefore digest the non-hybridized, single-stranded DNA regions at each end of the DNA–RNA hybrid, along with any looped-out introns (Figure 11.4b). Those parts of the DNA polynucleotide that were protected from S1 nuclease digestion can then be recovered if the RNA strand is degraded by treatment with alkali.

Unfortunately, the manipulations shown in Figure 11.4 are not very informative. The sizes of the protected DNA fragments could be measured by gel electrophoresis, but this does not allow their order or relative positions in the DNA sequence to be determined. However, a few subtle modifications to the technique allow the precise start and end points of the transcript and of any introns it contains to be mapped onto the DNA sequence. This procedure is called S1 nuclease mapping, and an example of how it is used to locate the start point of a transcript is shown in Figure 11.5. Here, a Sau3A fragment that contains 100 bp of coding region, along with 300 bp of the leader sequence preceding the gene, has been cloned into an M13 vector and obtained as a single-stranded molecule. A sample of the RNA transcript is added and allowed to anneal to the DNA molecule. The DNA molecule is still primarily single-stranded but now has a small region protected by the RNA transcript. All but this protected region is digested by S1 nuclease, and the RNA is degraded with alkali to leave a short, single-stranded DNA fragment. If these manipulations are examined closely, it will become clear that the size of this single-stranded fragment corresponds to the distance between the transcription start point and the right-hand Sau3A site. The size of the single-stranded fragment is therefore determined by gel electrophoresis and this information is used to locate the transcription start point on the DNA sequence. Exactly the same strategy could be used to locate the end point of transcription and the junction points between introns and exons. The only difference would be the position of the restriction site chosen to delimit one end of the protected single-stranded DNA fragment.

11.1.3 Transcript analysis by primer extension

S1 nuclease analysis is a powerful technique that allows both the 5′ and 3′ termini of a transcript, as well as the positions of intron–exon boundaries, to be identified. The second method of transcript analysis that we will consider – primer extension – is less adaptable, because it can only identify the 5′ end of an RNA. It is, nonetheless, an important technique that is frequently used to check the results of S1 analyses.

Primer extension can only be used if at least part of the sequence of the transcript is known. This is because a short oligonucleotide primer must be annealed to the RNA at a known position, ideally within 100–200 nucleotides of the 5′ end of the transcript. Once annealed, the primer is extended by reverse transcriptase (p. 52). This is a cDNA synthesis reaction, but one that is very likely to proceed to completion as only a short segment of RNA has to be copied (Figure 11.6). The 3′ end of the newly synthesized strand of DNA will therefore correspond with the 5′ terminus of the transcript. Locating the position of this terminus on the DNA sequence is achieved simply by determining

Figure 11.5

Locating a transcription start point by S1 nuclease mapping.

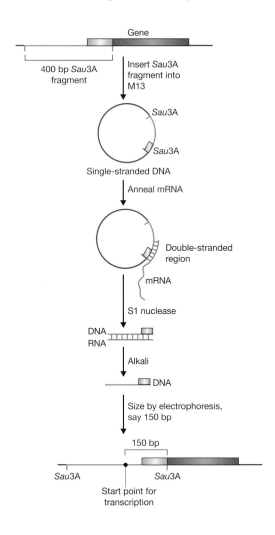

the length of the single-stranded DNA molecule and correlating this information with the annealing position of the primer.

11.1.4 Transcript analysis by PCR

We have already studied the way in which reverse transcriptase PCR is used to quantify the amount of a particular RNA in an extract (p. 171). The standard reverse transcriptase PCR procedure provides a copy of the internal region of an RNA molecule, but does not provide any information about the ends of the molecule. However, a modified version termed rapid amplification of cDNA ends (RACE) can be used to identify the 5′ and 3′ termini of RNA molecules and, like S1 analysis, can be used to map the ends of transcripts.

There are several variations to the RACE method, but here we will consider how the 5′ end of an RNA molecule can be mapped (Figure 11.7). This procedure uses a primer that is specific for an internal region of the RNA molecule. The primer attaches to the RNA and directs the first, reverse transcriptase-catalysed stage of the process,

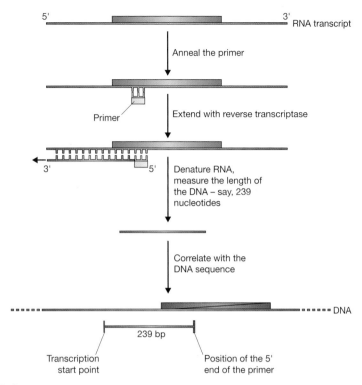

Figure 11.6

Locating a transcription start point by primer extension.

during which a single-stranded cDNA is made. As in the primer extension method, the 3′ end of the cDNA corresponds with the 5′ end of the RNA. Terminal deoxynucleotidyl transferase (p. 53) is now used to attach a series of A nucleotides to the 3′ end of the cDNA, forming the priming site for a second PCR primer, which is made up entirely of Ts and hence anneals to the poly(A) tail created by terminal transferase. Now the standard PCR begins, first converting the single-stranded cDNA into a double-stranded molecule, and then amplifying this molecule as the PCR proceeds. The PCR product is then sequenced to reveal the precise position of the start of the transcript.

11.2 Studying the regulation of gene expression

Few genes are expressed all the time. In fact, most are subject to regulation and are switched on only when their gene product is required by the cell. The simplest gene regulation systems are found in bacteria such as *E. coli*, which can regulate the expression of genes for biosynthetic and metabolic processes, so that gene products that are not needed are not synthesized. For instance, the genes coding for the enzymes involved in

Figure 11.7

One version of RACE.

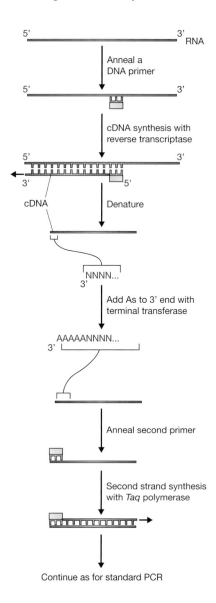

tryptophan biosynthesis can be switched off when there are abundant amounts of tryptophan in the cell, and switched on again when tryptophan levels drop. Similarly, genes for the utilization of sugars such as lactose are activated only when the relevant sugar is there to be metabolized. In higher organisms, gene regulation is more complex because there are many more genes to control. The differentiation of cells involves wholesale changes in gene expression patterns, and the process of development from fertilized egg cell to adult requires coordination between different cells as well as time-dependent changes in gene expression.

Many of the problems in gene regulation require a classical genetic approach. Genetics enables genes that control regulation to be located, allows the biochemical signals that influence gene expression to be identified, and can explore the interactions between different genes and gene families. It is for this reason that most of the breakthroughs

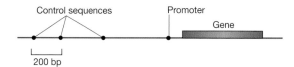

Figure 11.8

Possible positions for control sequences in the region upstream of a gene.

Control sequences

Promoter

Gene

200 bp

in understanding gene regulation in higher organisms have started with genetic studies of model species such as the fruit fly *Drosophila melanogaster*. Gene cloning and DNA analysis complement classical genetics as they provide much more detailed information on the molecular events involved in regulating the expression of a single gene.

We now know that a gene subject to regulation has one or more control sequences in its upstream region (Figure 11.8), and that the gene is switched on and off by the attachment of regulatory proteins to these sequences. A regulatory protein may repress gene expression, in which case the gene is switched off when the protein is bound to the control sequence or, alternatively, the protein may have a positive or enhancing role, switching on or increasing expression of the target gene. In this section we will examine methods for locating control sequences and determining their roles in regulating gene expression.

11.2.1 Identifying protein binding sites on a DNA molecule

A control sequence is a region of DNA that can bind a regulatory protein. It should therefore be possible to identify control sequences upstream of a gene by searching the relevant region for protein binding sites. There are three different ways of doing this.

Gel retardation of DNA–protein complexes

Proteins are quite substantial structures, and attachment of a protein to a DNA molecule results in a large increase in the overall molecular mass. If this increase can be detected, a DNA fragment containing a protein binding site will have been identified. In practice a DNA fragment carrying a bound protein is identified by gel electrophoresis, as it has a lower mobility than the uncomplexed DNA molecule (Figure 11.9). This procedure is referred to as gel retardation.

In a gel retardation experiment (Figure 11.10), the region of DNA upstream of the gene being studied is digested with a restriction endonuclease and then mixed with the regulatory protein or, if the protein has not yet been purified, with an unfractionated extract of nuclear protein (remember that gene regulation occurs in the nucleus). The

Lane 1: DNA fragment
Lane 2: DNA fragment +
      bound protein

Figure 11.9

A bound protein decreases the mobility of a DNA fragment during gel electrophoresis.

Figure 11.10

Carrying out a gel retardation experiment.

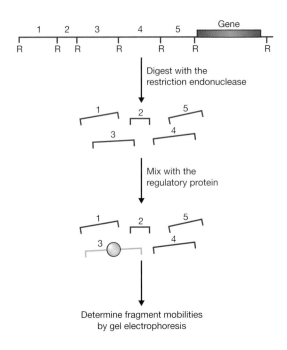

restriction fragment containing the control sequence forms a complex with the regulatory protein, and all the other fragments remain as 'naked' DNA. The location of the control sequence is then determined by finding the position on the restriction map of the fragment that is retarded during gel electrophoresis. The precision with which the control sequence can be located depends on how conveniently placed the restriction sites are. A single control sequence may be less than 10 bp in size, and so gel retardation is rarely able to pinpoint it exactly. More precise techniques are therefore needed to delineate the position of the control sequence within the fragment identified by gel retardation.

Footprinting with DNase I

The procedure generally called footprinting enables a control region to be positioned within a restriction fragment that has been identified by gel retardation. Footprinting works on the basis that the interaction with a regulatory protein protects the DNA in the region of a control sequence from the degradative action of an endonuclease such as DNase I (Figure 11.11). This phenomenon can be used to locate the protein binding site on the DNA molecule.

The DNA fragment being studied is first labelled at one end, and then complexed with the regulatory protein (Figure 11.12a). DNase I is then added, though the amount used is limited so that complete degradation of the DNA fragment does not occur. Instead, the aim is to cut each molecule at just a single phosphodiester bond (Figure 11.12b). If the DNA fragment has no protein attached to it, the result of this treatment is a family of labelled fragments that differ in size by just one nucleotide each.

After removal of the bound protein and separation on a polyacrylamide gel, the family of labelled fragments appears as a ladder of bands (Figure 11.12c). However, the bound protein protected certain phosphodiester bonds from being cut by DNase I, which means that in this case the family of fragments is not complete, as the fragments

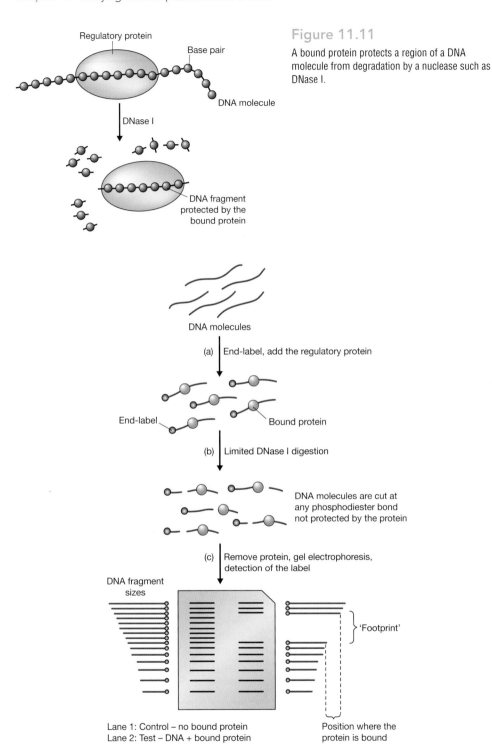

Regulatory protein

Base pair

DNA molecule

DNase I

DNA fragment protected by the bound protein

Figure 11.11

A bound protein protects a region of a DNA molecule from degradation by a nuclease such as DNase I.

DNA molecules

(a) End-label, add the regulatory protein

End-label

Bound protein

(b) Limited DNase I digestion

DNA molecules are cut at any phosphodiester bond not protected by the protein

(c) Remove protein, gel electrophoresis, detection of the label

DNA fragment sizes

'Footprint'

Lane 1: Control – no bound protein
Lane 2: Test – DNA + bound protein

Position where the protein is bound

Figure 11.12

DNase I footprinting.

Figure 11.13

A bound protein can protect a region of DNA that is much longer than the control sequence.

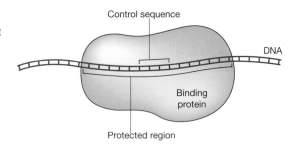

resulting from cleavage within the control sequence are absent. Their absence shows up as a 'footprint', clearly seen in Figure 11.12c. The region of the DNA molecule containing the control sequence can now be worked out from the sizes of the fragments on either side of the footprint.

Modification interference assays

Gel retardation analysis and footprinting enable control sequences to be located, but do not provide any information about the interaction between the binding protein and the DNA molecule. The more precise of these two techniques – footprinting – only reveals the region of DNA that is protected by the bound protein. Proteins are relatively large compared to a DNA double helix, and can protect several tens of base pairs when bound to a control sequence that is just a few base pairs in length (Figure 11.13). Footprinting therefore does not delineate the control region itself, only the region within which it is located.

Nucleotides that actually form attachments with a bound protein can be identified using the modification interference assay. As in footprinting, the DNA fragments must first be labelled at one end. Then they are treated with a chemical that modifies specific nucleotides; an example is dimethyl sulphate, which attaches methyl groups to guanine nucleotides (Figure 11.14). This modification is carried out under limiting conditions, so an average of just one nucleotide per DNA fragment is modified. The DNA is then mixed with the protein extract. The key to the assay is that the binding protein will probably not attach to the DNA if one of the guanines within the control region is modified, because methylation of a nucleotide interferes with the specific chemical reaction that enables it to form an attachment with a protein.

How is the absence of protein binding monitored? If the DNA and protein mixture is examined using agarose gel electrophoresis, two bands will be seen – one comprising the DNA–protein complex and one containing DNA with no bound protein. In essence, this part of the procedure is a gel retardation assay (Figure 11.14). The band made up of unbound DNA is purified from the gel and treated with piperidine, a chemical which cleaves DNA molecules at methylated nucleotides. The products of piperidine treatment are now separated on a polyacrylamide gel and the labelled bands visualized. The sizes of the bands that are seen indicate the position in the DNA fragment of guanines whose methylation prevented protein binding. These guanines lie within the control sequence. The modification assay can now be repeated with chemicals that target A, T, or C nucleotides to determine the precise position of the control sequence.

11.2.2 Identifying control sequences by deletion analysis

Gel retardation, footprinting and modification interference assays are able to locate possible control sequences upstream of a gene, but they do not provide any information

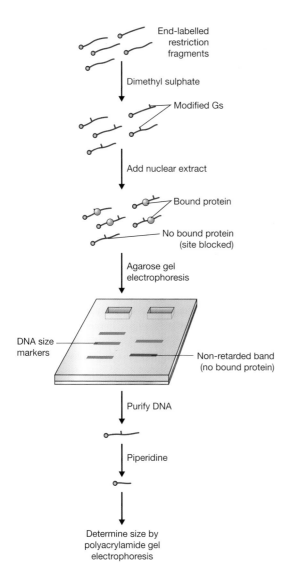

Figure 11.14
A modification interference assay.

on the function of the individual sequences. Deletion analysis is a totally different approach that not only can locate control sequences (though only with the precision of gel retardation) but more importantly can also indicate the function of each sequence.

The technique depends on the assumption that deletion of the control sequence will result in a change in the way in which expression of a cloned gene is regulated (Figure 11.15). For instance, the deletion of a sequence that represses the expression of a gene should result in that gene being expressed at a higher level. Similarly, tissue-specific control sequences can be identified as their deletion results in the target gene being expressed in tissues other than the correct one.

Reporter genes
Before carrying out a deletion analysis, a way must be found to assay the effect of a deletion on expression of the cloned gene. The effect will probably only be observed

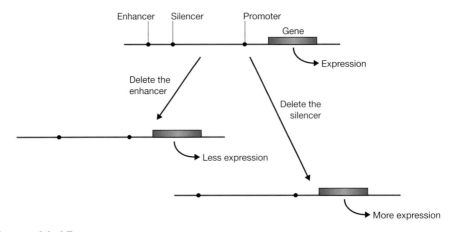

Figure 11.15

The principle behind deletion analysis.

when the gene is cloned into the species from which it was originally obtained – it would be no use attempting to assay the light regulation of a plant gene if the gene is cloned in a bacterium.

As cloning vectors have now been developed for most organisms (Chapter 7), cloning the gene under study back into its host should not cause a problem. The difficulty is that in most cases the host already possesses a copy of the gene within its chromosomes. So, how can changes in the expression pattern of the cloned gene be distinguished from the normal pattern of expression displayed by the chromosomal copy of the gene? The answer is to use a reporter gene. This is a test gene that is been fused to the upstream region of the cloned gene (Figure 11.16), replacing the latter gene. When cloned into the host organism, the expression pattern of the reporter gene should exactly mimic that of the original gene, as the reporter gene is under the influence of exactly the same control sequences as the original gene.

The reporter gene must be chosen with care. The first criterion is that the reporter gene must code for a phenotype not already displayed by the host organism. The phenotype of the reporter gene must be relatively easy to detect after it has been cloned into the host, and ideally it should be possible to assay the phenotype quantitatively. These criteria have not proved difficult to meet and a variety of different reporter genes have been used in studies of gene regulation. A few examples are listed in Table 11.1.

Figure 11.16

A reporter gene.

Table 11.1
A few examples of reporter genes used in studies of gene regulation in higher organisms.

GENE[a]	GENE PRODUCT	ASSAY
lacZ	β-Galactosidase	Histochemical test
neo	Neomycin phosphotransferase	Kanamycin resistance
cat	Chloramphenicol acetyltransferase	Chloramphenicol resistance
dhfr	Dihydrofolate reductase	Methotrexate resistance
aphIV	Hygromycin phosphotransferase	Hygromycin resistance
lux	Luciferase	Bioluminescence
GFP	Green fluorescent protein	Fluoresence
uidA	β-Glucuronidase	Histochemical test

[a] All of these genes are obtained from *E. coli*, except for: *lux* which has three sources, the luminescent bacteria *Vibrio harveyii* and *V. fischeri*, and the firefly *Photinus pyralis*; and GFP, which is obtained from the jellyfish *Aequorea victoria*.

Carrying out a deletion analysis

Once a reporter gene has been chosen and the necessary construction made, carrying out a deletion analysis is fairly straightforward. Deletions can be made in the upstream region of the construct by any one of several strategies, a simple example being shown in Figure 11.17. The effect of the deletion is then assessed by cloning the deleted construct into the host organism and determining the pattern and extent of expression of the reporter gene. An increase in expression will imply that a repressing or silencing sequence has been removed, a decrease will indicate the removal of an activator or enhancer, while a change in tissue specificity (as shown in Figure 11.17) will pinpoint a tissue-responsive control sequence.

The results of a deletion analysis project must be interpreted very carefully. Complications may arise if a single deletion removes two closely linked control sequences with different regulatory functions or, as is fairly common, two different control sequences that are some distance apart cooperate to produce a single response. Despite these potential difficulties, deletion analyses – in combination with studies of protein-binding sites – have provided important information about how the expression of individual genes is regulated, and have supplemented and extended the more broadly based genetic analyses of differentiation and development.

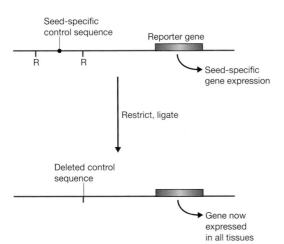

Figure 11.17

Deletion analysis. A reporter gene has been attached to the upstream region of a seed-specific gene from a plant. Removal of the restriction fragment bounded by the sites R deletes the control sequence that mediates seed-specific gene expression, so that the reporter gene is now expressed in all tissues of the plant.

11.3 Identifying and studying the translation product of a cloned gene

Over the years gene cloning has become increasingly useful in the study not only of genes themselves but also of the proteins coded by cloned genes. Investigations into protein structure and function have benefited greatly from the development of techniques that allow mutations to be introduced at specific points in a cloned gene, resulting in directed changes in the structure of the encoded protein.

Before considering these procedures, we should first look at the more mundane problem of how to isolate the protein coded by a cloned gene. In many cases this analysis will not be necessary, as the protein will have been characterized long before the gene cloning experiment is performed and pure samples of the protein will already be available. However, there are occasions when the translation product of a cloned gene has not been identified, and a method for isolating the protein is then needed.

11.3.1 HRT and HART can identify the translation product of a cloned gene

Two related techniques, hybrid-release translation (HRT) and hybrid-arrest translation (HART), are used to identify the translation product encoded by a cloned gene. Both depend on the ability of purified mRNA to direct the synthesis of proteins in cell-free translation systems. These are cell extracts, usually prepared from germinating wheat seeds or from rabbit reticulocyte cells, both of which are each exceptionally active in protein synthesis. The extracts contain ribosomes, tRNAs, and all the other molecules needed for protein synthesis. The mRNA sample is added to the cell-free translation system, along with a mixture of the 20 amino acids found in proteins, one of which is labelled (often, ^{35}S-methionine is used). The mRNA molecules are translated into a mixture of radioactive proteins (Figure 11.18), which can be separated using gel electrophoresis and visualized by autoradiography. Each band represents a single protein coded by one of the mRNA molecules present in the sample.

Both, HRT and HART function best when the gene being studied has been obtained as a cDNA clone. For HRT, the cDNA is denatured, immobilized on a nitrocellulose or nylon membrane, and then incubated with the mRNA sample (Figure 11.19). The specific mRNA counterpart of the cDNA hybridizes and remains attached to the membrane. After discarding the unbound molecules, the hybridized mRNA is recovered and translated in a cell-free system. This provides a pure sample of the protein coded by the cDNA.

Hybrid-arrest translation is slightly different in that the denatured cDNA is added directly to the mRNA sample (Figure 11.20). Hybridization again occurs between the cDNA and its mRNA counterpart, but in this case the unbound mRNA is not discarded. Instead, the entire sample is translated in the cell-free system. As the hybridized mRNA is unable to direct translation, all of the proteins except the one coded by the cloned gene are synthesized. The cloned gene's translation product is therefore identified as the protein that is absent from the autoradiograph.

11.3.2 Analysis of proteins by in vitro mutagenesis

Although HRT and HART can identify the translation product of a cloned gene, these techniques tell us little about the protein itself. Some of the major questions asked by the

Pure mRNA

Mix with
^{35}S met

Cell-free
translation
system

Labelled protein

mRNA

Gel electrophoresis,
autoradiography

Labelled
translation
products

Protein molecular
weight size markers

Figure 11.18
Cell-free translation.

Specific cDNA

Membrane

Apply mRNA
mixture

Specific mRNA
hybridizes

Remove
hybridized
mRNA

Cell-free translation,
electrophoresis and
autoradiography

Protein coded
by the cDNA

Figure 11.19
Hybrid-release translation.

Figure 11.20

Hybrid-arrest translation.

molecular biologist today centre on the relationship between the structure of a protein and its mode of activity. The best way of tackling these problems is to induce a mutation in the gene coding for the protein and then to determine what effect the change in amino acid sequence has on the properties of the translation product (Figure 11.21). Under normal circumstances, mutations occur randomly and a large number may have to be screened before one that gives useful information is found. A solution to this problem is provided by *in vitro* mutagenesis, a technique that enables a directed mutation to be made at a specific point in a cloned gene.

Different types of in vitro mutagenesis techniques

An almost unlimited variety of DNA manipulations can be used to introduce mutations into cloned genes. The following are the simplest:

- A restriction fragment can be deleted (Figure 11.22a).
- The gene can be opened at a unique restriction site, a few nucleotides removed with a double-strand-specific endonuclease such as Bal31 (p. 48), and the gene religated (Figure 11.22b).

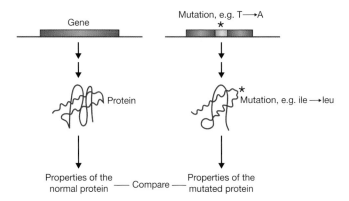

Figure 11.21

A mutation may change the amino acid sequence of a protein, possibly affecting its properties.

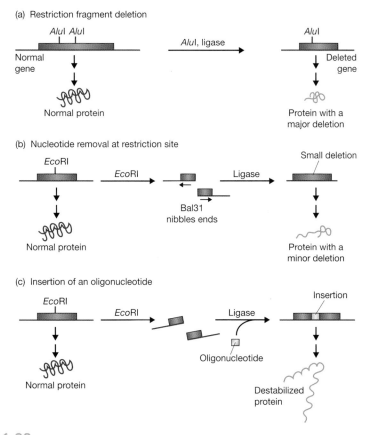

Figure 11.22

Various *in vitro* mutagenesis techniques.

- A short, double-stranded oligonucleotide can be inserted at a restriction site (Figure 11.22c). The sequence of this oligonucleotide can be such that the additional stretch of amino acids inserted into the protein produces, for example, a new structure such as an α-helix, or destabilizes an existing structure.

Although potentially useful, these manipulations depend on the fortuitous occurrence of a restriction site at the area of interest in the cloned gene. Oligonucleotide-directed mutagenesis is a more versatile technique that can create a mutation at any point in the gene.

Using an oligonucleotide to create a point mutation in a cloned gene

There are a number of different ways of carrying out oligonucleotide-directed mutagenesis. We will consider one of the simplest of these methods. The gene to be mutated must be obtained in a single-stranded form and so is generally cloned into an M13 or phagemid vector. A short oligonucleotide is synthesized that is complementary to the region to be mutated, and which contains the desired nucleotide alteration (Figure 11.23a). Despite this mismatch, the oligonucleotide will anneal to the single-stranded DNA and act as a primer for complementary strand synthesis by a DNA polymerase (Figure 11.23b). This strand synthesis reaction is continued until an entire new strand is made and the recombinant molecule is completely double-stranded.

After introduction, by transfection, into competent *E. coli* cells, DNA replication produces numerous copies of the recombinant DNA molecule. The semi-conservative nature of DNA replication means that half the double-stranded molecules that are produced are unmutated in both strands, whereas half are mutated in both strands. Similarly, half the resulting phage progeny carry copies of the unmutated molecule and half carry the mutation. The phages produced by the transfected cells are plated onto solid agar so that plaques are produced. Half of these plaques should contain the original recombinant molecule, and half the mutated version. Which are which is determined by plaque hybridization, using the oligonucleotide as the probe, and employing very strict conditions so that only the completely base-paired hybrid is stable.

Cells infected with M13 vectors do not lyse, but instead continue to divide (p. 19). The mutated gene can therefore be expressed in the host *E. coli* cells, resulting in the production of a recombinant protein. The protein coded by the mutated gene can be purified from the recombinant cells and its properties studied. The effect of a single base pair mutation on the activity of the protein can therefore be assessed.

Other methods of creating a point mutation in a cloned gene

The oligonucleotide-directed procedure illustrated in Figure 11.23 is an effective way of creating a single point mutation in a cloned gene. But what if the objective is to make a number of changes to the sequence of the gene? The oligonucleotide procedure could, of course, be repeated several times, introducing a new mutation at each stage, but this would be a very lengthy process.

An alternative for short genes (up to about 1.2 kb) would be to construct the gene in the test tube, placing mutations at all the desired positions. This approach is feasible now that oligonucleotides of 150 units and longer can be made by chemical synthesis (p. 149). The gene is constructed by synthesizing a series of partially overlapping oligonucleotides, the sequences of these oligonucleotides making up the desired sequence for the gene. The overlaps can be partial rather than complete because the gaps can be filled in with a DNA polymerase, and the final phosphodiester

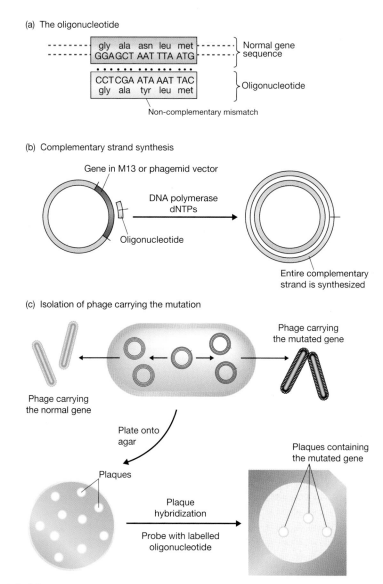

Figure 11.23

One method for oligonucleotide-directed mutagenesis.

bonds synthesized with DNA ligase to create the completed, double-stranded gene (Figure 11.24). If restriction sites are included in the end-sequences of the gene, then treatment with the appropriate enzyme will produce sticky ends that allow the gene to be ligated into a cloning vector. This procedure is called artificial gene synthesis.

PCR can also be used to create mutations in cloned genes, although like oligonucleotide-directed mutagenesis, only one mutation can be created per experiment. Various procedures have been devised, one of which is shown in Figure 11.25. In this example, the starting DNA molecule is amplified by two PCRs. In each of these, one primer is normal and forms a fully base-paired hybrid with the template DNA, but the

Figure 11.24

Artificial gene synthesis.

second is mutagenic, as it contains a single base-pair mismatch corresponding to the mutation that we wish to introduce into the DNA sequence. This mutation is therefore present in the two PCR products, each of which represents one half of the starting DNA molecule. The two PCR products are mixed together and a final PCR cycle carried out. In this cycle, complementary strands from the two products anneal to one another and are then extended by the polymerase, producing the full-length, mutated DNA molecule.

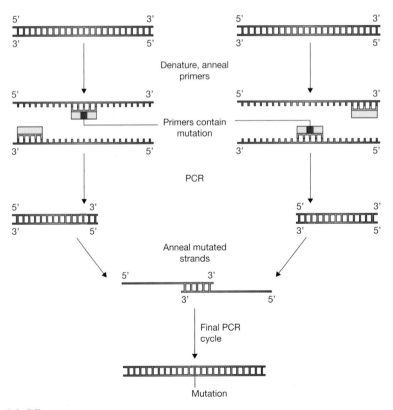

Figure 11.25

One method for using PCR to create a directed mutation.

Although this technique (and related ones using PCR) is very quick and easy to carry out, a major problem is caused by the high error rate of the *Taq* polymerase used in PCR (p. 167). This error rate makes it likely that not only the desired mutation but also random mutations will be present at the end of the experiment. The PCR product therefore has to be cloned and the sequences of individual clones checked to find one that has just the desired mutation.

The potential of in vitro mutagenesis

In vitro mutagenesis has remarkable potential, both for pure research and for applied biotechnology. For example, the biochemist can now ask very specific questions about the way that protein structure affects the action of an enzyme. In the past, it has been possible through biochemical analysis to gain some idea of the identity of the amino acids that provide the substrate binding and catalytic functions of an enzyme molecule. Mutagenesis techniques provide a much more detailed picture by enabling the role of each individual amino acid to be assessed by replacing it with an alternative residue and determining the effect this has on the enzymatic activity.

The ability to manipulate enzymes in this way has resulted in dramatic advances in our understanding of biological catalysis, and has led to the new field of protein engineering, in which mutagenesis techniques are used to develop new enzymes for biotechnological purposes. For example, careful alterations to the amino acid sequence of subtilisin, an enzyme used in biological washing powders, have resulted in engineered versions with greater resistances to the thermal and bleaching (oxidative) stresses encountered in washing machines.

Further reading

Berk, A.J. (1989) Characterization of RNA molecules by S1 nuclease analysis. *Methods in Enzymology*, **180**, 334–347.

Carrigan, P.E., Ballar, P., and Tuzman, S. (2011) Site-directed mutagenesis. *Methods in Molecular Biology*, **700**, 107–124.

Fried, M. and Crothers, D.M. (1981) Equilibria and kinetics of *lac* repressor–operator interactions by polyacrylamide gel electrophoresis. *Nucleic Acids Research*, **9**, 6505–6525. [Gel retardation.]

Frohman, M.A., Dush, D.K., and Martin, G.R. (1988) Rapid production of full length cDNAs from rare transcripts: amplification using a single gene-specific oligonucleotide primer. *Proceedings of the National Academy of Sciences of the United States of America*, **85**, 8998–9002. [An example of RACE.]

Galas, D.J. and Schmitz, A. (1978) DNase footprinting: a simple method for the detection of protein–DNA binding specificity. *Nucleic Acids Research*, **5**, 3157–3170.

Garner, M.M. and Rezvin, A. (1981) A gel electrophoretic method for quantifying the binding of proteins to specific DNA regions: application to components of the *Escherichia coli* lactose operon regulatory system. *Nucleic Acids Research*, **9**, 3047–3060. [Gel retardation.]

Hendrickson, W. and Schleif, R. (1985) A dimer of AraC protein contacts three adjacent major groove regions at the Ara I DNA site. *Proceedings of the National Academy of Sciences of the USA*, **82**, 3129–3133. [An example of the modification interference assay.]

Kunkel, T.A. (1985) Rapid and efficient site-specific mutagenesis without phenotypic selection. *Proceedings of the National Academy of Sciences of the United States of America*, **82**, 488–492. [Oligonucleotide-directed mutagenesis.]

Matzke, A.J.M., Stöger, E.M., Schernthaner, J.P., and Matzke, M.A. (1990) Deletion analysis of a zein gene promoter in transgenic tobacco plants. *Plant Molecular Biology*, **14**, 323–332.

Paterson, B.M., Roberts, B.E., and Kuff, E.L. (1977) Structural gene identification and mapping by DNA.mRNA hybrid-arrested cell-free translation. *Proceedings of the National Academy of Sciences of the United States of America*, **74**, 4370–4374.

Pellé, R. and Murphy, N.B. (1993) Northern hybridization: rapid and simple electrophoretic conditions. *Nucleic Acids Research*, **21**, 2783–2784.

Tsien, R. (1998) The green fluorescent protein. *Annual Reviews of Biochemistry*, **67**, 509–544. [A reporter gene system.]

Chapter 12

Studying Genomes

Chapter contents

At the start of the 21st century the emphasis in molecular biology shifted from the study of individual genes to the study of entire genomes. This change in emphasis was prompted by the development during the 1990s of methods for sequencing large genomes. Genome sequencing predates the 1990s – we saw in Chapter 10 how the first genome, that of the phage φX174, was completed in 1975 – but it was not until 20 years later, in 1995, that the first genome of a free-living organism, the bacterium *Haemophilus influenzae*, was completely sequenced. The next five years were a watershed with the genome sequences of almost 50 other bacteria published, along with complete sequences for the much larger genomes of yeast, fruit fly, *Caenorhabditis elegans* (a nematode worm), *Arabidopsis thaliana* (a plant), and humans. Today, the sequencing of bacterial and archaeal genomes has become routine, with several thousand completed, and almost 300 eukaryotic genomes have also been sequenced.

Genome sequencing has led to the development of a new area of DNA research, loosely called post-genomics or functional genomics. Post-genomics includes the use of computer systems in genome annotation, the process by which the genes, control sequences and other interesting features are identified in a genome sequence, as well as computer-based and experimental techniques aimed at determining the functions of any unknown genes that are discovered. Post-genomics also encompasses techniques designed to identify which genes are expressed in a particular type of cell or tissue, and how this pattern of genome expression changes over time.

12.1 Genome annotation

Once a genome sequence has been completed, the next step is to locate all the genes and determine their functions. It is in this area that bioinformatics, sometimes referred to

Gene Cloning and DNA Analysis: An Introduction, Seventh Edition. T.A. Brown.
© 2016 John Wiley & Sons, Ltd. Published 2016 by John Wiley & Sons, Ltd.
Companion Website: www.wiley.com/go/brown/cloning

as molecular biology *in silico*, is proving of major value as an adjunct to conventional experiments.

Genome annotation is a far from trivial process, even with genomes that have been extensively studied by genetic analysis and gene cloning techniques prior to complete sequencing. For example, the sequence of the yeast *Saccharomyces cerevisiae*, one of the best studied of all organisms, revealed that this genome contains about 6000 genes. When the sequence was completed in 1996, some 3600 of these genes could immediately be assigned a function, either on the basis of previous studies that had been carried out with yeast, or because the yeast gene had a similar sequence to a gene that had been studied in another organism. This left 2400 genes whose functions were not known. Despite a massive amount of work since 1996, the functions of some of these orphans have still not been determined.

12.1.1 Identifying the genes in a genome sequence

Locating a gene in a genome sequence is easy if the amino acid sequence of the protein product is known, allowing the nucleotide sequence of the gene to be predicted, or if the corresponding cDNA has been previously sequenced. But for many genes there is no prior information that enables the correct DNA sequence to be recognized. How can these genes be located in a genome sequence?

Searching for open reading frames

The DNA sequence of a gene is an open reading frame (ORF), a series of nucleotide triplets beginning with an initiation codon (usually, but not always, ATG) and ending in a termination codon (TAA, TAG or TGA in most genomes). Searching a genome sequence for ORFs, by eye or more usually by computer, is therefore the first step in gene location. When carrying out the search it is important to remember that each DNA sequence has six reading frames, three in one direction and three in the reverse direction on the complementary strand (Figure 12.1).

The key to the success of ORF scanning is the frequency with which termination codons appear in the DNA sequence. If the DNA has a random sequence and a GC content of 50%, then each of the three termination codons will appear, on average, once every $4^3 = 64$ bp. This means that there should not be many ORFs longer than 30–40 codons in random DNA, and not all of these ORFs will start with ATG. Most genes are much longer than this: the average lengths are 300–350 codons for bacterial genes and approximately 450 codons for humans. ORF scanning, in its simplest form, therefore takes a figure of 100 codons as the shortest length of a putative gene and records positive hits for all ORFs longer than this.

Figure 12.1

A double-stranded DNA molecule has six reading frames.

Figure 12.2

The typical result of a search for ORFs in a prokaryotic genome. The arrows indicate the directions in which the genes and spurious ORFs run.

With prokaryotic genomes, simple ORF scanning is an effective way of locating most of the genes in a DNA sequence. Most prokaryotic genes are much longer than 100 codons in length and so are easily recognized (Figure 12.2). With prokaryotes, the analysis is further simplified by the fact that most genes are closely spaced with very little intergenic DNA between them. If we assume that the real genes do not overlap, which is true for most prokaryotic genomes, then it is only in these short intergenic regions that there is a possibility of mistaking a spurious ORF for a real gene.

Simple ORF scans are less effective at locating genes in eukaryotic genomes

Although ORF scans work well for prokaryotic genomes, they are less effective for locating genes in eukaryotic genomes. This is partly because there is substantially more intergenic DNA in a eukaryotic genome, increasing the chances of finding spurious ORFs, but the main problem is the presence of introns (see Figure 11.1). If a gene contains one or more introns, then it does not appear as a continuous ORF in the genome sequence. Many exons are shorter than 100 codons, and some have fewer than 50 codons. Continuing the reading frame into an intron usually leads to a termination sequence that appears to close the ORF (Figure 12.3). In other words, many of the genes in a eukaryotic genome are not full-length, uninterrupted ORFs, and simple ORF scanning cannot locate them.

Finding ways of locating genes by inspection of a eukaryotic sequence is a major challenge in bioinformatics. Two approaches are being followed:

- Codon bias can be taken into account. Not all codons are used equally frequently in the genes of a particular organism. For example, leucine is specified by six codons (TTA, TTG, CTT, CTC, CTA, and CTG), but in human genes leucine is most frequently coded by CTG and is only rarely specified by TTA or CTA. Similarly, of the four valine codons, human genes use GTG four times more frequently than

Figure 12.3

ORF scans are complicated by introns. The nucleotide sequence of a short gene containing a single intron is shown. The correct amino acid sequence of the protein translated from the gene is given immediately below the nucleotide sequence, using the single-letter amino acid abbreviations. In this sequence the intron has been left out, because it is removed from the transcript before the mRNA is translated into protein. In the lower line, the sequence has been translated without recognizing that an intron is present. As a result of this error, the amino acid sequence appears to terminate within the intron.

Figure 12.4

The consensus sequences for the upstream and downstream exon–intron boundaries of vertebrate introns. Py = pyrimidine nucleotide (C or T); N = any nucleotide. The arrows indicate the boundary positions.

GTA. The biological reason for codon bias is not understood, but all organisms have a bias, which is different in different species. Real exons display this bias, whereas chance series of triplets usually do not.

- Exon–intron boundaries can be searched for as these have distinctive sequence features, although unfortunately the distinctiveness of these sequences is not so great as to make their location a trivial task. In vertebrates, the sequence of the upstream exon–intron boundary is usually described as 5′–AG↓GTAAGT–3′ and the downstream one as 5′–PyPyPyPyPyPyNCAG↓–3′, where 'Py' means one of the pyrimidine nucleotides (T or C), 'N' is any nucleotide, and the arrow shows the precise location of the boundary (Figure 12.4). These are consensus sequences, the averages of a large number of related but non-identical sequences, so the search has to include not only the sequences shown but also at least the most common variants. Despite these problems, this type of search can predict the locations of the exon–intron boundaries with an accuracy of 60–70%.

These two extensions of simple ORF scanning, despite their limitations, are generally applicable to all higher eukaryotic genomes. Additional strategies are also possible with individual organisms based on the special features of their genomes. For example, vertebrate genomes contain CpG islands upstream of many genes, these being sequences of approximately 1 kb in which the GC content is greater than the average for the genome as a whole. Some 40–50% of human genes have an upstream CpG island. These sequences are distinctive, and when one is located in vertebrate DNA a strong assumption can be made that a gene begins in the region immediately downstream.

Gene location is aided by homology searching

Tentative identification of a gene is usually followed by a homology search. This is an analysis, carried out by computer, in which the sequence of the gene is compared with all the gene sequences present in the international DNA databases, not only the known genes of the organism under study but also genes from all other species. The rationale is that two genes from different organisms which have similar functions have similar sequences, reflecting their common evolutionary histories (Figure 12.5).

In order to carry out a homology search, the nucleotide sequence of the tentative gene is usually translated into an amino acid sequence, as this allows a more sensitive search. This is because there are 20 different amino acids but only four nucleotides, so there is less chance of two amino acid sequences appearing to be similar purely by chance (Figure 12.6). The analysis is carried out through the internet, by logging on to the web site of one of the DNA databases and using a search program such as BLAST (Basic Local Alignment Search Tool). If the test sequence is over 200 amino acids in

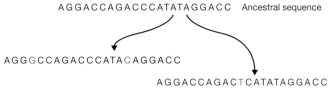

AGGACCAGACCCATATAGGACC   Ancestral sequence

AGGGCCAGACCCATACAGGACC

AGGACCAGACTCATATAGGACC

Two modern sequences

Figure 12.5

Homology between two sequences that share a common ancestor. The two sequences have acquired mutations during their evolutionary histories but their sequence similarities indicate that they are homologues.

length and has 30% or greater identity with a sequence in the database (i.e., at 30 out of 100 positions the same amino acid occurs in both sequences), then the two can be considered possible homologues and the ORF under study looked on as a possible gene.

Comparing the sequences of related genomes

A more precise version of homology searching is possible when genome sequences are available for two or more related species. Related species have genomes that share similarities inherited from their common ancestor, overlaid with species-specific differences that have arisen since the two species began to evolve independently. Because of natural selection, the sequence similarities between related genomes is greatest within the genes and least in the intergenic regions. Therefore, when related genomes are compared, homologous genes are easily identified because they have high sequence similarity, and any ORF that does not have a clear homologue in the related genome can be discounted as almost certainly being a chance sequence and not a genuine gene.

This type of homology analysis – called comparative genomics – has proved very valuable for locating genes in the *S. cerevisiae* genome, as complete sequences are available not only for this yeast but also for several related species, such as *Saccharomyces paradoxus*, *Saccharomyces mikatae*, and *Saccharomyces bayanus*. Comparisons between these genomes confirmed the authenticity of a number of *S. cerevisiae* ORFs, and also enabled almost 500 putative ORFs to be discounted on the grounds that they have no equivalents in the related genomes. The gene maps of these yeasts are very similar, and although each genome has undergone its own species-specific rearrangements, there are still substantial regions where the gene order in the *S. cerevisiae* genome is the same as in one or more of the related genomes. The conservation of gene order is termed synteny,

```
            G   A   P   G   M   W   L   R   L   A   A   G   S   F   E   H   A   G
Sequence 1  GGTGCACCCGGTATGTGACTGCGATTAGCAGCGGGATCATTTCAGCATGCAGGG
            *  * *****  * **** **** ** *** **** ***** ** ** **** ** *
Sequence 2  GATACACCCCGTATTTGACAGCAATTTGCAGGGGGATTATTGCACCATGGAGCG
            D   T   P   R   I   W   E   E   P   A   G   G   L   L   H   H   G   A
```

Figure 12.6

Lack of homology between two sequences is often more apparent when comparisons are made at the amino acid level. Two nucleotide sequences are shown, with nucleotides that are identical in the two sequences given in red and non-identities given in blue. The two nucleotide sequences are 76% identical, as indicated by the asterisks. This might be taken as evidence that the sequences are homologous. However, when the sequences are translated into amino acids, the identity decreases to 28%, suggesting that the genes are not homologous, and that the similarity at the nucleotide level was fortuitous. Identical amino acids are shown in brown, and non-identities in green. The amino acid sequences have been written using the one-letter abbreviations.

Figure 12.7

Using comparisons between the genomes of related species to test the authenticity of a short ORF. In this example, the questionable ORF is not present at the expected position in the related genome and so is probably not a real gene.

and this makes it very easy to identify homologous genes. More importantly, a spurious ORF – especially a short one – can be discarded with confidence because its expected location in a related genome can be searched in detail to ensure that no equivalent is present (Figure 12.7).

Identifying the binding sites for regulatory proteins in a genome sequence

Genome annotation can be enhanced by the identification of regulatory sequences that act as binding sites for proteins such as transcription factors. These regulatory sequences are usually positioned upstream of the genes whose expression they control, so identifying them can indicate regions of the genome where genes are present, and can also help to confirm the authenticity of a questionable ORF.

Regulatory sequences have distinctive features that they possess in order to carry out their role as recognition signals for the DNA-binding proteins involved in gene expression. As with exon–intron boundaries, the regulatory sequences are variable, and it is difficult to locate them simply by scanning a genome sequence. These sequences are more easily located by ChIP-seq (chromatin immunoprecipitation sequencing), which identifies the positions where individual DNA-binding proteins attach to a genome. To perform a ChIP-seq experiment, chromatin – the complex of DNA and bound proteins present in the nucleus – is first extracted and then treated with formaldehyde to crosslink the proteins to the DNA (Figure 12.8). The DNA is then sonicated to break it

Figure 12.8

ChIP-seq (chromatin immunoprecipitation sequencing).

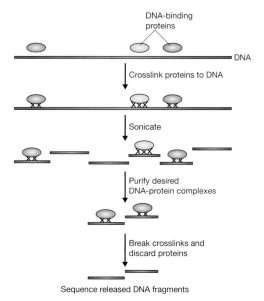

into fragments suitable for the later sequencing stage. Because of the crosslinking, the bound proteins are not lost and remain attached to their DNA fragments. Those fragments attached to a particular transcription factor or other interesting DNA-binding protein can therefore be purified with an antibody specific for that protein. The purified fragments are then heated at 65 °C overnight to break the crosslinks and detach the proteins, after which the released DNA fragments are sequenced using a next-generation method. The reads are then mapped onto the genome sequence in order to locate the binding sites for the protein being studied.

Only those binding sites occupied by the protein will be represented in the reads obtained by a ChIP-seq experiment. In any particular tissue, some sites will be occupied and others empty, depending on which genes are active in that tissue. ChIP-seq therefore provides information not only on the locations of the binding sites for a transcription factor but also on the expression profiles of the genes controlled by that factor. This information, in turn, enables genome annotation to move beyond the simple mapping of gene positions within a genome, and to encompass information on the expression of the genome.

12.1.2 Determining the function of an unknown gene

Homology searching serves two purposes. As well as testing the veracity of a tentative gene, it can also give an indication of the function of the gene, presuming that the function of the homologous gene is known. Almost 2000 of the genes in the yeast genome were assigned functions in this way. Frequently, however, the matches found by homology searching are to other genes whose functions have yet to be determined. These unassigned genes are called 'orphans', and determining their function is one of the key objectives of post-genomics research.

Structural bioinformatics, which attempts to predict the structure of a protein from its amino acid sequence, can provide at least an insight into the function of an orphan gene. It is possible to use the nucleotide sequence of a gene to predict the positions of α-helices and β-sheets in the encoded protein (albeit with limited accuracy), and the resulting structural information can sometimes be used to make inferences about the function of the protein. Proteins that attach to membranes can often be identified because they possess α-helical arrangements that span the membrane, and DNA-binding motifs such as zinc fingers can be recognized. A greater scope and accuracy to this aspect of bioinformatics will be possible when more information is obtained about the relationship between the structure of a protein and its function. In the meantime, the functional analysis of orphans depends largely on conventional experiments.

Assigning gene function by experimental analysis requires a reverse approach to genetics

Experimental identification of the function of an unknown gene is proving to be one of the biggest challenges in genomics research. The problem is that the objective – to plot a course from gene to function – is the reverse of the route normally taken by genetic analysis, in which the starting point is a phenotype and the objective is to identify the underlying gene or genes (Figure 12.9). In conventional genetic analysis, the genetic basis of a phenotype is usually studied by searching for mutant organisms in

Figure 12.9

Forward and reverse genetics.

which the phenotype has become altered. The mutants might be obtained experimentally, for example by treating a population (e.g., a culture of bacteria) with ultraviolet radiation or a mutagenic chemical, or the mutants might be present in a natural population. The gene or genes that have been altered in the mutant organism are then studied by genetic crosses, which can locate the position of a gene in a genome and also determine if the gene is the same as one that has already been characterized. The gene can then be studied further by molecular biology techniques such as cloning and sequencing.

The general principle of this conventional analysis – forward genetics – is that the genes responsible for a phenotype can be identified by determining which genes are inactivated in organisms that display a mutant version of the phenotype. If the starting point is the gene, rather than the phenotype, then the equivalent strategy – reverse genetics – would be to mutate the gene and identify the phenotypic change that results. This is the basis of most of the techniques used to assign functions to unknown genes.

Specific genes can be inactivated by homologous recombination

The easiest way to inactivate a specific gene is to disrupt it with an unrelated segment of DNA (Figure 12.10). This can be achieved by homologous recombination between the chromosomal copy of the gene and a second piece of DNA that shares some sequence identity with the target gene.

How is gene inactivation carried out in practice? We will consider two examples, the first with *S. cerevisiae*. Once the genome sequence had been completed in 1996, yeast molecular biologists embarked on a coordinated, international effort to determine the functions of as many of the unknown genes as possible. One technique makes use of a deletion cassette, which carries a gene for antibiotic resistance (Figure 12.11). This gene is not a normal component of the yeast genome but it will work if transferred into a yeast chromosome, giving rise to a transformed yeast cell that is resistant to the antibiotic geneticin. Before using the deletion cassette, new segments of DNA are attached as tails to either end. These segments have sequences identical to parts of the yeast gene that is going to be inactivated. After the modified cassette is introduced into a yeast cell, homologous recombination occurs between the DNA tails and the chromosomal copy

Figure 12.10

Gene disruption by homologous recombination.

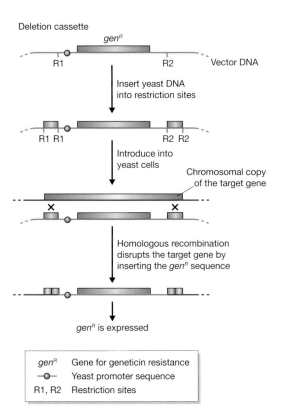

Figure 12.11

The use of a yeast deletion cassette.

of the yeast gene, replacing the latter with the antibiotic resistance gene. The target gene therefore becomes inactivated. Cells which have undergone the replacement are selected by plating the culture onto agar medium containing geneticin, and their phenotypes examined to gain some insight into the function of the gene.

The second example of gene inactivation uses an analogous process, but with mice rather than yeast. The mouse is frequently used as a model organism for humans because the mouse genome contains many of the same genes. Identifying the functions of unknown human genes is therefore being carried out largely by inactivating the equivalent genes in mice. The homologous recombination part of the procedure is identical to that described for yeast and once again results in a cell in which the target gene has been inactivated. The problem is that we do not want just one mutant cell, we want a whole mutant mouse, as only with the complete organism can we make a full assessment of the effect of the gene inactivation on the phenotype. To achieve this, the vector carrying the deletion cassette is initially microinjected into an embryonic stem cell (p. 132). The eventual result is a knockout mouse, whose phenotype will, with luck, provide the desired information on the function of the gene being studied.

12.2 Studies of the transcriptome and proteome

So far, we have considered those aspects of post-genomics research that are concerned with studies of individual genes. The change in emphasis from genes to the genome has resulted in new types of analysis that are aimed at understanding the activity of the genome as a whole. This work has led to the invention of two new terms:

- The transcriptome, which is the RNA content of a cell, and which reflects the overall pattern of gene expression in that cell.
- The proteome, which is the protein content of a cell and which reflects its biochemical capability.

12.2.1 Studying the transcriptome

Transcriptomes can have highly complex compositions, containing hundreds or thousands of different RNAs, each making up a different fraction of the overall population. To characterize a transcriptome it is therefore necessary to identify the RNAs that it contains and, ideally, to determine their relative abundances.

Studying transcriptomes by microarray or chip analysis

Techniques that enabled accurate comparisons to be made of the amounts of individual mRNAs in a transcriptome were first developed as part of the yeast post-genomics project. In essence, these techniques involve a sophisticated type of hybridization analysis. Every yeast gene – all 6000 of them – was obtained as an individual clone and samples spotted onto glass slides in arrays of 80 × 80 spots. This is called a microarray. To determine which genes are active in yeast cells grown under particular conditions, mRNA was extracted from the cells, converted to cDNA, and the cDNA labelled and hybridized to the microarrays (Figure 12.12). Fluorescent labels were used and hybridization was detected by examining the microarrays by confocal microscopy. Those spots that gave a signal indicated genes that were active under the conditions being studied, and the intensities of the hybridization signals revealed the relative amounts of the mRNAs in the transcriptome. Changes in gene expression, when the yeast were transferred to different growth conditions (e.g., oxygen starvation) could be monitored by repeating the experiment with a second cDNA preparation.

Currently, microarrays are being used to monitor changes in the transcriptomes of many organisms. In some cases, the strategy is the same as used with yeast, with the microarray representing all of the genes in the genome, but this is possible only for those organisms that have relatively few genes. A microarray for all the human genes

Figure 12.12

Microarray analysis. The microarray shown here has been hybridized to two different cDNA preparations, each labelled with a fluorescent marker. The clones which hybridize with the cDNAs are identified by confocal microscopy.

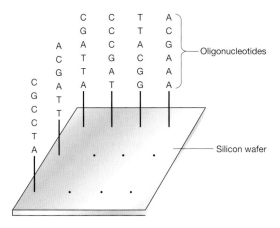

Figure 12.13

A DNA chip. A real chip would carry many more oligonucleotides than those shown here, and each oligonucleotide would usually be 20–30 nucleotides in length.

could be carried by just 10 glass slides of 18 × 18 mm, but preparing clones of every one of the 20 000–30 000 human genes would be a massive task. Fortunately this is not necessary, because in many applications only those genes expressed in a particular tissue are of interest. For example, to study changes in the transcriptome occurring as a result of cancer, a microarray could be prepared with a cDNA library from the normal tissue. This cDNA library will contain only that subset of human genes that are expressed in the tissue being studied. Hybridization with labelled cDNAs from the normal and cancerous tissues would then reveal which genes are upregulated or downregulated in response to the cancerous state.

An alternative to microarrays is provided by DNA chips, which are thin wafers of silicon that carry many different oligonucleotides (Figure 12.13). These oligonucleotides are synthesized directly on the surface of the chip and can be prepared at a density of 1 million per cm², substantially higher than is possible with a conventional microarray. Hybridization between an oligonucleotide and the probe is detected electronically. Because the oligonucleotides are synthesized *de novo*, using special automated procedures, a chip carrying 20 000–30 000 oligonucleotides – each one specific for a different human gene – is relatively easy to prepare.

Studying a transcriptome by SAGE

The most direct way to characterize a transcriptome is to convert its RNA into cDNA, and then to sequence the cDNA. One possibility would be to sequence the inserts of every clone in a cDNA library. Comparisons between the cDNA sequences and the genome sequence will then reveal the identities of the genes whose RNAs are present in the transcriptome. Whilst this approach is feasible it is laborious, with many different cDNA sequences being needed before a near-complete picture of the composition of the transcriptome begins to emerge. Can any shortcuts be used to obtain the vital sequence information more quickly?

Serial analysis of gene expression (SAGE) was the first method to be developed for the rapid identification of the mRNAs in a transcriptome. Rather than studying complete cDNAs, SAGE yields short sequences (as little as 12 bp in length), each of which represents a single mRNA. The basis of the technique is that these 12-bp sequences, despite their shortness, are sufficient to enable the gene that codes for the mRNA to be identified.

The first step in generating the 12-bp sequences is to immobilize the mRNA in a chromatography column by annealing the poly(A) tails present at the 3' ends of these

Figure 12.14

Serial analysis of gene expression (SAGE).

molecules to oligo(dT) strands that have been attached to cellulose beads (Figure 12.14). The mRNA is converted into double-stranded cDNA and then treated with a restriction enzyme that recognizes a 4-bp target site, such as *Alu*I, and so cuts frequently in each cDNA. The terminal restriction fragment of each cDNA remains attached to the cellulose beads, enabling all the other fragments to be eluted and discarded. A short linker is then attached to the free end of each of the terminal cDNA fragments, this linker containing a recognition sequence for *Bsm*FI. This is an unusual restriction enzyme which, rather than cutting within its recognition sequence, cuts 10–14 nucleotides downstream. Treatment with *Bsm*FI therefore removes a fragment with an average length of 12 bp from the end of each cDNA. The fragments are collected, ligated head-to-tail to produce a catenane, and sequenced. In the original method, the catenanes were cloned and sequenced by the chain-termination method, but they can also be sequenced using a next-generation approach. The individual mRNA sequences can be identified within each catenane, because they are separated by *Bsm*FI sites.

Sequencing a transcriptome by RNA-seq

The next-generation sequencing techniques developed during the late 2000s can easily be adapted for sequencing RNA as well as DNA. All that is necessary is to convert the RNA into cDNA prior to preparation of the sequencing library. For example, to obtain sequences of the mRNAs in a transcriptome, the RNA extract is mixed with magnetic beads coated with oligo(dT), to which the mRNAs hybridize via their poly(A) tails (Figure 12.15). After purification, the mRNAs are released from the beads, resuspended in a buffer containing magnesium or some other divalent cation, and heated to 94 °C for

Figure 12.15

Capture of mRNA with oligo(dT)-coated magnetic beads.

a few minutes. This treatment results in random breakage of the mRNA, producing fragments of an appropriate size for next-generation sequencing. The RNA fragments are then mixed with short oligonucleotides of random sequences, some of which will anneal at various positions on the RNA, priming synthesis of the first cDNA strand by reverse transcriptase. The RNA is partially degraded with RNase H and the second cDNA strand is synthesized as in the standard method for preparing cDNA (see Figure 8.7). Once the double-stranded cDNA has been created, library preparation can proceed exactly as for DNA sequencing (p. 182).

The next-generation sequencing of RNA is termed RNA-seq. As is the case when DNA is sequenced with a next-generation method, many millions of individual RNA sequences can be obtained. This means that every mRNA in the transcriptome is likely to be sequenced many times, so even mRNAs that are rare in a particular tissue will be detected. As the numbers of reads obtained for each mRNA will be proportional to the relative abundance of that mRNA in the transcriptome, RNA-seq can provide both quantitative and qualitative information on the transcriptome. Modified versions of the initial purification steps can direct RNA-seq at other RNA types such as miRNAs.

Advantages of the different methods for transcriptome analysis

RNA-seq is such a powerful technique that it might be imagined that other methods for transcriptome analysis are now superseded. This is, arguably, true for SAGE, which was devised initially as a means of obtaining sequences from as many mRNA molecules as possible, at a time when sequence acquisition was limited by reliance on the chain-termination method. SAGE products can now be sequenced using a next-generation method, and so can yield as much information as RNA-seq, though the preparation of an RNA-seq library is less laborious. SAGE also presents an additional computation burden as the resulting reads must be deconstructed to produce the sequences of the individual *Bsm*FI fragments that were ligated together.

RNA-seq is less likely to replace microarray analysis as the latter is quicker and cheaper, and so more appropriate for high-throughput studies. There is a limit to the number of mRNAs that can be included on a single microarray, but this is not an issue for applications such as the screening of clinical samples for gene expression abnormalities associated with disease, which only require detection of particular mRNAs. Microarrays diagnostic for specific diseases are available commercially or can be custom-made. The great value of RNA-seq lies with its ability to provide information on a transcriptome that has not been previously characterized because, unlike microarray analysis, an RNA-seq experiment can be carried out without any *a priori* knowledge of the RNAs that are present.

12.2.2 Studying the proteome

The proteome is the entire collection of proteins in a cell. Proteome studies (also called proteomics) provide additional information that is not obtainable simply by examining the transcriptome. Examination of the transcriptome gives an accurate indication of which genes are active in a particular cell, but provides a less accurate indication of the proteins that are present. This is because the factors that influence protein content include not only the amount of each mRNA that is available, but also the rate at which an mRNA is translated into protein and the rate at which the protein is degraded.

The method used to identify the constituents of a proteome, termed protein profiling, is based on two techniques – protein electrophoresis and mass spectrometry – both of which have long pedigrees but which were rarely applied together in the pre-genomics era.

Separating the proteins in a proteome

In order to characterize a proteome, it is first necessary to prepare pure samples of its constituent proteins. A mammalian cell may contain 10 000–20 000 different proteins, so a highly discriminating separation system is needed.

Polyacrylamide gel electrophoresis is the standard method for separating the proteins in a mixture. Depending on the composition of the gel and the conditions under which the electrophoresis is carried out, the different chemical and physical properties of proteins can be used as the basis for their separation. The most popular technique makes use of the detergent sodium dodecyl sulphate (SDS), which denatures proteins and confers a negative charge that is roughly equivalent to the length of the unfolded polypeptide. Under these conditions, the proteins separate according to their molecular masses, with the smallest proteins migrating more quickly toward the positive electrode. Alternatively, proteins can be separated by isoelectric focusing in a gel that contains chemicals that establish a pH gradient when an electrical charge is applied. In this type of gel, a protein migrates to its isoelectric point, the position in the gradient where its net charge is zero.

In protein profiling, these methods are combined in two-dimensional gel electrophoresis. In the first dimension the proteins are separated by isoelectric focusing. The gel is then soaked in SDS solution, rotated by 90° and a second electrophoresis, separating the proteins according to their sizes, carried out at right angles to the first (Figure 12.16). This approach can separate several thousand proteins in a single gel.

After electrophoresis, staining of the gel reveals a complex pattern of spots, each of which contains a different protein. When two gels are compared, differences in the pattern and intensities of the spots indicate differences in the identities and relative amounts of individual proteins in the two proteomes that are being studied. Interesting spots can therefore be targeted for the second stage of profiling in which actual protein identities are determined.

Load the protein sample

First electrophoresis   Rotate   Second electrophoresis

Figure 12.16

Two-dimensional gel electrophoresis.

Identifying the individual proteins after separation

The second stage of protein profiling is to identify the individual proteins that have been separated from the starting mixture. Originally, this was a difficult proposition but peptide mass fingerprinting has provided a rapid and accurate identification procedure.

Peptide mass fingerprinting was made possible by advances in mass spectrometry, a technique which originally was designed for identifying compounds based on the mass-to-charge (m/z) ratios of the ionized forms produced when molecules of the compound are exposed to a high-energy field. Unfortunately, the standard MS technique could not be used with proteins because they are too large to be ionized effectively, but the development of matrix-assisted laser desorption ionization time-of-flight (MALDI-TOF) mass spectrometry overcame this problem, at least for peptides up to 50 amino acids in length. Of course, most proteins are much longer than 50 amino acids, and it is therefore necessary to break them into fragments before examining them by MALDI-TOF. The usual approach is to digest the protein with a sequence-specific protease such as trypsin, which cleaves proteins immediately after arginine or lysine residues. For most proteins, this results in a series of peptides ranging from five to 75 amino acids in length.

Once ionized, the m/z ratio of a peptide is determined from its 'time of flight' within the mass spectrometer as it passes from the ionization source to the detector (Figure 12.17). The m/z ratio enables the molecular mass of the peptide to be determined, which in turn allows its amino acid composition to be deduced. If a number of

(a) MALDI-TOF mass spectrometry

(b) MALDI-TOF spectrum

Figure 12.17

The use of MALDI-TOF mass spectrometry. (a) Ionized peptides are injected into the mass spectrometer. (b) The mass-to-charge ratios of the peptides are measured and displayed as a spectrum.

peptides from a single protein spot in the two-dimensional gel are analysed, then the resulting compositional information can be related to the genome sequence to identify the gene that specifies that protein.

If two proteomes are being compared, then a key requirement is that proteins present in different amounts can be identified. If these differences are relatively large they will be apparent simply by looking at the stained gels after two-dimensional electrophoresis. However, important changes in the biochemical properties of a proteome can result from relatively minor changes in the amounts of individual proteins, and methods for detecting small-scale changes are therefore essential. One possibility is to label the constituents of two proteomes with different fluorescent markers, and then to run them together in a single two-dimensional gel. Visualization of the gel at different wavelengths enables the intensities of equivalent spots to be judged more easily than is possible when two separate gels are compared.

A more accurate alternative is to label each proteome with an isotope-coded affinity tag (ICAT). These markers can be obtained in two forms, one containing normal hydrogen atoms and the other containing deuterium, the heavy isotope of hydrogen. The normal and heavy versions can be distinguished by mass spectrometry, enabling the relative amounts of a protein in two proteomes that have been mixed together to be determined during the MALDI-TOF stage of the profiling procedure (Figure 12.18).

12.2.3 Studying protein–protein interactions

Important information on the activity of a genome can also be obtained by identifying pairs or groups of proteins that function together. Within living cells, few – if any – proteins act in total isolation. Instead they interact with one another in biochemical pathways and in multiprotein complexes. Two important techniques – phage display and the yeast two-hybrid system – enable these protein–protein interactions to be examined.

Figure 12.18

Analysing two proteomes by ICAT. In the MALDI-TOF spectrum, peaks resulting from peptides containing normal hydrogen atoms are shown in red, and those from peptides containing deuterium are shown in blue. The protein under study is approximately 1.5-fold more abundant in the proteome that has been labelled with normal hydrogen.

(a) Protein display on the surface of a phage

Displayed proteins

(b) Fusion between the cloned gene and a coat protein gene

Phage gene   Cloned gene

Gene
expression

Phage coat
protein   Displayed
protein

(c) Using a phage display library

Microtitre well

Phage display
library

Retained
phage

Washes

Test protein

Figure 12.19
Phage display. (a) Display of proteins on the surface of a recombinant filamentous phage. (b) The gene fusion used to display a protein. (c) One way of detecting interactions between a test protein and phages from within a display library.

Phage display
This technique is called phage display because it involves the 'display' of proteins on the surface of a bacteriophage, usually M13 (Figure 12.19a). This is achieved by cloning the gene for the protein in a special type of M13 vector, one that results in the cloned gene becoming fused with a gene for a phage coat protein (Figure 12.19b). After transfection of *E. coli*, this gene fusion directs the synthesis of a hybrid protein, composed partly of the coat protein and partly of the product of the cloned gene. With luck, this hybrid protein will be inserted into the phage coat so that the product of the cloned gene is now located on the surface of the phage particles.

Normally, this technique is carried out with a phage display library composed of many recombinant phages, each displaying a different protein. Large libraries can be prepared by cloning a mixture of cDNAs from a particular tissue or, less easily, by cloning genomic DNA fragments. The library consists of phages displaying a range of different proteins, and is used to identify those that interact with a test protein. This test protein could be a pure protein or one that is itself displayed on a phage surface. The protein is immobilized in the wells of a microtitre tray or on particles that can be used in an affinity chromatography column, and then mixed with the phage display

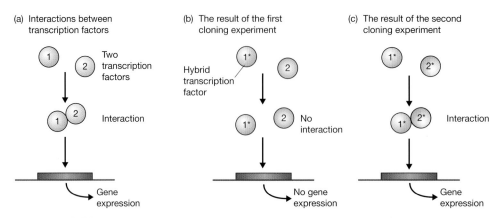

Figure 12.20

The yeast two-hybrid system. (a) A pair of transcription factors that must interact in order for a yeast gene to be expressed. (b) Replacement of transcription factor 1 with the hybrid protein 1* abolishes gene expression because 1* cannot interact with transcription factor 2. (c) Replacement of transcription factor 2 with the hybrid protein 2* restores gene expression if the hybrid parts of 1* and 2* are able to interact.

library (Figure 12.19c). The phages that are retained in the microtitre tray or within the column after a series of washes are those that display proteins which interact with the immobilized test protein.

The yeast two-hybrid system

The yeast two-hybrid system is very different to phage display. This procedure is based on the discovery that gene expression in *S. cerevisiae* depends on interactions between pairs of transcription factors (Figure 12.20a). In the two-hybrid system, a pair of transcription factors responsible for the expression of a yeast gene is replaced by hybrid proteins, each one composed partly of transcription factor and partly of test protein. The ability of this pair of hybrids to direct expression of the yeast target gene is then tested.

To use the system, two yeast cloning experiments must be carried out. The first experiment involves the gene whose protein product is being studied. This gene is ligated to the gene for one of the pair of transcription factors and the construct inserted into a yeast vector. The recombinant yeasts that are produced are unable to express the target gene, because this modified transcription factor cannot interact with its partner (Figure 12.20b).

In the second experiment, a hybrid version of the partner is made and cloned into the yeast cells. Restoration of expression of the target gene indicates that the two hybrid transcription factors can interact. The fusions are designed in such a way that this can only happen if the interactions occur between the test protein components of the hybrids, and not between the transcription factor segments (Figure 12.20c). Pairs of interacting test proteins are therefore identified. The second cloning experiment can involve a library of recombinants representing different proteins, so that one protein can be tested against many others.

Further reading

Altschul, S.F., Gish, W., Miller, W., Myers, E.W., and Lipman, D.J. (1990) Basic local alignment search tool. *Journal of Molecular Biology*, **215**, 403–410. [The BLAST program.]

Görg, A., Weiss, W., and Dunn, M.J. (2004) Current two-dimensional electrophoresis technology for proteomics. *Proteomics*, **4**, 3665–3685.

Guigó, R., Flicek, P., Abril, J.F., *et al.* (2006) EGASP: the human ENCODE Genome Annotation Assessment Project. *Genome Biology*, **7**, S2, doi:10.1186/gb-2006-7-s1-s2. [Comparison of the accuracy of different computer programs for gene location.]

Kellis, M., Patterson, N., Birren, B., and Lander, E.S. (2003) Sequencing and comparison of yeast species to identify genes and regulatory elements. *Nature*, **423**, 241–254. [Using comparative genomics to annotate the yeast genome sequence.]

Pande, J., Szewczyk, M.M., and Grover, A.K. (2010) Phage display: concept, innovations, applications and future. *Biotechnology Advances*, **28**, 849–858.

Parrish, J.R., Gulyas, K.D., and Finley, R.L. (2006) Yeast two-hybrid contributions to interactome mapping. *Current Opinion in Biotechnology*, **17**, 387–393.

Park, P.J. (2009) ChIP-seq: advantages and challenges of a maturing technology. *Nature Reviews Genetics*, **10**, 669–680.

Phizicky, E., Bastiaens, P.I.H., Zhu, H., Snyder, M., and Fields, S. (2003) Protein analysis on a proteomics scale. *Nature*, **422**, 208–215. [Reviews all aspects of proteomics].

Ramsay, G. (1998) DNA chips: state of the art. *Nature Biotechnology*, **16**, 40–44.

Stoughton, R.B. (2005) Applications of DNA microarrays in biology. *Annual Review of Biochemistry*, **74**, 53–82.

Velculescu, V.E., Vogelstein, B., and Kinzler, K.W. (2000) Analysing uncharted transcriptomes with SAGE. *Trends in Genetics*, **16**, 423–425.

Wach, A., Brachat, A., Pohlmann, R., and Philippsen, P. (1994) New heterologous modules for classical or PCR-based gene disruptions in *Saccharomyces cerevisiae*. *Yeast*, **10**, 1793–1808. [Gene inactivation by homologous recombination.]

Wang, Z., Gerstein, M., and Snyder, M. (2009) RNA-Seq: a revolutionary tool for transcriptomics. *Nature Reviews Genetics*, **10**, 57–63.

Tandell, M. and Ence, D. (2012) A beginner's guide to eukaryotic genome annotation. *Nature Reviews Genetics*, **13**, 329–342.

PART III

The Applications of Gene Cloning and DNA Analysis in Biotechnology

Chapter 13

Production of Protein from Cloned Genes

Chapter contents

Now that we have covered the basic techniques involved in gene cloning and DNA analysis and examined how these techniques are used in research, we can move on to consider how recombinant DNA technology is being applied in biotechnology. This is not a new subject, although biotechnology has received far more attention during recent years than it ever has in the past. Biotechnology can be defined as the use of biological processes in industry and technology. According to archaeologists, the British biotechnology industry dates back 4000 years, to the late Neolithic period, when fermentation processes that make use of living yeast cells to produce ale and mead were first introduced into this country. Certainly brewing was well established by the time of the Roman invasion.

During the 20th century, biotechnology expanded with the development of a variety of industrial uses for microorganisms. The discovery by Alexander Fleming in 1929 that the mould *Penicillium* synthesizes a potent antibacterial agent led to the use of fungi and bacteria in the large-scale production of antibiotics. At first, the microorganisms were grown in large culture vessels from which the antibiotic was purified after the cells had been removed (Figure 13.1a), but this batch culture method has been largely supplanted by continuous culture techniques, making use of a fermenter, from which samples of medium can be continuously drawn off, providing a non-stop supply of the product (Figure 13.1b). This type of process is not limited to antibiotic production, and has also been used to obtain large amounts of other compounds produced by microorganisms (Table 13.1).

Gene Cloning and DNA Analysis: An Introduction, Seventh Edition. T.A. Brown.
© 2016 John Wiley & Sons, Ltd. Published 2016 by John Wiley & Sons, Ltd.
Companion Website: www.wiley.com/go/brown/cloning

Figure 13.1

Two different systems for the growth of microorganisms. (a) Batch culture. (b) Continuous culture.

(a) Batch culture

Centrifuge → Prepare product from medium or cells

Closed culture vessel

Pellet of cells

(b) Continuous culture

Stirrer

Fresh medium in → → Medium + cells out → Prepare product

Large-volume culture

One of the reasons why biotechnology has received so much attention during the past three decades is because of gene cloning. Although many useful products can be obtained from microbial culture, the list in the past has been limited to those compounds naturally synthesized by microorganisms. Many important pharmaceuticals, which are produced not by microbes but by higher organisms, could not be obtained in this way.

Table 13.1

Some of the compounds produced by industrial scale culture of microorganisms.

COMPOUND	MICROORGANISM
Antibiotics	
Cephalosporins	*Cephalosporium* spp.
Chloramphenicol, streptomycin	*Streptomyces* spp.
Gramicidins, polymixins	*Bacillus* spp.
Penicillins	*Penicillium* spp.
Enzymes	
Invertase	*Saccharomyces cerevisiae*
Proteases, amylases	*Bacillus* spp., *Aspergillus* spp.
Other compounds	
Acetone, butanol	*Clostridium* spp.
Alcohol	*S. cerevisiae, Saccharomyces carlsbergensis*
Butyric acid	Butyric acid bacteria
Citric acid	*Aspergillus niger*
Dextran	*Leuconostoc* spp.
Glycerol	*S. cerevisiae*
Vinegar	*S. cerevisiae*, acetic acid bacteria

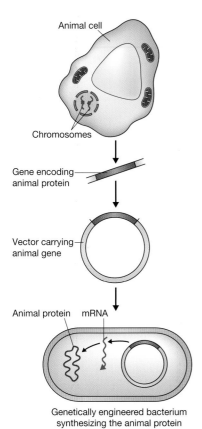

Figure 13.2

A possible scheme for the production of an animal protein by a bacterium.

Animal cell

Chromosomes

Gene encoding animal protein

Vector carrying animal gene

Animal protein   mRNA

Genetically engineered bacterium synthesizing the animal protein

This has been changed by the application of gene cloning to biotechnology. The ability to clone genes means that a gene for an important animal or plant protein can now be taken from its normal host, inserted into a cloning vector, and introduced into a bacterium (Figure 13.2). If the manipulations are performed correctly, the gene will be expressed and the recombinant protein synthesized by the bacterial cell. It may then be possible to obtain large amounts of the protein.

Of course, in practice the production of recombinant protein is not as easy as it sounds. Special types of cloning vector are needed, and satisfactory yields of recombinant protein are often difficult to obtain. In this chapter we will look at cloning vectors for recombinant protein synthesis and examine some of the problems associated with their use.

13.1 Special vectors for the expression of foreign genes in *E. coli*

If a foreign (i.e., non-bacterial) gene is simply ligated into a standard vector and cloned in *E. coli*, it is very unlikely that a significant amount of recombinant protein will be synthesized. This is because expression is dependent on the gene being surrounded by a collection of signals that can be recognized by the bacterium. These signals, which are

Figure 13.3

The three most important signals for gene expression in *E. coli.*

short sequences of nucleotides, advertise the presence of the gene and provide instructions for the transcriptional and translational apparatus of the cell. The three most important signals for *E. coli* genes are as follows (Figure 13.3):

- The promoter, which marks the point at which transcription of the gene should start. In *E. coli*, the promoter is recognized by the σ subunit of the transcribing enzyme RNA polymerase.
- The terminator, which marks the point at the end of the gene where transcription should stop. A terminator is usually a nucleotide sequence that can base pair with itself to form a stem–loop structure.
- The ribosome-binding site, a short nucleotide sequence recognized by the ribosome as the point at which it should attach to the mRNA molecule. The initiation codon of the gene is always a few nucleotides downstream of this site.

The genes of higher organisms are also surrounded by expression signals, but their nucleotide sequences are not the same as the *E. coli* versions. This is illustrated by comparing the promoters of *E. coli* and human genes (Figure 13.4). There are similarities, but it is unlikely that an *E. coli* RNA polymerase would be able to attach to a human promoter. A foreign gene is inactive in *E. coli*, simply because the bacterium does not recognize its expression signals.

A solution to this problem would be to insert the foreign gene into the vector in such a way that it is placed under control of a set of *E. coli* expression signals. If this can be achieved, then the gene should be transcribed and translated (Figure 13.5). Cloning vectors that provide these signals, and can therefore be used in the production of recombinant protein, are called expression vectors.

Figure 13.4

Typical promoter sequences for *E. coli* and animal genes.

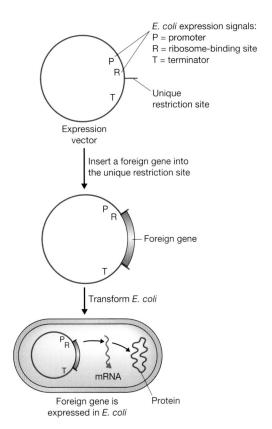

E. coli expression signals:
P = promoter
R = ribosome-binding site
T = terminator

Unique restriction site

Expression vector

Insert a foreign gene into the unique restriction site

Foreign gene

Transform E. coli

mRNA

Foreign gene is expressed in E. coli

Protein

Figure 13.5

The use of an expression vector to achieve expression of a foreign gene in *E. coli*.

13.1.1 The promoter is the critical component of an expression vector

The promoter is the most important component of an expression vector. This is because the promoter controls the very first stage of gene expression (attachment of an RNA polymerase enzyme to the DNA) and determines the rate at which mRNA is synthesized. The amount of recombinant protein obtained therefore depends to a great extent on the nature of the promoter carried by the expression vector.

The promoter must be chosen with care

The two sequences shown in Figure 13.4a are consensus sequences, averages of all the *E. coli* promoter sequences that are known. Although most *E. coli* promoters do not differ much from these consensus sequences (e.g., TTTACA instead of TTGACA), a small variation may have a major effect on the efficiency with which the promoter can direct transcription. Strong promoters are those that can sustain a high rate of transcription, and usually control genes whose translation products are required in large amounts by the cell (Figure 13.6a). In contrast, weak promoters, which are relatively inefficient, direct transcription of genes whose products are needed in only small amounts (Figure 13.6b). Clearly, an expression vector should carry a strong promoter, so that the cloned gene is transcribed at the highest possible rate.

A second factor to be considered when constructing an expression vector is whether it will be possible to regulate the promoter in any way. Two major types of gene

Figure 13.6

Strong and weak promoters.

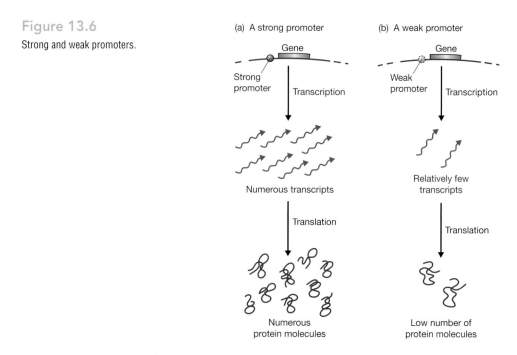

(a) A strong promoter

Gene

Strong promoter

Transcription

Numerous transcripts

Translation

Numerous protein molecules

(b) A weak promoter

Gene

Weak promoter

Transcription

Relatively few transcripts

Translation

Low number of protein molecules

regulation are recognized in *E. coli* – induction and repression. An inducible gene is one whose transcription is switched on by addition of a chemical to the growth medium. Often, this chemical is one of the substrates for the enzyme coded by the inducible gene (Figure 13.7a). In contrast, a repressible gene is switched off by addition of the regulatory chemical (Figure 13.7b).

Gene regulation is a complex process that only indirectly involves the promoter itself. However, many of the sequences important for induction and repression lie in

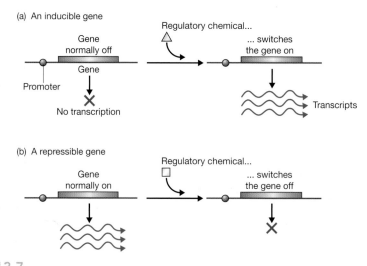

(a) An inducible gene

Gene normally off

Promoter

Gene

No transcription

Regulatory chemical...

... switches the gene on

Transcripts

(b) A repressible gene

Gene normally on

Regulatory chemical...

... switches the gene off

Figure 13.7

Examples of the two major types of gene regulation that occur in bacteria. (a) An inducible gene. (b) A repressible gene.

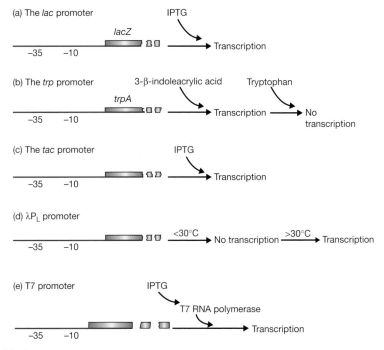

Figure 13.8

Five promoters frequently used in expression vectors. The *lac* and *trp* promoters are shown upstream of the genes that they normally control in *E. coli*.

the region surrounding the promoter and are therefore also present in an expression vector. It may therefore be possible to extend the regulation to the expression vector, so that the chemical which induces or represses the gene that is normally controlled by the promoter is also able to regulate expression of the cloned gene. This can be a distinct advantage in the production of recombinant protein. For example, if the recombinant protein has a harmful effect on the bacterium, then its synthesis must be carefully monitored to prevent accumulation of toxic levels. This can be achieved by a judicious use of the regulatory chemical to control expression of the cloned gene. Even if the recombinant protein has no harmful effects on the host cell, regulation of the cloned gene is still desirable, as a continuously high level of transcription may affect the ability of the recombinant plasmid to replicate, leading to its eventual loss from the culture.

Examples of promoters used in expression vectors

Several *E. coli* promoters combine the desired features of strength and ease of regulation. Those most frequently used in expression vectors are as follows:

- The *lac* **promoter** (Figure 13.8a) is the sequence that controls transcription of the *lacZ* gene coding for β-galactosidase (and also the *lacZ'* gene fragment carried by the pUC and M13mp vectors; p. 81). The *lac* promoter is induced by isopropylthiogalactoside (IPTG, p. 82), so addition of this chemical into the growth medium switches on transcription of a gene inserted downstream of the *lac* promoter carried by an expression vector.

- The *trp* promoter (Figure 13.8b) is normally upstream of the cluster of genes coding for several of the enzymes involved in biosynthesis of the amino acid tryptophan. The *trp* promoter is repressed by tryptophan, but is more easily induced by 3-β-indoleacrylic acid.
- The *tac* promoter (Figure 13.8c) is a hybrid between the *trp* and *lac* promoters. It is stronger than either, but still induced by IPTG.
- The λP_L promoter (Figure 13.8d) is one of the promoters responsible for transcription of the λ DNA molecule. λP_L is a very strong promoter that is recognized by the *E. coli* RNA polymerase, which is subverted by λ into transcribing the bacteriophage DNA. The promoter is repressed by the product of the λcI gene. Expression vectors that carry the λP_L promoter are used with a mutant *E. coli* host that synthesizes a temperature-sensitive form of the cI protein (p. 41). At a low temperature (less than 30 °C) this mutant cI protein is able to repress the λP_L promoter, but at higher temperatures the protein is inactivated, resulting in transcription of the cloned gene.
- The T7 promoter (Figure 13.8e) is specific for the RNA polymerase coded by T7 bacteriophage. This RNA polymerase is much more active than the *E. coli* RNA polymerase (p. 98), which means that a gene inserted downstream of the T7 promoter will be expressed at a high level. The gene for the T7 RNA polymerase is not normally present in the *E. coli* genome, so a special strain of *E. coli* is needed, one which is lysogenic for T7 phage. Remember that a lysogen contains an inserted copy of the phage DNA in its genome (p. 19). In this particular strain of *E. coli*, the phage DNA has been altered by placing a copy of the *lac* promoter upstream of its gene for the T7 RNA polymerase. The addition of IPTG to the growth medium therefore switches on synthesis of the T7 RNA polymerase, which in turn leads to activation of the gene carried by the T7 expression vector.

13.1.2 Cassettes and gene fusions

An efficient expression vector requires not only a strong, regulatable promoter, but also an *E. coli* ribosome-binding sequence and a terminator. In most vectors these expression signals form a cassette, so-called because the foreign gene is inserted into a unique restriction site present in the middle of the expression signal cluster (Figure 13.9). Ligation of the foreign gene into the cassette therefore places it in the ideal position relative to the expression signals.

With some cassette vectors the cloning site is not immediately adjacent to the ribosome-binding sequence, but instead is preceded by a segment from the beginning of an *E. coli* gene (Figure 13.10). Insertion of the foreign gene into this restriction site must be performed in such a way as to fuse the two reading frames, producing a hybrid

Figure 13.9
A typical cassette vector and the way it is used. P = promoter; R = ribosome-binding site; T = terminator.

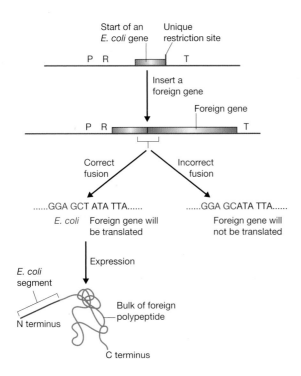

Figure 13.10

The construction of a hybrid gene and the
synthesis of a fusion protein.

gene that starts with the *E. coli* segment and progresses without a break into the codons
of the foreign gene. The product of gene expression is therefore a hybrid or fusion protein, consisting of the short peptide coded by the *E. coli* reading frame fused to the
amino-terminus of the foreign protein. This fusion system has four advantages:

- Efficient translation of the mRNA produced from the cloned gene depends not only
  on the presence of a ribosome-binding site, but is also affected by the nucleotide
  sequence at the start of the coding region. This is probably because secondary
  structures resulting from intrastrand base pairs could interfere with attachment of
  the ribosome to its binding site (Figure 13.11). This possibility is avoided if the
  pertinent region is made up entirely of natural *E. coli* sequences.
- The presence of the bacterial peptide at the start of the fusion protein may stabilize
  the molecule and prevent it from being degraded by the host cell. In contrast,
  foreign proteins that lack a bacterial segment are often destroyed by the host.
- The bacterial segment may constitute a signal peptide, responsible for directing the
  E. coli protein to its correct position in the cell. If the signal peptide is derived from
  a protein that is exported by the cell (e.g., the products of the *ompA* or *malE*
  genes), the recombinant protein may itself be exported, either into the culture
  medium or into the periplasmic space between the inner and outer cell membranes.

Figure 13.11

A problem caused by secondary structure at the start of an mRNA.

Figure 13.12

The use of affinity chromatography to purify a glutathione-S-transferase fusion protein.

Export is desirable because it simplifies the problem of purification of the recombinant protein from the culture.

- The bacterial segment may also aid purification by enabling the fusion protein to be recovered by affinity chromatography. For example, fusions involving the E. coli glutathione-S-transferase protein can be purified by adsorption onto agarose beads carrying bound glutathione (Figure 13.12).

The disadvantage with a fusion system is that the presence of the E. coli segment may alter the properties of the recombinant protein. Methods for removing the bacterial segment are therefore needed. Usually, this is achieved by treating the fusion protein with a chemical or enzyme that cleaves the polypeptide chain at or near the junction between the two components. For example, if a methionine is present at the junction the fusion protein can be cleaved with cyanogen bromide, which cuts polypeptides specifically at methionine residues (Figure 13.13). Alternatively, enzymes such as thrombin (which

Figure 13.13

One method for the recovery of the foreign polypeptide from a fusion protein. The methionine residue at the fusion junction must be the only one present in the entire polypeptide. If others are present, cyanogen bromide will cleave the fusion protein into more than two fragments.

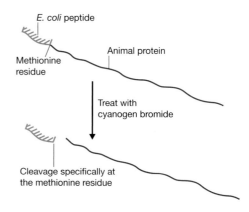

cleaves adjacent to arginine residues) or factor Xa (which cuts after the arginine of Gly–Arg) can be used. The important consideration is that recognition sequences for the cleavage agent must not occur within the recombinant protein.

13.2 General problems with the production of recombinant protein in *E. coli*

Despite the development of sophisticated expression vectors, there are still numerous difficulties associated with the production of protein from foreign genes cloned in *E. coli*. These problems can be grouped into two categories: (i) those that are due to the sequence of the foreign gene; and (ii) those that are due to the limitations of *E. coli* as a host for recombinant protein synthesis.

13.2.1 Problems resulting from the sequence of the foreign gene

There are three ways in which the nucleotide sequence might prevent efficient expression of a foreign gene cloned in *E. coli*:

- The foreign gene might contain introns. This would be a major problem, as *E. coli* genes do not contain introns and therefore the bacterium does not possess the necessary machinery for removing introns from the transcripts (Figure 13.14a).
- The foreign gene might contain sequences that act as termination signals in *E. coli* (Figure 13.14b). These sequences are perfectly innocuous in the normal host cell, but in the bacterium the result is premature termination and a loss of gene expression.
- The codon bias of the gene may not be ideal for translation in *E. coli*. As described on p. 227, although virtually all organisms use the same genetic code, each organism has a bias toward preferred codons. This bias reflects the efficiency with which the tRNA molecules in the organism are able to recognize the different codons. If a cloned gene contains a high proportion of disfavoured codons, the *E. coli* tRNAs may encounter difficulties in translating the gene, reducing the amount of protein that is synthesized (Figure 13.14c).

These problems can usually be solved, although the necessary manipulations may be time-consuming and costly (these are important considerations in an industrial project). If the gene contains introns then its complementary DNA (cDNA), prepared from the mRNA (p. 142) and so lacking introns, might be used as an alternative. *In vitro* mutagenesis could then be employed to change the sequences of possible terminators and to replace disfavoured codons with those preferred by *E. coli*. With genes that are less than 1.2 kb in length, an alternative would be to make an artificial version (p. 221), which involves synthesizing a set of overlapping oligonucleotides that are ligated together. The sequences of these oligonucleotides are designed to ensure that the resulting gene contains preferred *E. coli* codons, and that terminators are absent.

Figure 13.14

Three of the problems that could be encountered when foreign genes are expressed in *E. coli*. (a) Introns are not removed in *E. coli*. (b) Premature termination of transcription. (c) A problem with codon bias.

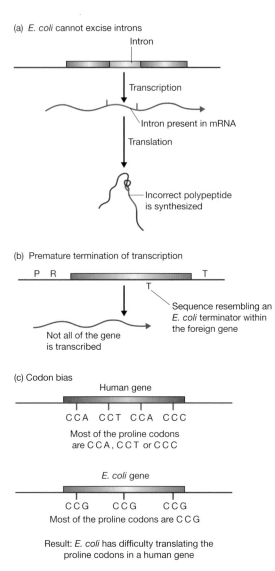

(a) *E. coli* cannot excise introns

Intron

Transcription

Intron present in mRNA

Translation

Incorrect polypeptide is synthesized

(b) Premature termination of transcription

P   R   T

T

Sequence resembling an *E. coli* terminator within the foreign gene

Not all of the gene is transcribed

(c) Codon bias

Human gene

CCA   CCT   CCA   CCC

Most of the proline codons are CCA, CCT or CCC

E. coli gene

CCG   CCG   CCG

Most of the proline codons are CCG

Result: *E. coli* has difficulty translating the proline codons in a human gene

13.2.2 Problems caused by E. coli

Some of the difficulties encountered when using *E. coli* as the host for recombinant protein synthesis stem from inherent properties of the bacterium. For example:

- *E. coli* might not process the recombinant protein correctly. The proteins of most organisms are processed after translation, by chemical modification of amino acids within the polypeptide. Often, these processing events are essential for the correct biological activity of the protein, but unfortunately the proteins of bacteria and higher organisms are not processed identically. In particular, some animal proteins are glycosylated, which means they have sugar groups attached to them after translation. Glycosylation is extremely uncommon in bacteria and recombinant proteins synthesized in *E. coli* are never glycosylated correctly.

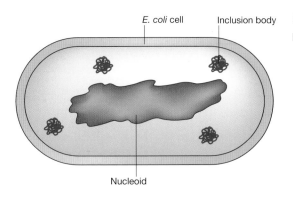

Figure 13.15

Inclusion bodies.

- *E. coli* might not fold the recombinant protein correctly, and generally is unable to synthesize the disulphide bonds present in many animal proteins. If the protein does not take up its correctly folded tertiary structure, then usually it is insoluble and forms an inclusion body within the bacterium (Figure 13.15). Recovery of the protein from the inclusion body is not a problem, but converting the protein into its correctly folded form can be difficult or even impossible in the test tube. Under these circumstances the protein is, of course, inactive.
- *E. coli* might degrade the recombinant protein. Exactly how *E. coli* can recognize the foreign protein, and thereby subject it to preferential turnover, is not known.

These problems are less easy to solve than the sequence problems described in the previous section, but can be alleviated to a certain extent by using special *E. coli* strains. The degradation of recombinant proteins can be reduced by using a mutant *E. coli* strain that is deficient in one or more of the proteases responsible for protein degradation. Correct folding of recombinant proteins can also be promoted by choosing a special host strain, in this case one that oversynthesizes the chaperone proteins thought to be responsible for protein folding in the cell. The main drawback, however, is the absence of glycosylation. Attempts have been made to solve this problem with *E. coli* strains that contain cloned genes for enzymes that carry out glycosylation in other organisms. These include *Campylobacter jejuni*, one of the few bacteria that has any glycosylation activity. So far, however, this approach has had limited success, and *E. coli* is generally looked on as only suitable for the synthesis of those animal proteins that do not need to be glycosylated.

13.3 Production of recombinant protein by eukaryotic cells

The problems associated with obtaining high yields of active recombinant proteins from genes cloned in *E. coli* have led to the development of expression systems for other organisms. A few attempts have been made to use other bacteria as hosts for recombinant protein synthesis, and some progress has been made with *Lactococcus lactis* and *Pseudomonas* species, but the main alternatives to *E. coli* are microbial eukaryotes. The argument here is that a microbial eukaryote, such as a yeast or filamentous fungus, is more closely related to an animal, and so may be able to deal with recombinant protein synthesis more efficiently than *E. coli*. Yeasts and fungi can be grown just as easily as bacteria in continuous culture, and might express a cloned gene from a higher organism, and process the resulting protein in a manner more akin to that occurring in the higher organism itself.

13.3.1 Recombinant protein from yeasts and filamentous fungi

To a large extent, the potential of microbial eukaryotes has been realized and these organisms are now being used for the routine production of several animal proteins. Expression vectors are still required because it turns out that the promoters and other expression signals for animal genes do not, in general, function efficiently in these lower eukaryotes. The vectors themselves are based on those described in Chapter 7.

Saccharomyces cerevisiae as the host for recombinant protein synthesis

The yeast *Saccharomyces cerevisiae* is currently the most popular microbial eukaryote for recombinant protein production. Cloned genes are often placed under the control of the *GAL* promoter (Figure 13.16a), which is normally upstream of the gene coding for galactose epimerase, an enzyme involved in the metabolism of galactose. The *GAL* promoter is induced by galactose, providing a straightforward system for regulating the expression of a cloned foreign gene. Other useful promoters are *PHO5*, which is regulated by the phosphate level in the growth medium, and *CUP1*, which is induced by copper. Most yeast expression vectors also carry a termination sequence from an *S. cerevisiae* gene, because animal termination signals do not function effectively in yeast.

Although yields of recombinant protein are relatively high, *S. cerevisiae* is unable to glycosylate animal proteins correctly, often adding too many sugar units ('hyperglycosylation'). This can be prevented – or at least reduced – by using a mutant host strain. *S. cerevisiae* also lacks an efficient system for secreting proteins into the growth medium. In the absence of secretion, recombinant proteins are retained in the cell and consequently are less easy to purify. Codon bias (p. 227) can also be a problem.

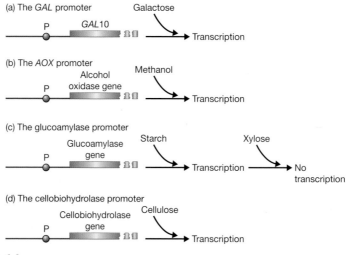

Figure 13.16
Four promoters frequently used in expression vectors for microbial eukaryotes.

Figure 13.17

Comparison between a typical glycosylation structure found on an animal protein and the structures synthesized by *P. pastoris* and *S. cerevisiae*.

Sugars

– Asn–   – Asn–   – Asn–

Human   *P. pastoris*   *S. cerevisiae*

Despite these drawbacks, *S. cerevisiae* remains the most frequently used microbial eukaryote for recombinant protein synthesis. This is partly because *S. cerevisiae* is accepted as a safe organism for the production of proteins to be used in medicines or in foods, and partly because of the wealth of knowledge built up over the years regarding the biochemistry and genetics of *S. cerevisiae*, which means that it is relatively easy to devise strategies for minimizing any difficulties that arise.

Other yeasts and fungi

Although *S. cerevisiae* retains the loyalty of many molecular biologists, there are other microbial eukaryotes that might be equally, if not more, effective in recombinant protein synthesis. In particular, *Pichia pastoris*, a second species of yeast, is able to synthesize large amounts of recombinant protein (up to 30% of the total cell protein), and its glycosylation abilities are very similar to those of animal cells. The sugar structures that it synthesizes are not precisely the same as the animal versions (Figure 13.17), but the differences are relatively trivial and would probably not have any significant effect on the activity of a recombinant protein. Importantly, the glycosylated proteins made by *P. pastoris* are unlikely to induce an antigenic reaction if injected into the bloodstream, a problem that is frequently encountered with the overglycosylated proteins synthesized by *S. cerevisiae*. Expression vectors for *P. pastoris* make use of the alcohol oxidase (*AOX*) promoter (Figure 13.16b), which is induced by methanol. The only significant problem with *P. pastoris* is that it sometimes degrades recombinant proteins before they can be purified, though this can be controlled by using special growth media. Other yeasts that have been used for recombinant protein synthesis include *Hansenula polymorpha*, *Yarrowia lipolytica*, and *Kluveromyces lactis*. The last of these has the attraction that it can be grown on waste products from the food industry.

The two most popular filamentous fungi are *Aspergillus nidulans* and *Trichoderma reesei*. The advantages of these organisms are their good glycosylation properties and their ability to secrete proteins into the growth medium. The latter is a particularly strong feature of the wood rot fungus *T. reesei*, which in its natural habitat secretes cellulolytic enzymes that degrade the wood it lives on. The secretion characteristics mean that these fungi are able to produce recombinant proteins in a form that aids purification. Expression vectors for *A. nidulans* usually carry the glucoamylase promoter (Figure 13.16c), induced by starch and repressed by xylose. Those for *T. reesei* make use of the cellobiohydrolase promoter (Figure 13.16d), which is induced by cellulose.

13.3.2 Using animal cells for recombinant protein production

Difficulties inherent in the synthesis of a fully active animal protein in a microbial host have prompted biotechnologists to explore the possibility of using animal cells for recombinant protein synthesis. For proteins with complex and essential glycosylation structures, an animal cell might be the only type of host within which the active protein can be synthesized.

Protein production in mammalian cells

Culture systems for animal cells have been available since the early 1960s, but only during the past 20 years have methods for large-scale continuous culture been developed. A problem with some animal cell lines is that they require a solid surface on which to grow, which complicates the design of the culture vessels. One solution to this is to fill the inside of the vessel with plates so as to provide a large surface area, but this has the disadvantage that complete and continuous mixing of the medium within the vessel becomes very difficult. A second possibility is to use a standard vessel but to provide the cells with small inert particles (e.g., cellulose beads) on which to grow. Rates of growth and maximum cell densities are much less for animal cells compared to microorganisms, limiting the yield of recombinant protein, but this drawback can be tolerated if it is the only way of obtaining the active protein.

Of course, gene cloning may not be necessary in order to obtain an animal protein from an animal cell culture. Nevertheless, expression vectors and cloned genes are still used to maximize yields, by placing the gene under control of a promoter that is stronger than the one to which it is normally attached. This promoter is often obtained from viruses such as SV40 (p. 130), cytomegalovirus (CMV), or Rous sarcoma virus (RSV). Mammalian cell lines derived from humans or hamsters have been used in the synthesis of several recombinant proteins, and in most cases these proteins have been processed correctly and are indistinguishable from the non-recombinant versions. However, this is the most expensive approach to recombinant protein production, especially as the possible copurification of viruses with the protein means that rigorous quality control procedures must be employed to ensure that the product is safe.

Protein production in insect cells

Insect cells provide an alternative to mammalian cells for animal protein production. Although, in culture, insect cells behave no differently from mammalian cells, they have a major advantage that, because of a natural expression system, they can provide high yields of recombinant protein.

The expression system is based on the baculoviruses, a group of viruses that are common in insects but do not normally infect vertebrates. The baculovirus genome includes the polyhedrin gene, whose product accumulates in the insect cell as large nuclear inclusion bodies toward the end of the infection cycle (Figure 13.18). The product of this single gene frequently comprises over 50% of the total cell protein. Similar levels of protein

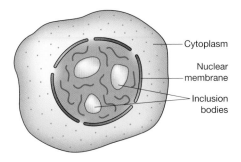

Figure 13.18

Crystalline inclusion bodies in the nucleus of an insect cell infected with a baculovirus.

Cytoplasm

Nuclear membrane

Inclusion bodies

production also occur if the polyhedrin gene is replaced by one for a recombinant protein. Baculovirus vectors have been used successfully in the production of a number of mammalian proteins, but unfortunately the resulting proteins are not glycosylated correctly. In this regard, the baculovirus system does not offer any advantages compared to *S. cerevisiae* or *P. pastoris*. However, the deficiencies in the insect glycosylation process can be circumvented by using a modified baculovirus that carries a mammalian promoter to express genes directly in mammalian cells. The infection is not productive, which means that the virus genome is unable to replicate, but genes cloned into one of the BacMam vectors (as they are called) are maintained stably in mammalian cells for enough time for expression to occur. This expression is accompanied by the mammalian cell's own post-translational processing activities, so the recombinant protein will be correctly glycosylated and therefore should be fully active.

Of course, in nature baculoviruses infect living insects rather than cell cultures. For example, one of the most popular baculoviruses used in cloning is the *Bombyx mori* nucleopolyhedrovirus (BmNPV), which is a natural pathogen of the silkworm. There is a huge conventional industry based on the culturing of silkworms for silk production, and this expertise is now being harnessed for production of recombinant proteins, using expression vectors based on the BmNPV genome. As well as providing an easy and cheap means of obtaining proteins, silkworms have the additional advantage of not being infected by viruses that are pathogenic to humans. The possibility that dangerous viruses are copurified with the recombinant protein is therefore avoided.

13.3.3 Pharming: Recombinant protein from live animals and plants

The use of silkworms for recombinant protein production is an example of the process often referred to as pharming, where a transgenic organism acts as the host for protein synthesis. Pharming is a recent and controversial innovation in gene cloning.

Pharming in animals

A transgenic animal is one that contains a cloned gene in all of its cells. Knockout mice (p. 233), used to study the function of human and other mammalian genes, are examples of transgenic animals. With mice, a transgenic animal can be produced by microinjection of the gene to be cloned into a fertilized egg cell (p. 90). Although this technique works well with mice, the injection of fertilized cells is inefficient or impossible with many other mammals, and the generation of transgenic animals for recombinant protein production usually involves a more sophisticated procedure called nuclear transfer (Figure 13.19).

Figure 13.19

Transfer of the nucleus from a transgenic somatic cell to an oocyte.

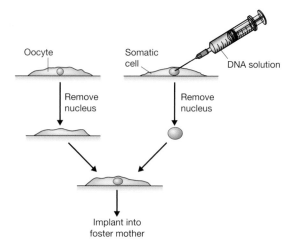

This involves microinjection of the recombinant protein gene into a somatic cell, which is a more efficient process than injection into a fertilized egg. Because the somatic cell will not itself differentiate into an animal, its nucleus – after microinjection – must be transferred to an oocyte whose own nucleus has been removed. After implantation into a foster mother, the engineered cell retains the ability of the original oocyte to divide and differentiate into an animal, one that will contain the transgene in every cell. This is a lengthy procedure and transgenic animals are therefore expensive to produce. However, the technology is cost-effective because once a transgenic animal has been made it can reproduce and pass its cloned gene to its offspring according to standard Mendelian principles.

Although proteins have been produced in the blood of transgenic animals, and in the eggs of transgenic chickens, the most successful approach has been to use farm animals such as sheep or pigs, with the cloned gene attached to the promoter for the animal's β-lactoglobulin gene. This promoter is active in the mammary tissue, which means that the recombinant protein is secreted in the milk (Figure 13.20). Milk production can be continuous during the animal's adult life, resulting in a high yield of the protein. For example, the average cow produces some 8000 litres of milk per year, yielding 40–80 kg

Figure 13.20

Recombinant protein production in the milk of a transgenic sheep.

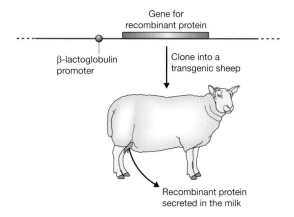

of protein. Because the protein is secreted its purification is relatively easy. Most importantly, sheep and pigs are mammals, and consequently human proteins produced in this way are correctly modified. The production of pharmaceutical proteins in farm animals therefore offers considerable promise for the synthesis of correctly modified human proteins for use in medicine.

Recombinant proteins from plants

Plants provide the final possibility for the production of recombinant protein. Plants and animals have similar protein processing activities, although there are slight differences in their glycosylation pathways. Plant cell culture is a well-established technology that is already used widely in the commercial synthesis of natural plant products. Alternatively, intact plants can be grown to a high density in fields. The latter approach to recombinant protein production has been used with a variety of crops such as maize, tobacco, rice, and sugarcane. One possibility is to place the transgene next to the promoter of a seed-specific gene such as β-phaseolin, which codes for the main seed protein of the bean *Phaseolus vulgaris*. The recombinant protein is therefore synthesized specifically in the seeds, which naturally accumulate large quantities of proteins and are easy to harvest and to process. Recombinant proteins have also been synthesized in the leaves of tobacco and alfalfa, as well as the tubers of potatoes. In all of these cases the protein has to be purified from the complex biochemical mixture that is produced when the seeds, leaves or tubers are crushed. One way to avoid this problem is to express the recombinant protein as a fusion with a signal peptide that directs secretion of the protein by the roots. Although this requires the plants to be grown in hydroponic systems rather than in fields, the decrease in yield is at least partly offset by the low cost of purification.

Whichever production system is used, plants offer a cheap and low-technology means for the mass production of recombinant proteins. A range of proteins have been produced in experimental systems, including important pharmaceuticals such as interleukins and antibodies. This is an area of intensive research at the present time, with a number of plant biotechnology companies developing systems that have reached or are nearing commercial production. One very promising possibility is that plants could be used to synthesize vaccines, providing the basis to a cheap and efficient vaccination program (Chapter 14).

Ethical concerns raised by pharming

With our discussion of pharming we have entered one of the areas of gene cloning that causes concern among the public. No student of gene cloning and DNA analysis should ignore the controversies raised by the genetic manipulation of animals and plants but, equally, no textbook on the subject should attempt to teach the 'correct' response to these ethical concerns. You must make up your own mind on such matters.

With transgenic animals, one of the fears is that the procedures used might cause suffering. These concerns do not centre on the recombinant protein, but rather on the manipulations that result in the production of a transgenic animal. Animals produced by nuclear transfer suffer a relatively high frequency of birth defects, and some of those that do survive do not synthesize the required protein adequately, which means that this type of pharming is accompanied by a high 'wastage'. Even the healthy animals appear to suffer from premature aging, as was illustrated most famously by 'Dolly the sheep' who, although not transgenic, was the first animal to be produced by nuclear transfer. Although most sheep of her breed live for up to 12 years, Dolly developed arthritis at the age of 5 years and was humanely killed one year later when she was found to be suffering from a terminal lung disease that normally occurs only in old sheep. It was speculated that such premature aging was related to the age of the somatic cell whose nucleus gave rise to Dolly. As this cell was from a six-year-old sheep, Dolly was effectively aged six when she was born. Although the technology has moved on dramatically since Dolly's birth in 1997, the welfare issues regarding transgenic animals have not been resolved, and the broader issues concerning the use of nuclear transfer to clone animals (i.e., to produce identical offspring rather than to clone individual genes) remain at the forefront of public awareness.

Pharming in plants raises a completely different set of ethical concerns, relating in part to the impact that genetically modified (GM) crops might have on the environment. These concerns apply to all GM crops, and not just to those used for pharming, and we will return to them in Chapter 15, after we have examined the more general uses of gene cloning in agriculture.

Further reading

Çelik, E. and Çelik, P. (2012) Production of recombinant proteins by yeast cells. *Biotechnology Advances*, **30**, 1108–1118.

Chen. R. (2012) Bacterial expression systems for recombinant protein production: *E. coli* and beyond. *Biotechnology Advances*, **30**, 1102–1107. [Includes information on recombinant protein production in bacteria other than *E. coli.*]

de Boer, H.A., Comstock, L.J., and Vasser, M. (1983) The *tac* promoter: a functional hybrid derived from the trp and lac promoters. *Proceedings of the National Academy of Sciences of the United States of America*, **80**, 21–25.

Drugmand, J.-C., Schneider, Y.-J., and Agathos, S.N. (2012) Insect cells as factories for biomanufacturing. *Biotechnology Advances*, **30**, 1140–1157.

Egelkrout, E., Rajan, V., and Howard, J.A. (2012) Overproduction of recombinant proteins in plants. *Plant Science*, **184**, 83–101.

Houdebine, L.-M. (2009) Production of pharmaceutical proteins by transgenic animals. *Comparative Immunology, Microbiology and Infectious Diseases*, **32**, 107–121.

Huang, C.-J., Lin, H., and Yang, X. (2012) Industrial production of recombinant therapeutics in *Escherichia coli* and its recent advancements. *Journal of Industrial Microbiology and Biotechnology*, **39**, 383–399.

Kind, A. and Schnieke, A. (2008) Animal pharming, two decades on. *Transgenic Research*, **17**, 1025–1033. [Reviews the progress and controversies with animal pharming.]

Kost, T.A., Condreay, J.P., and Jarvis, D.L. (2005) Baculovirus as versatile vectors for protein expression in insect and mammalian cells. *Nature Biotechnology*, **23**, 567–575.

Remaut, E., Stanssens, P., and Fiers, W. (1981) Plasmid vectors for high-efficiency expression controlled by the P_L promoter of coliphage. *Gene*, **15**, 81–93. [Construction of an expression vector.]

Thomson, A.J. and McWhir, J. (2004) Biomedical and agricultural applications of animal transgenesis. *Molecular Biotechnology*, **27**, 231–244.

Ward, O.P. (2012) Production of recombinant proteins by filamentous fungi. *Biotechnology Advances*, **30**, 1119–1139.

Wilmut, I., Schnieke, A.E., McWhir, J., *et al.* (1997) Viable offspring derived from fetal and adult mammalian cells. *Nature*, **385**, 810–813. [The method used to produce Dolly the sheep.]

Xu, J., Dolan, M., Medrano, G., *et al.* (2012) Green factory: plants as bioproduction platforms for recombinant proteins. *Biotechnology Advances*, **30**, 1171–1184.

Zhu, J. (2012) Mammalian cell protein expression for biopharmaceutical production. *Biotechnology Advances*, **30**, 1158–1170.

Chapter 14

Gene Cloning and DNA Analysis in Medicine

Chapter contents

Medicine has been, and will continue to be, a major beneficiary of the recombinant DNA revolution, and an entire book could be written on this topic. Later in this chapter we will see how recombinant DNA techniques are being used to identify genes responsible for inherited diseases and to devise new therapies for these disorders. First, we will continue the theme developed in the last chapter and examine the ways in which cloned genes are being used in the production of recombinant pharmaceuticals.

14.1 Production of recombinant pharmaceuticals

A number of human disorders can be traced to the absence or malfunction of a protein normally synthesized in the body. Most of these disorders can be treated by supplying the patient with the correct version of the protein, but for this to be possible the relevant protein must be available in relatively large amounts. If the defect can be corrected only by administering the human protein, then obtaining sufficient quantities will be a major problem unless donated blood can be used as the source. Animal proteins are therefore used whenever these are effective, but there are not many disorders that can be treated with animal proteins and there is always the possibility of side effects, such as an allergenic response.

Gene Cloning and DNA Analysis: An Introduction, Seventh Edition. T.A. Brown.
© 2016 John Wiley & Sons, Ltd. Published 2016 by John Wiley & Sons, Ltd.
Companion Website: www.wiley.com/go/brown/cloning

We learned in Chapter 13 that gene cloning can be used to obtain large amounts of recombinant human proteins. How are these techniques being applied to the production of proteins for use as pharmaceuticals?

14.1.1 Recombinant insulin

Insulin, synthesized by the β-cells of the islets of Langerhans in the pancreas, controls the level of glucose in the blood. An insulin deficiency manifests itself as diabetes mellitus, a complex of symptoms which may lead to death if untreated. Fortunately, many forms of diabetes can be alleviated by a continuing programme of insulin injections, thereby supplementing the limited amount of hormone synthesized by the patient's pancreas. The insulin used in this treatment was originally obtained from the pancreas of pigs and cows slaughtered for meat production. Although animal insulin is generally satisfactory, problems may arise in its use to treat human diabetes. One problem is that the slight differences between the animal and human proteins can lead to side effects in some patients. Another is that the purification procedures are difficult, and potentially dangerous contaminants cannot always be completely removed.

Insulin displays two features that facilitate its production by recombinant DNA techniques. The first is that the human protein is not modified after translation by the addition of sugar molecules (p. 258), which means that recombinant insulin synthesized by a bacterium should be active. The second feature concerns the size of the molecule. Insulin is a relatively small protein, comprising two polypeptides, one of 21 amino acids (the A chain) and one of 30 amino acids (the B chain; Figure 14.1). In humans these chains are synthesized as a precursor called preproinsulin, which contains the A and B segments linked by a third chain (C) and preceded by a leader sequence. The leader sequence is removed after translation and the C chain is excised, leaving the A and B polypeptides linked to each other by two disulphide bonds.

Several strategies have been used to obtain recombinant insulin. One of the first projects, involving the synthesis of artificial genes for the A and B chains followed by the production of fusion proteins in *E. coli*, illustrates a number of the general techniques used in recombinant protein production.

Synthesis and expression of artificial insulin genes

During the late 1970s, the idea of making an artificial gene was extremely innovative. Oligonucleotide synthesis was in its infancy at that time, and the available methods for creating artificial DNA molecules were much more cumbersome than the present-day automated techniques. Nevertheless, genes coding for the A and B chains of insulin were synthesized as early as 1978.

The procedure used was to synthesize trinucleotides representing all the possible codons and then to join these together in the order dictated by the amino acid sequences of the A and B chains. The artificial genes would not necessarily have the same nucleotide sequences as the real gene segments coding for the A and B chains, but they would still specify the correct polypeptides. Two recombinant plasmids were constructed – one carrying the artificial gene for the A chain, and one the gene for the B chain.

In each case the artificial gene was ligated to a *lacZ'* reading frame present in a pBR322-type vector (Figure 14.2a). The insulin genes were therefore under the control of the strong *lac* promoter (p. 253), and were expressed as fusion proteins, consisting of the first few amino acids of β-galactosidase followed by the A or B polypeptides (Figure 14.2b). Each gene was designed so that its β-galactosidase and insulin segments were separated by a methionine residue, so that the insulin polypeptides could be cleaved

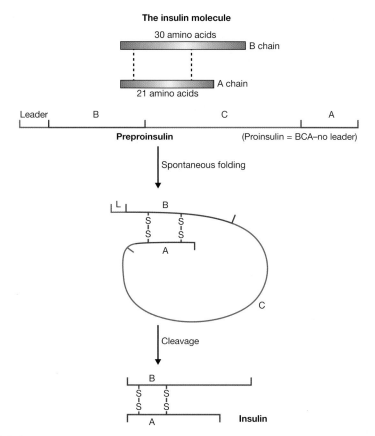

Figure 14.1

The structure of the insulin molecule and a summary of its synthesis by processing from preproinsulin.

from the β-galactosidase segments by treatment with cyanogen bromide (p. 256). The purified A and B chains were then attached to each other by disulphide bond formation in the test tube.

The final step, involving disulphide bond formation, is actually rather inefficient, and a subsequent improvement was to synthesize not the individual A and B genes but rather the entire proinsulin reading frame, specifying B chain–C chain–A chain (see Figure 14.1). Although this is a more daunting proposition in terms of DNA synthesis, the prohormone has the major advantage of folding spontaneously into the correct disulphide-bonded structure. The C chain segment can then be excised relatively easily by proteolytic cleavage.

14.1.2 Synthesis of human growth hormones in E. coli

At about the same time that recombinant insulin was first being made in *E. coli*, other research groups were working on similar projects with the human growth hormones somatostatin and somatotrophin. These two proteins act in conjunction to control growth processes in the human body, their malfunction leading to painful and disabling disorders such as acromegaly (uncontrolled bone growth) and dwarfism.

Somatostatin was the first human protein to be synthesized in *E. coli*. Being a very short protein of only 14 amino acids in length, it was ideally suited for artificial

Figure 14.2

The synthesis of recombinant insulin from artificial A and B chain genes.

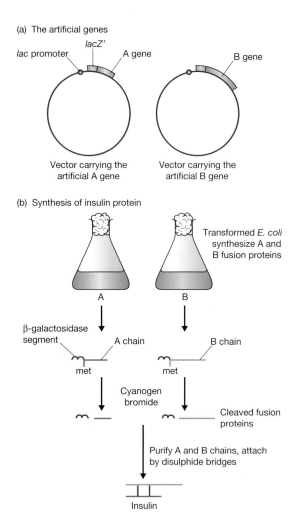

(a) The artificial genes

Vector carrying the artificial A gene

Vector carrying the artificial B gene

(b) Synthesis of insulin protein

Transformed *E. coli* synthesize A and B fusion proteins

β-galactosidase segment

A chain

B chain

met

met

Cyanogen bromide

Cleaved fusion proteins

Purify A and B chains, attach by disulphide bridges

Insulin

gene synthesis. The strategy used was the same as described for recombinant insulin, involving insertion of the artificial gene into a *lacZ'* vector (Figure 14.3), synthesis of a fusion protein, and cleavage with cyanogen bromide.

Somatotrophin presented a more difficult problem. This protein is 191 amino acids in length, equivalent to almost 600 bp, and was an impossible prospect for the DNA synthesis capabilities of the late 1970s. Instead, a combination of artificial gene synthesis and complementary DNA (cDNA) cloning was used to obtain a somatotrophin-producing *E. coli* strain. Messenger RNA was obtained from the pituitary gland, which produces somatotrophin in the human body, and a cDNA library prepared. The somatotrophin cDNA contained a single site for the restriction endonuclease *Hae*III, which therefore cuts the cDNA into two segments (Figure 14.4a). The longer segment, consisting of codons 24–191, was retained for use in construction of the recombinant plasmid. The smaller segment was replaced by an artificial DNA molecule that reproduced the start of the somatotrophin gene and provided the correct signals for translation in *E. coli* (Figure 14.4b). The modified gene was then ligated into an expression vector carrying the *lac* promoter.

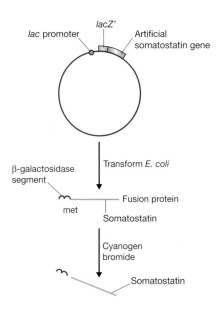

Figure 14.3

Production of recombinant somatostatin.

(a) Preparation of the somatotrophin cDNA fragment

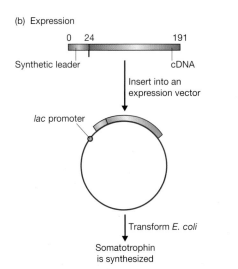

Figure 14.4

Production of recombinant somatotrophin.

(b) Expression

14.1.3 Recombinant factor VIII

Although a number of important pharmaceutical compounds have been obtained from genes cloned in *E. coli*, the general problems associated with using bacteria to synthesize foreign proteins (p. 257) have led in many cases to these organisms being replaced by eukaryotes. An example of a recombinant pharmaceutical produced in eukaryotic cells is human factor VIII, a protein that plays a central role in blood clotting. The commonest form of haemophilia in humans results from an inability to synthesize factor VIII, leading to a breakdown in the blood-clotting pathway and the well-known symptoms associated with the disease.

Until recently, the only way to treat haemophilia was by the injection of purified factor VIII protein, obtained from human blood provided by donors. The purification of factor VIII is a complex procedure and the treatment is expensive. More critically, the purification is beset with difficulties, in particular in removing virus particles that may be present in the blood. Hepatitis and human immunodeficiency viruses can be – and have been – passed on to haemophiliacs via factor VIII injections. Recombinant factor VIII, free from contamination problems, would be a significant achievement for biotechnology.

The factor VIII gene is very large, over 186 kb in length, and is split into 26 exons and 25 introns (Figure 14.5a). The mRNA codes for a large polypeptide (2351 amino acids), which undergoes a complex series of post-translational processing events, eventually resulting in a dimeric protein consisting of a large subunit, derived from the upstream region of the initial polypeptide, and a small subunit from the downstream segment (Figure 14.5b). The two subunits contain a total of 17 disulphide bonds and a number of glycosylated sites. As might be anticipated for such a large and complex protein, it has not been possible to synthesize an active version in *E. coli*.

Initial attempts to obtain recombinant factor VIII therefore involved mammalian cells. In the first experiments to be carried out, the entire cDNA was cloned in hamster cells, but the yields of protein were disappointingly low. This was probably because the post-translational events, although carried out correctly in hamster cells, did not convert all of the initial product into an active form, thus limiting the overall yield. As

Figure 14.5

The factor VIII gene and its translation product.

Figure 14.6

The expression signals used in production of recombinant factor VIII. The promoter is an artificial hybrid of the chicken β-actin and rabbit β-globin sequences, and the polyadenylation signal (needed for correct processing of the mRNA before translation into protein) is obtained from SV40 virus.

an alternative, two separate fragments from the cDNA were used, one fragment coding for the large subunit polypeptide, and the second fragment for the small subunit. Each cDNA fragment was ligated into an expression vector, downstream of the Ag promoter (a hybrid between the chicken β-actin and rabbit β-globin sequences) and upstream of a polyadenylation signal from SV40 virus (Figure 14.6). The plasmid was introduced into a hamster cell line and recombinant protein obtained. The yields were over tenfold greater than those from cells containing the complete cDNA, and the resulting factor VIII protein was indistinguishable from the native form in terms of function.

Pharming (p. 263) has also been used for the production of recombinant factor VIII. The complete human cDNA has been attached to the promoter for the whey acidic protein gene of pig, leading to the synthesis of human factor VIII in pig mammary tissue and subsequent secretion of the protein in milk. The factor VIII produced in this way appears to be exactly the same as the native protein and is fully functional in blood-clotting assays. Similar results have been obtained with transgenic rabbits.

14.1.4 Synthesis of other recombinant human proteins

The list of human proteins synthesized by recombinant technology continues to grow (Table 14.1). As well as proteins used to treat disorders by replacement or supplementation of the dysfunctional versions, the list includes a number of growth factors (e.g., interferons and interleukins) with potential uses in cancer therapy. As these proteins are synthesized in very limited amounts in the body, recombinant technology is the only viable means of obtaining them in the quantities needed for clinical purposes. Other proteins, such as serum albumin, are more easily obtained, but are needed in such large quantities that production in microorganisms is still a more attractive option.

14.1.5 Recombinant vaccines

The final category of recombinant protein is slightly different from the examples given in Table 14.1. A vaccine is an antigenic preparation that, after injection into the blood-stream, stimulates the immune system to synthesize antibodies that protect the body against infection. The antigenic material present in a vaccine is normally an inactivated form of the infectious agent. For example, antiviral vaccines often consist of virus particles that have been attenuated by heating or a similar treatment. In the past, two problems have hindered the preparation of attenuated viral vaccines:

- The inactivation process must be 100% efficient, as the presence in a vaccine of just one live virus particle could result in infection. This has been a problem with vaccines for the cattle disease foot-and-mouth.

Table 14.1

Some of the human proteins that have been synthesized from genes cloned in bacteria and/or eukaryotic cells or by pharming.

PROTEIN	USED IN THE TREATMENT OF
α_1-Antitrypsin	Emphysema
Deoxyribonuclease	Cystic fibrosis
Epidermal growth factor	Ulcers
Erythropoietin	Anaemia
Factor IX	Christmas disease
Factor VIII	Haemophilia
Fibroblast growth factor	Ulcers
Follicle-stimulating hormone	Infertility treatment
Granulocyte colony-stimulating factor	Cancers
Insulin	Diabetes
Insulin-like growth factor 1	Growth disorders
Interferon-β	Cancers, AIDS
Interferon-γ	Cancers, rheumatoid arthritis
Interferon-α	Leukaemia and other cancers
Interleukins	Cancers, immune disorders
Lung surfactant protein	Respiratory distress
Relaxin	Used to aid childbirth
Serum albumin	Used as a plasma supplement
Somatostatin	Growth disorders
Somatotrophin	Growth disorders
Superoxide dismutase	Free radical damage in kidney transplants
Tissue plasminogen activator	Heart attack
Tumour necrosis factor	Cancers

- The large amounts of virus particles needed for vaccine production are usually obtained from tissue cultures. Unfortunately, some viruses – notably hepatitis B virus – do not grow in tissue culture.

Producing vaccines as recombinant proteins

The use of gene cloning in this field centres on the discovery that virus-specific antibodies are sometimes synthesized in response not only to the whole virus particle, but also to isolated components of the virus. This is particularly true of purified preparations of the proteins present in the virus coat (Figure 14.7). If the genes coding for the antigenic proteins of a particular virus could be identified and inserted into an expression vector, the methods described above for the synthesis of animal proteins could be employed in the production of recombinant proteins that might be used as vaccines. These vaccines would have the advantages that they would be free from intact virus particles and could be obtained in large quantities.

The first success with this approach was with hepatitis B virus. Hepatitis B is endemic in many tropical parts of the world and leads to liver disease and possibly, after chronic infection, to cancer of the liver. A person who recovers from hepatitis B is immune to future infection because their blood contains antibodies to the hepatitis B surface antigen (HBsAg), which is one of the virus coat proteins. This protein has been synthesized in both *Saccharomyces cerevisiae*, using a vector based on the 2 μm plasmid (p. 111), and in Chinese hamster ovary (CHO) cells. In both cases, the protein was obtained in reasonably high quantities and, when injected into test animals, provided protection against hepatitis B.

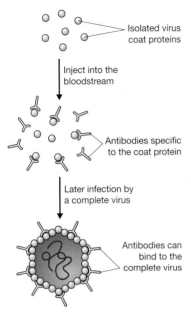

Figure 14.7
The principle behind the use of a preparation of isolated virus coat proteins as a vaccine.

The key to the success of recombinant HBsAg as a vaccine is provided by an unusual feature of the natural infection process for the virus. The bloodstream of infected individuals contains not only intact hepatitis B virus particles, which are 42 nm in diameter, but also smaller, 22-nm spheres composed entirely of HBsAg protein molecules. The assembly of these 22-nm spheres occurs during HBsAg synthesis in both yeast and hamster cells, and it is almost certainly these spheres, rather than single HBsAg molecules, that are the effective component of the recombinant vaccine. The recombinant vaccine therefore mimics part of the natural infection process and stimulates antibody production, although as the spheres are not viable viruses the vaccine does not itself cause the disease. Both, the yeast and hamster cell vaccines have been approved for use in humans, and the World Health Organization is promoting their use in national vaccination programmes.

Recombinant vaccines have also been successfully developed for human papillomavirus (HPV), different subtypes of which are responsible for a variety of cancers. The L1 coat protein gene has been expressed in *S. cerevisiae*, giving rise to virus-like particles composed of aggregates of the L1 protein. Just like the HBsAg vaccines, these particles lack any nucleic acid and so are not infectious. In order to broaden the effectiveness of recombinant HPV vaccines, a variety of recombinant yeast strains have been produced, each synthesizing the L1 protein from a different HPV subtype. The virus-like particles from these strains are then combined to make a divalent or quadrivalent vaccine (Figure 14.8). These vaccines provide protection against cervical cancer in women, as well as a range of sexually transmitted infections in men.

Recombinant vaccines in transgenic plants
The advent of pharming (p. 263) has led to the possibility of using transgenic plants as the hosts for synthesis of recombinant vaccines. The ease with which plants can be grown and harvested means that this technology might be applicable for developing

Figure 14.8

Synthesis of a quadrivalent HPV vaccine. The four recombinant yeast strains synthesize L1 proteins from HPV subtypes 6, 11, 16, and 18.

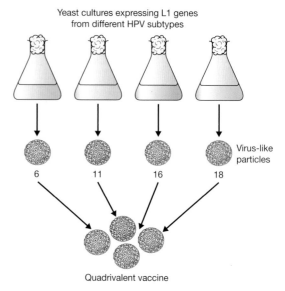

Yeast cultures expressing L1 genes from different HPV subtypes

Virus-like particles

Quadrivalent vaccine

parts of the world, where more expensive approaches to recombinant protein production might be difficult to sustain. If the recombinant vaccine is effective after oral administration, then immunity could be acquired simply by eating part or all of the transgenic plant. A simpler and cheaper means of carrying out a mass vaccination programme would be hard to imagine.

The feasibility of this approach has been demonstrated by trials with vaccines such as HBsAg and the coat proteins of measles virus and respiratory syncytial virus. In each case, immunity was conferred by feeding the transgenic plant to test animals. Attempts are also being made to engineer plants that express a variety of vaccines, so that immunity against a range of diseases can be acquired from a single source. The main problem currently faced by companies developing this technology is that the amount of recombinant protein synthesized by the plant is often insufficient to stimulate complete immunity against the target disease. In order to be completely effective, the vaccine should comprise 8–10% of the soluble protein content of the part of the plant which is eaten, but in practice yields are much less than this, usually not more than 0.5%. Variability in the yields between different plants in a single crop is also a concern. A possible solution would be to place the cloned gene in the chloroplast genome rather than the plant nucleus (p. 125), as this generally results in much higher yields of recombinant protein. Chloroplast synthesis of various vaccine proteins has been achieved, including ones active against anthrax, plague, smallpox and the parasitic protozoan *Entamoeba histolytica*. Tobacco and lettuce chloroplasts have also been used to synthesize fusions between surface proteins from *Vibrio cholerae* and *Plasmodium falciparum*. When fed to mice, these plants provided protection against cholera toxin and also stimulated the synthesis of antibodies that reduced the ability of *P. falciparum* parasites to invade human red blood cells (effectiveness against malaria could not be directed tested in mice because *P. falciparum* does not infect mice). Cholera and malaria are major diseases in many parts of the world, and the development of an easily administered vaccine that confers dual protection would be an important advance for human health.

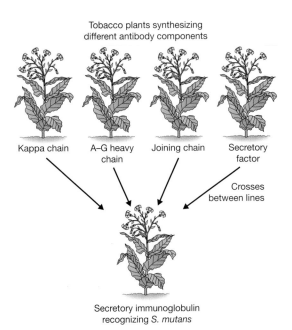

Tobacco plants synthesizing
different antibody components

Kappa chain | A–G heavy chain | Joining chain | Secretory factor

Crosses
between lines

Secretory immunoglobulin
recognizing *S. mutans*

Figure 14.9

Transgenic tobacco that synthesizes an antibody
active against *S. mutans*.

Plants are also being used in a radically different approach to disease protection, where the goal is not to make a protein that will stimulate an immune response but rather to use the plant to synthesize the antibody that is the product of the immune response. To this end, four transgenic tobacco lines have been engineered, each producing a different component of an antibody specific for the surface protein antigen A gene of *Streptococcus mutans*. After crosses between these lines, a single variety of transgenic tobacco containing all four cloned genes was obtained (Figure 14.9). The plants assembled functional antibodies in their cells. *S. mutans* inhabits the human oral cavity and is a major cause of dental caries, which affects over 60% of schoolchildren in most industrialized countries. A cheap source of *S. mutans* antibodies which might provide protection against caries, and which could be administered by rubbing on to the surface of the teeth rather than by injection, has been sought for many years.

Live recombinant virus vaccines

The use of live vaccinia virus as a vaccine for smallpox dates back to 1796, when Edward Jenner first realized that this virus, which is harmless to humans, could stimulate immunity against the related and much more dangerous smallpox virus. The term 'vaccine' comes from vaccinia. Its use resulted in the worldwide eradication of smallpox in 1980.

A more recent idea is that recombinant vaccinia viruses could be used as live vaccines against other diseases. If a gene coding for a virus coat protein, for example HBsAg, is ligated into the vaccinia genome under control of a vaccinia promoter, then the gene will be expressed (Figure 14.10). After injection into the bloodstream, replication of the recombinant virus results not only in new vaccinia particles but also in significant quantities of the major surface antigen. Immunity against both smallpox and hepatitis B would result.

Figure 14.10

The rationale behind the use of a recombinant vaccinia virus.

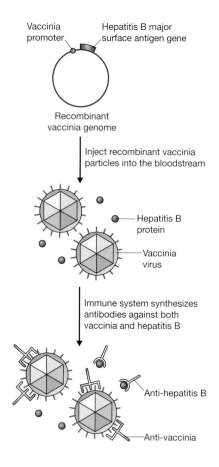

This remarkable technique has considerable potential. Recombinant vaccinia viruses expressing a number of foreign genes have been constructed and shown to confer immunity against the relevant diseases in experimental animals (Table 14.2). The possibility of broad-spectrum vaccines is raised by the demonstration that a single recombinant vaccinia, expressing the genes for influenza virus haemagglutinin, HBsAg, and herpes simplex virus glycoprotein, confers immunity against each disease in monkeys. Studies of vaccinia viruses expressing the rabies glycoprotein have shown that deletion of the vaccinia gene for the enzyme thymidine kinase prevents the virus from replicating. This avoids the possibility that animals treated with the live vaccine will develop any form of cowpox, the disease caused by normal vaccinia. This particular live virus vaccine is now being used in rabies control in Europe and North America.

14.2 Identification of genes responsible for human diseases

A second major area of medical research in which gene cloning is having an impact is in the identification and isolation of genes responsible for human diseases. A genetic or inherited disease is one that is caused by a defect in a specific gene (Table 14.3),

Table 14.2
Some of the foreign genes that have been expressed in recombinant vaccinia viruses.

GENE
Human diseases
Plasmodium falciparum (malaria parasite) surface antigen
Influenza virus coat proteins
Hepatitis B surface antigen
Herpes simplex glycoproteins
Human immunodeficiency virus (HIV) envelope proteins
Vesicular somatitis coat proteins
Sindbis virus proteins
Animal diseases
Rabies virus G protein
Rinderpest virus F and H proteins
Feline leukaemia virus envelope protein
Newcastle disease virus F glycoprotein

with individuals carrying the defective gene being predisposed towards developing the disease at some stage of their lives. With some inherited diseases, such as haemophilia, the gene is present on the X chromosome and so all males carrying the gene will express the disease state. Females with one defective gene and one correct gene are healthy but can pass the disease on to their male offspring. Genes for other diseases are present on autosomes and in most cases are recessive, so both chromosomes of the pair must carry a defective version for the disease to occur. A few diseases, including Huntington's chorea, are autosomal dominant, so a single copy of the defective gene is enough to lead to the disease state.

With some genetic diseases, the symptoms manifest themselves early in life, but with others the disease may not be expressed until the individual is middle-aged or elderly. Cystic fibrosis is an example of the former, and neurodegenerative diseases such as Alzheimer's and Huntington's are examples of the latter. With a number of diseases that appear to have a genetic component, cancers in particular, the overall syndrome is complex with the disease remaining dormant until triggered by some metabolic or environmental stimulus. If predisposition to these diseases can be diagnosed, the risk

Table 14.3
Some of the most common genetic diseases in the UK.

DISEASE	SYMPTOMS	FREQUENCY (BIRTHS PER YEAR)
Inherited breast cancer	Cancer	1 in 300 females
Cystic fibrosis	Lung disease	1 in 2000
Huntington's chorea	Neurodegeneration	1 in 2000
Duchenne muscular dystrophy	Progressive muscle weakness	1 in 3000 males
Haemophilia A	Blood disorder	1 in 4000 males
Sickle cell anaemia	Blood disorder	1 in 10 000
Phenylketonuria	Mental retardation	1 in 12 000
β-Thalassaemia	Blood disorder	1 in 20 000
Retinoblastoma	Cancer of the eye	1 in 20 000
Haemophilia B	Blood disorder	1 in 25 000 males
Tay–Sachs disease	Blindness, loss of motor control	1 in 200 000

factor can be reduced by careful management of the patient's lifestyle to minimize the chances of the disease being triggered.

Genetic diseases have always been present in the human population, but their importance has increased during recent decades. This is because vaccination programmes, antibiotics and improved sanitation have reduced the prevalence of infectious diseases such as smallpox, tuberculosis and cholera, which were major killers up to the mid-20th century. The result is that a greater proportion of the population now dies from a disease that has a genetic component, especially the late-onset diseases that are now more common because of increased life expectancies. Medical research has been successful in controlling many infectious diseases, but can it be equally successful with genetic disease?

There are a number of reasons why identifying the gene responsible for a genetic disease is important:

- Gene identification may provide an indication of the biochemical basis to the disease, enabling therapies to be designed.
- Identification of the mutation present in a defective gene can be used to devise a screening programme so that the mutant gene can be identified in individuals who are carriers or who have not yet developed the disease. Carriers can receive counselling regarding the chances of their children inheriting the disease. The early identification of individuals who have not yet developed the disease allows appropriate precautions to be taken to reduce the risk of the disease becoming expressed.
- Identification of the gene is a prerequisite for gene therapy (p. 286).

14.2.1 How to identify a gene for a genetic disease

There is no single strategy for the identification of genes that cause diseases, the best approach depending on the information that is available about the disease. To gain an understanding of the principles of this type of work, we will consider the most common and most difficult scenario, when all that is known about the disease is that certain people have it. Even with such an unpromising starting point, DNA techniques can locate the relevant gene.

Locating the approximate position of the gene in the human genome

If there is no information about the desired gene, how can it be located in the human genome? The answer is to use basic genetics to determine the approximate position of the gene on the human genetic map. Genetic mapping is usually carried out by linkage analysis, in which the inheritance pattern for the target gene is compared with the inheritance patterns for genetic loci whose map positions are already known. If two loci are inherited together, they must be very close on the same chromosome (Figure 14.11a). If they are on different chromosomes then random segregation during meiosis will result in the loci displaying different inheritance patterns (Figure 14.11b). If they are on the same chromosome but not close together, then recombination events will occasionally unlink them (Figure 14.11c). The demonstration of linkage with one or more mapped genetic loci is therefore the key to understanding the chromosomal position of an unmapped gene.

With humans, it is not possible to carry out directed breeding programmes aimed at determining the map position of a desired gene. Instead, the mapping of disease genes

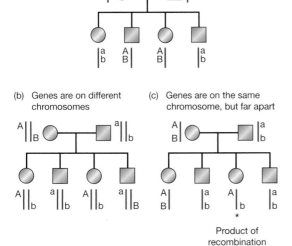

(a) Genes are linked

(b) Genes are on different chromosomes

(c) Genes are on the same chromosome, but far apart

Product of recombination

Figure 14.11

Inheritance patterns for linked and unlinked genes. Three families are shown, circles representing females and squares representing males. (a) Two closely linked genes are almost always inherited together. (b) Two genes on different chromosomes display random segregation. (c) Two genes that are far apart on a single chromosome are often inherited together, but recombination may unlink them.

must make use of data available from pedigree analysis, in which inheritance of the gene is examined in families with a high incidence of the disease being studied. It is important to be able to obtain DNA samples from at least three generations of each family, and the more family members there are the better, but unless the disease is very uncommon it is usually possible to find suitable pedigrees. Linkage is studied between the presence/absence of the disease and the inheritance of DNA markers. A DNA marker is simply a DNA sequence that is variable and so exists in two or more allelic forms, and whose location in the genome is known. To illustrate how linkage analysis is used with DNA markers, we will examine the way in which one of the genes conferring susceptibility to human breast cancer was mapped.

Linkage analysis of the human BRCA1 gene

The first breakthrough in mapping the human breast cancer susceptibility gene *BRCA1* occurred in 1990 as a result of restriction fragment length polymorphism (RFLP) linkage analyses. This study showed that in families with a high incidence of breast cancer, a significant number of the women who suffered from the disease all possessed the same version of an RFLP called *D17S74*. An RFLP results from a DNA sequence variation that causes a change in a restriction site (Figure 14.12a). When digested with a restriction endonuclease, a loss of the site is revealed because two fragments remain joined together. Originally, RFLPs were typed by Southern hybridization of restricted genomic DNA, but this is a time-consuming process and so nowadays the presence or absence of the restriction site is usually determined by PCR (Figure 14.12b). The *D17S74* RFLP had previously been mapped to the long arm of chromosome 17: the gene being sought – *BRCA1* – must therefore also be located on the long arm of chromosome 17.

This initial linkage result was extremely important, as it indicated in which region of the human genome the *BRCA1* gene was to be found, but it was far from the end of the story. In fact, over 1000 genes are thought to lie in this particular stretch of chromosome 17. The next objective was therefore to carry out more linkage studies to try to pinpoint *BRCA1* more accurately. This was achieved by examining the region containing *BRCA1*

Figure 14.12

Restriction fragment length polymorphisms. (a) An RFLP is a sequence variation that changes a restriction site. (b) Typing an RFLP by PCR. In the middle lane the PCR product gives two bands because it is cut by treatment with the restriction enzyme. In the right-hand lane there is just one band because the template DNA lacks the restriction site.

(a) An RFLP

EcoRI site

...AGTGAATTCGTAC...

...AGTGTATTCGTAC...

Sequence variation results in loss of EcoRI site

(b) Typing an RFLP by PCR

Restriction site

PCR primers

Markers   PCR, restrict   PCR, restrict

for short tandem repeats (STRs). These sequences, also called microsatellites, are made up of short repetitive sequences of between one and 13 nucleotides in length, linked head to tail (Figure 14.13a). The number of repeats present in a particular STR varies, usually between 5 and 20. The number can be determined by carrying out a PCR using primers that anneal either side of the STR, and then examining the size of the resulting product by agarose or polyacrylamide gel electrophoresis (Figure 14.13b). STRs are useful for fine-scale mapping because many of them exist in three or more allelic forms, rather than just the two alleles that are possible for an RFLP. Several alleles of an STR might therefore be present within a single pedigree, enabling more detailed mapping to be carried out. STR linkage mapping placed *BRCA1* between two STRs called *D17S1321* and *D17S1325*, reducing the size of the *BRCA1* region from 20 Mb down to just 600 kb (Figure 14.14). This approach to locating a gene is called positional cloning.

Identification of candidates for the disease gene

One might imagine that once the map location of the disease gene has been determined, the next step would be simply to refer to the genome sequence in order to identify the gene. Unfortunately a great deal of work still has to be done. Genetic mapping, even in its most precise form, gives only an approximate indication of the location of a gene. In the breast cancer project, the investigators were fortunate in being able to narrow the search area down to just 600 kb, as often 10 Mb or more of DNA sequence needs to be examined. Such large stretches of DNA could contain many genes: the 600 kb breast cancer region contained over 60 genes, any one of which could have been *BRCA1*.

(a) Two versions of an STR

...TCGCACACAGTG...
...TCGCACACACAGTG...

(b) Typing an STR by PCR

Figure 14.13

Short tandem repeats. (a) An STR is a repetitive sequence made up of short units 1–13 nucleotides in length. (b) Typing an STR by PCR. The PCR product in the right-hand lane is slightly longer than that in the middle lane, because the template DNA from which it is generated contains an additional CA unit.

A variety of approaches can be used to identify which of the genes in a mapped region is the disease gene:

- The expression profiles of the candidate genes can be examined by hybridization analysis or RT–PCR of RNA from different tissues. For example, *BRCA1* would be expected to hybridize to RNA prepared from breast tissue, and also to ovary tissue RNA, as ovarian cancer is frequently associated with inherited breast cancer.

Figure 14.14

Mapping the breast cancer susceptibility gene *BRCA1*. Initially, the gene was mapped to a segment of the left arm of chromosome 17, between map positions q12 and q23 (highlighted region in the left-hand drawing). Additional mapping experiments narrowed this down to a 600 kb region flanked by two STR loci, *D17S1321* and *D17S1325* (middle drawing). After examination of expressed sequences, a strong candidate for *BRCA1* was eventually identified (right-hand drawing).

Chromosome 17
(83 Mb)

- The presence of homologues of the candidate gene in other mammalian species can be tested by searching their genome sequences using BLAST (p. 228). The rationale is that a human gene which is important enough to cause disease when mutated will almost certainly have homologues in a range of other mammals. The same analysis can be carried out by Southern hybridization of the DNA from the related species (these are called zoo blots). If a homologue is present, it will be detectable by hybridization with a suitable probe, even though the homologue has a slightly different sequence from the human version.
- The gene sequences could be examined in individuals with and without the disease to see if the genes from affected individuals contain mutations that might explain why they have the disease.
- To confirm the identity of a candidate gene, it might be possible to prepare a knockout mouse (p. 233) that has an inactive version of the equivalent mouse gene. If the knockout mouse displays symptoms compatible with the human disease, then the candidate gene is almost certainly the correct one.

When applied to the region between *D17S1321* and *D17S1325*, these analyses resulted in identification of an approximately 100 kb gene, composed of 22 exons and coding for a 1863-amino acid protein, that was a strong candidate for *BRCA1*. Transcripts of the gene were detectable in breast and ovary tissues, and homologues were present in mice, rats, rabbits, sheep and pigs, but not in chickens. Most importantly, the genes from five susceptible families contained mutations (such as frameshift and nonsense mutations) likely to lead to a non-functioning protein. Although circumstantial, the evidence in support of the candidate was sufficiently overwhelming for this gene to be identified as *BRCA1*. Subsequent investigations have shown that both this gene and *BRCA2*, a second gene associated with susceptibility to breast cancer, are involved in transcription regulation and DNA repair, and that both act as tumour suppressor genes, inhibiting abnormal cell division.

14.3 Gene therapy

The final application of recombinant DNA technology in medicine that we will consider is gene therapy. This is the name originally given to methods that aim to cure an inherited disease by providing the patient with a correct copy of the defective gene. Gene therapy has now been extended to include attempts to cure any disease by the introduction of a cloned gene into the patient. First, we will examine the techniques used in gene therapy, and then we will attempt to address the ethical issues.

14.3.1 Gene therapy for inherited diseases

There are two basic approaches to gene therapy, namely germline therapy and somatic cell therapy. In germline therapy, a fertilized egg is provided with a copy of the correct version of the relevant gene and reimplanted into the mother. If successful, the gene is present and expressed in all cells of the resulting individual. Germline therapy is usually carried out by microinjection of a somatic cell, followed by nuclear transfer into an oocyte (p. 263) and, in theory, could be used to treat any inherited disease.

Somatic cell therapy involves the manipulation of cells, which either can be removed from the organism, transfected, and then placed back in the body, or transfected *in situ*

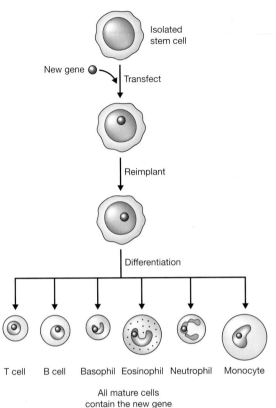

Figure 14.15

Differentiation of a transfected stem cell leads to the new gene being present in all the mature blood cells.

without removal. The technique has most promise for inherited blood diseases (e.g., haemophilia and thalassaemia), with genes being introduced into stem cells from the bone marrow, which give rise to all the specialized cell types in the blood. The strategy is to prepare a bone extract containing several billion cells, transfect these with a retrovirus-based vector, and then reimplant the cells. Subsequent replication and differentiation of transfectants leads to the added gene being present in all the mature blood cells (Figure 14.15). The advantage of a retrovirus is that this type of vector has an extremely high transfection frequency, enabling a large proportion of the stem cells in a bone marrow extract to receive the new gene.

Somatic cell therapy also has potential in the treatment of lung diseases such as cystic fibrosis, as DNA cloned in adenovirus vectors (p. 131) or contained in liposomes (p. 88) is taken up by the epithelial cells in the lungs after introduction into the respiratory tract via an inhaler. However, gene expression occurs for only a few weeks, and as yet this has not been developed into an effective means of treating cystic fibrosis.

With those genetic diseases where the defect arises because the mutated gene does not code for a functional protein, all that is necessary is to provide the cell with the correct version of the gene. Removal of the defective gene is unnecessary. The situation is less easy with dominant genetic diseases (p. 281), as with these it is the defective gene product itself that is responsible for the disease state, and so the therapy must include not only addition of the correct gene but also removal of the defective version. This requires a gene delivery system that promotes recombination between the chromosomal and vector-borne versions of the gene, so that the defective chromosomal copy is replaced by the gene from the vector. The technique is complex and unreliable, and broadly applicable procedures have not yet been developed.

14.3.2 Gene therapy and cancer

The clinical uses of gene therapy are not limited to treatment of inherited diseases. There have also been attempts to use gene cloning to disrupt the infection cycles of pathogens such as human immunodeficiency virus (HIV). However, the most intensive area of current research concerns the potential use of gene therapy as a treatment for cancer.

Most cancers result from the activation of an oncogene that leads to tumour formation, or the inactivation of a gene that normally suppresses tumour formation. In both cases, a gene therapy could be envisaged to treat the cancer. Inactivation of a tumour suppressor gene could be reversed by the introduction of a correct version of the gene, using one of the methods described above for inherited disease. Inactivation of an oncogene would, however, require a more subtle approach, as the objective would be to prevent expression of the oncogene, not to replace it with a non-defective copy. One possible way of doing this would be to introduce into a tumour a gene specifying an antisense version of the mRNA transcribed from the oncogene (Figure 14.16a). An antisense RNA is the reverse complement of a normal RNA, and can prevent synthesis of the protein coded by the gene it is directed against, probably by hybridizing to the mRNA producing a double-stranded RNA molecule that is rapidly degraded by cellular ribonucleases (Figure 14.16b). The target is therefore inactivated.

An alternative would be to introduce a gene that selectively kills cancer cells or promotes their destruction by drugs administered in a conventional fashion. This is called suicide gene therapy and is looked on as an effective general approach to cancer treatment because it does not require a detailed understanding of the genetic basis of the particular disease being treated. Many genes that code for toxic proteins are known, and there are also examples of enzymes that convert non-toxic precursors of drugs into the toxic form. Introduction of the gene for one of these toxic proteins or enzymes into a tumour should result in death of the cancer cells, either immediately or after drug administration. It is obviously important that the introduced gene is targeted accurately at the cancer cells, so that healthy cells are not killed. This requires a very precise delivery system, such as direct inoculation into the tumour, or some other means of ensuring that the gene is expressed only in the cancer cells. One possibility is to place the gene under control of the human telomerase promoter, which is active only in cancerous tissues.

Another approach is to use gene therapy to improve the natural killing of cancer cells by the patient's immune system, perhaps with a gene that causes the tumour cells to synthesize strong antigens that are efficiently recognized by the immune system. All of these approaches, and many not based on gene therapy, are currently being tested in the fight against cancer.

14.3.3 The ethical issues raised by gene therapy

Should gene therapy be used to cure human disease? As with many ethical questions, there is no simple answer. On the one hand, there could surely be no justifiable objection to the routine application via a respiratory inhaler of correct versions of the cystic fibrosis gene as a means of managing this disease. Similarly, if bone marrow transplants are acceptable, then it is difficult to argue against gene therapies aimed at correction of

(a) Synthesis of an antisense RNA

Figure 14.16

Antisense RNA can be used to silence a cellular mRNA.

(b) Mode of action of antisense RNA

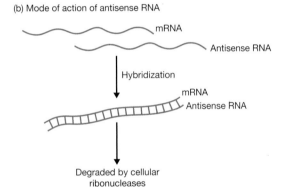

blood disorders via stem cell transfection. And cancer is such a terrible disease that the withholding of effective treatment regimens on moral grounds could itself be criticized as immoral.

Germline therapy is a more difficult issue, the problem being that the techniques used for germline correction of inherited diseases are exactly the same as those that could be used for the germline manipulation of other inherited characteristics. Indeed, the development of this technique in farm animals has not been prompted by a desire to cure genetic diseases, but to 'improve' the animals, perhaps by making genetic changes that would result in a lower fat content. This type of manipulation, where the genetic constitution of an organism is changed in a directed, heritable fashion, is clearly unacceptable with humans. At present, human germline manipulation is impossible because the cells resulting from somatic nuclear transfer do not divide to produce a blastocyst (the structure that comprises about 100 cells and attaches to the wall of the uterus during a normal pregnancy). Geneticists are uncertain if this problem can ever be solved, but many agree that a method for human germline therapy will eventually become a reality. But before that happens the ethical issues that germline therapy raises must be debated, so that when the procedure does become possible there is already an understanding of the rules and regulations under which it should be used.

Further reading

Brocher, B., Kieny, M.P., Costy, F., *et al.* (1991) Large-scale eradication of rabies using recombinant vaccinia-rabies vaccine. *Nature*, **354**, 520–522.

Broder, C.C. and Earl, P.L. (1999) Recombinant vaccinia viruses – design, generation, and isolation. *Molecular Biotechnology*, **13**, 223–245.

Davoodi-Semiromi, A., Schreiber, M., Nalapalli, S., *et al.* (2009) Chloroplast-derived vaccine antigens confer dual immunity against cholera and malaria by oral or injectable delivery. *Plant Biotechnology Journal*, **8**, 223–242.

Goeddel, D.V., Heyneker, H.L., Hozumi, T., *et al.* (1979) Direct expression in *Escherichia coli* of a DNA sequence coding for human growth hormone. *Nature*, **281**, 544–548. [Production of recombinant somatotrophin.]

Goeddel, D.V., Kleid, D.G., Bolivar, F., *et al.* (1979) Expression in *Escherichia coli* of chemically synthesized genes for human insulin. *Proceedings of the National Academy of Sciences of the United States of America*, **76**, 106–110.

Itakura, K., Hirose, T., Crea, R., *et al.* (1977) Expression in *Escherichia coli* of a chemically synthesized gene for the hormone somatostatin. *Science*, **198**, 1056–1063.

Kaufman, R.J., Wasley, L.C., and Dorner, A.J. (1988) Synthesis, processing, and secretion of recombinant human factor VIII expressed in mammalian cells. *Journal of Biological Chemistry*, **263**, 6352–6362.

Liu, M.A. (1998) Vaccine developments. *Nature Medicine*, **4**, 515–519. [Describes the development of recombinant vaccines.]

Ma, J.K.-C., Hiatt, A., Hein, M., *et al.* (1995) Generation and assembly of secretory antibodies in plants. *Science*, **268**, 716–719. [Antibodies against *Streptococcus mutans*.]

Miki, Y., Swensen, J., Shattuck Eidens, D., *et al.* (1994) A strong candidate for the breast and ovarian cancer susceptibility gene *BRCA1*. *Science*, **266**, 66–71.

Paleyanda, R.K., Velander, W.H., Lee, T.K., *et al.* (1997) Transgenic pigs produce functional human factor VIII in milk. *Nature Biotechnology*, **15**, 971–975.

Schepelmann, S., Ogilvie, L.M., Hedley, D., *et al.* (2007) Suicide gene therapy of human colon carcinoma xenografts using an armed oncolytic adenovirus expressing carboxypeptidase G2. *Cancer Research*, **67**, 4949–4955. [An example of suicide gene therapy using the telomerase promoter and an enzyme that converts a non-toxic pro-drug into the toxic form.]

Smith, K.R. (2003) Gene therapy: theoretical and bioethical concepts. *Archives of Medical Research*, **34**, 247–268.

Tiwari, S., Verma, P.C., Sing, P.K., and Tuli, R. (2009) Plants as bioreactors for the production of vaccine antigens. *Biotechnology Advances*, **27**, 449–467.

Chapter 15

Gene Cloning and DNA Analysis in Agriculture

Chapter contents

Agriculture, or more specifically the cultivation of plants, is the world's oldest biotechnology, with an unbroken history that stretches back at least 10 000 years. Throughout this period humans have constantly searched for improved varieties of their crop plants: varieties with better nutritional qualities, higher yields, or features that aid cultivation and harvesting. During the first few millennia, crop improvements occurred in a sporadic fashion, but in recent centuries new varieties have been obtained by breeding programmes of ever-increasing sophistication. However, the most sophisticated breeding programme still retains an element of chance, dependent as it is on the random merging of parental characteristics in the hybrid offspring that are produced. The development of a new variety of crop plant, displaying a precise combination of desired characteristics, is a lengthy and difficult process.

Gene cloning provides a new dimension to crop breeding by enabling directed changes to be made to the genotype of a plant, circumventing the random processes inherent in conventional breeding. Two general strategies have been used:

- Gene addition, in which cloning is used to alter the characteristics of a plant by providing it with one or more new genes.
- Gene subtraction, in which genetic engineering techniques are used to inactivate one or more of the plant's existing genes.

A number of projects are being carried out worldwide, many by biotechnology companies, aimed at exploiting the potential of gene addition and gene subtraction in crop improvement. In this chapter we will investigate a representative selection of these

Gene Cloning and DNA Analysis: An Introduction, Seventh Edition. T.A. Brown.
© 2016 John Wiley & Sons, Ltd. Published 2016 by John Wiley & Sons, Ltd.
Companion Website: www.wiley.com/go/brown/cloning

projects, and examine some of the problems that must be solved if plant genetic engineering is to gain widespread acceptance in agriculture.

15.1 The gene addition approach to plant genetic engineering

Gene addition involves the use of cloning techniques to introduce into a plant one or more new genes coding for a useful characteristic that the plant lacks. A good example of the technique is provided by the development of plants that resist insect attack by synthesizing insecticides coded by cloned genes.

15.1.1 Plants that make their own insecticides

Plants are subject to predation by virtually all other types of organism – viruses, bacteria, fungi, and animals – but in agricultural settings the greatest problems are caused by insects. To reduce losses, crops are regularly sprayed with insecticides. Most conventional insecticides (e.g., pyrethroids and organophosphates) are relatively non-specific poisons that kill a broad spectrum of insects, not just the ones eating the crop. Because of their high toxicity, several of these insecticides also have potentially harmful side effects for other members of the local biosphere, including in some cases humans. These problems are exacerbated by the need to apply conventional insecticides to the surfaces of plants by spraying, which means that subsequent movement of the chemicals in the ecosystem cannot be controlled. Furthermore, insects that live within the plant, or on the undersurfaces of leaves, can sometimes avoid the toxic effects altogether.

What features would be displayed by the ideal insecticide? Clearly, it must be toxic to the insects against which it is targeted, but if possible this toxicity should be highly selective, so that the insecticide is harmless to other insects and is not poisonous to animals and to humans. The insecticide should be biodegradable, so that any residues that remain after the crop is harvested, or which are carried out of the field by rainwater, do not persist long enough to damage the environment. And it should be possible to apply the insecticide in such a way that all parts of the crop, not just the upper surfaces of the plants, are protected against insect attack.

The ideal insecticide has not yet been discovered. The closest we have are the δ-endotoxins produced by the soil bacterium *Bacillus thuringiensis*.

The δ-endotoxins of Bacillus thuringiensis

Insects not only eat plants: bacteria also form an occasional part of their diet. In response, several types of bacteria have evolved defence mechanisms against insect predation, an example being *B. thuringiensis* which, during sporulation, forms intracellular crystalline bodies that contain an insecticidal protein called the δ-endotoxin. The activated protein is highly poisonous to insects, some 80 000-fold more toxic than organophosphate insecticides, and is relatively selective, with different strains of the bacterium synthesizing proteins effective against the larvae of different groups of insects (Table 15.1).

The δ-endotoxin protein that accumulates in the bacterium is an inactive precursor. After ingestion by the insect, this protoxin is cleaved by proteases, resulting in shorter versions of the protein that display toxic activity. The toxins bind to the inside of the insect's gut and damage the surface epithelium, so that the insect is unable to feed and

Table 15.1

The range of insects poisoned by the various types of *B. thuringiensis* δ-endotoxins.

δ-ENDOTOXIN TYPE	EFFECTIVE AGAINST
CryI	Lepidoptera (moth and butterfly) larvae
CryII	Lepidoptera and Diptera (two-winged fly) larvae
CryIII	Coleoptera (beetles)
CryIV	Diptera larvae
CryV	Nematode worms
CryVI	Nematode worms

consequently starves to death (Figure 15.1). Variation in the structure of the gut binding sites in different groups of insects is probably the underlying cause of the high specificities displayed by the different types of δ-endotoxin.

 B. thuringiensis toxins are not recent discoveries, the first patent for their use in crop protection having been granted in 1904. Over the years, several attempts have been made to market them as environmentally friendly insecticides, but their biodegradability acts as a disadvantage because it means they must be reapplied at regular intervals during the growing season, increasing the farmer's costs. Research has therefore been aimed at developing δ-endotoxins that do not require regular application. One approach is via protein engineering (p. 223), modifying the structure of the toxin so that it is more stable. A second approach is to engineer the crop so that it is capable of synthesizing its own toxin.

Cloning a δ-endotoxin gene in maize

Maize is an example of a crop plant that is not served well by conventional insecticides. One major pest of maize is the European corn borer (*Ostrinia nubilialis*), which tunnels into the plant after emerging from eggs laid on the undersurfaces of leaves, thereby

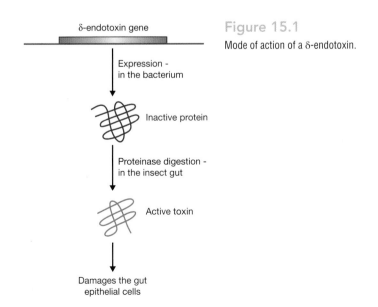

Figure 15.1

Mode of action of a δ-endotoxin.

evading the effects of insecticides applied by spraying. The first attempt at counter-ing this pest by engineering maize plants to synthesize δ-endotoxin was made by plant biotechnologists in 1993, working with the CryIA(b) version of the toxin. The CryIA(b) protein is 1155 amino acids in length, with the toxic activity residing in the segment from amino acids 29–607. Rather than isolating the natural gene, a shortened version con-taining the first 648 codons was made by artificial gene synthesis. This strategy enabled modifications to be introduced into the gene to improve its expression in maize plants. For example, the codons that were used in the artificial gene were those known to be pre-ferred by maize, and the overall GC content of the gene was set at 65%, compared to the 38% GC content of the native bacterial version of the gene (Figure 15.2a). The artificial gene was ligated into a cassette vector (p. 254) between a promoter and polyadenylation signal from cauliflower mosaic virus (Figure 15.2b), and introduced into maize embryos by bombardment with DNA-coated microprojectiles (p. 89). The embryos were grown into mature plants, and transformants identified by PCR analysis of DNA extracts, using primers specific for a segment of the artificial gene (Figure 15.2c).

The next step was to use an immunological test to determine if δ-endotoxin was being synthesized by the transformed plants. The results showed that the artificial gene was

(a) Synthesis of an artificial δ-endotoxin gene

1   1155
B. thuringiensis gene

1   Artificial gene   648
Preferred codons and
GC content for maize

(b) Attachment of a promoter and polyadenylation signal

Promoter sequence

Polyadenylation sequence

(c) PCR analysis of mature plants

1   2   3

1. DNA size markers

2. Result of PCR with DNA from
   a transformed plant

3. Result of PCR with DNA from
   a non-transformed plant

Figure 15.2
Important steps in the procedure used to obtain genetically engineered maize plants expressing an artificial δ-endotoxin gene.

Figure 15.3

Positional effects.

indeed active, but that the amounts of δ-endotoxin being produced varied from plant to plant, from about 250 to 1750 ng of toxin per mg total protein. These differences were probably due to positional effects, the level of expression of a gene cloned in a plant (or animal) often being influenced by the exact location of the gene in the host chromosomes (Figure 15.3).

Were the transformed plants able to resist the attentions of the corn borers? This point was assessed by conducting field trials in which transformed and normal maize plants were artificially infested with larvae, and the effects of predation measured over a period of six weeks. The criteria used were the amount of damage suffered by the foliage of the infested plants, and the lengths of the tunnels produced by the larvae when boring into the plants. In both respects, the transformed plants produced better results than their normal counterparts, with the average length of the larval tunnels in particular being reduced from 40.7 cm for control plants to just 6.3 cm for engineered plants. In real terms, this is a very significant level of resistance.

Cloning δ-endotoxin genes in chloroplasts

One objection that has been raised to the use of genetically modified (GM) crops is the possibility that the cloned gene might escape from the engineered plant and become established in a weed species. From a biological viewpoint, this is an unlikely scenario as the pollen produced by a plant is usually only able to fertilize the ovary of a plant of the same species, so the transfer of a gene from a crop to a weed is highly unlikely. However, one way of making such transfer totally impossible would be to place the cloned gene not in the nucleus but in the plant's chloroplasts. A transgene located in the chloroplast genome cannot escape via pollen for the simple reason that pollen does not contain chloroplasts.

The synthesis of δ-endotoxin protein in transgenic chloroplasts was first achieved in an experimental system with tobacco. The CryIIA(a2) gene was used, which codes for a protein that has a broader toxicity spectrum than the CryIA toxins, killing the larvae of two-winged flies as well as lepidopterans (see Table 15.1). In the *B. thuringiensis* genome, the CryIIA(a2) gene is the third gene in a short operon, the first two genes of which code for proteins that help to fold and process the δ-endotoxin (Figure 15.4). One advantage of using chloroplasts as the sites of recombinant protein synthesis is that the gene expression machinery of chloroplasts, being related to that of bacteria (because chloroplasts were once free-living prokaryotes), is able to express all of the genes in an operon. In contrast, each gene that is placed in a plant (or animal) nuclear genome

Figure 15.4

The CryIIA(a2) operon.

must be cloned individually, with its own promoter and other expression signals, which makes it very difficult to introduce two or more genes at the same time.

The CryIIA(a2) operon was attached to a kanamycin resistance gene along with flanking chloroplast DNA sequences, and biolistics then used to introduce the construct into tobacco leaf cells. Insertion into the chloroplast genome was ensured by rigorously selecting for kanamycin resistance, by placing leaf segments on agar containing kanamycin for up to 13 weeks (p. 126). Transgenic shoots growing out of the leaf segments were then placed on a medium that induced root formation, and the plants were grown.

The amounts of CryIIA(a2) protein produced in the tissues of these GM plants was quite remarkable, the toxin comprising over 45% of the total soluble protein, more than previously achieved in any plant cloning experiment. This high level of expression almost certainly resulted from the combined effects of the high copy number for the transgene (there being many chloroplast genomes per cell, compared to just two copies of the nuclear genome) and the presence in the chloroplasts of the two helper proteins coded by the other genes in the CryIIA(a2) operon. As might be anticipated, the plants proved to be extremely toxic to susceptible insect larvae. Five days after being placed on the GM plants, all cotton bollworm and beet armyworm larvae were dead, with appreciable damage being visible only on the leaves of the plants exposed to armyworms, which have a relatively high natural resistance to δ-endotoxins. The presence of large amounts of toxin in the leaf tissues appeared not to affect the plants themselves, and the GM tobacco was undistinguishable from non-GM plants when factors such as growth rates, chlorophyll content and level of photosynthesis were considered. Unfortunately, attempts to repeat this experiment with more useful crops have been hampered by difficulties in achieving chloroplast transformation with plants other than tobacco, though some success has been reported with cotton.

Countering insect resistance to δ-endotoxin crops

It has long been recognized that crops synthesizing δ-endotoxins might become ineffective after a few seasons due to a build-up of resistance among the insect populations feeding on the crops. This would be a natural consequence of exposing these populations to high amounts of toxins and, of course, could render the GM plants no better than the non-GM versions after a just a few years. Various strategies have been proposed to prevent the development of δ-endotoxin-resistant insects. One of the first to be suggested was to develop crops expressing both the CryI and CryII genes, the rationale being that as these toxins are quite different it would be difficult for an insect population to develop resistance to both types (Figure 15.5a). Whether or not this is a sound argument is not yet clear. Most of the documented examples of δ-endotoxin resistance have not been broad spectrum: for example, the CryIIA(a2) tobacco plants described above were equally poisonous to cotton budworms that were or were not resistant to CryIA(b). However, some strains of meal moth larvae exposed to plants containing the CryIA(c) toxin have acquired a resistance that also provides protection against the CryII toxins. In any case, it would be risky to base a counter-resistance strategy on assumed limitations to the genetic potential of the insect pests.

An alternative might be to engineer toxin production in such a way that synthesis occurs only in those parts of the plant that need protection. For example, in a crop such as maize some damage to the non-fruiting parts of the plant could be tolerated if this did not affect the production of cobs (Figure 15.5b). If expression of the toxin only

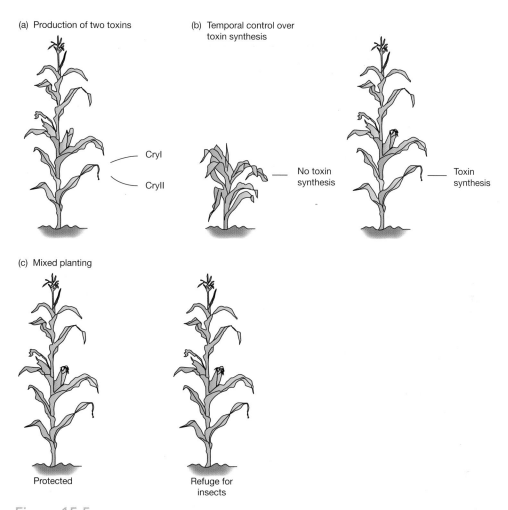

(a) Production of two toxins

Cryl

Cryll

(b) Temporal control over toxin synthesis

No toxin synthesis

Toxin synthesis

(c) Mixed planting

Protected

Refuge for insects

Figure 15.5

Three strategies for countering the development of insect resistance to δ-endotoxin crops.

occurred late in the plant life cycle, when the cobs were developing, then the overall exposure of the insects to the toxin might be reduced without any decrease in the value of the crop. This strategy might delay the onset of resistance, but it is unlikely to avoid it altogether.

A third strategy would be to mix GM plants with non-GM plants, so that each field contained plants on which the insects could feed without being exposed to the toxin produced by the engineered versions (Figure 15.5c). The non-GM plants would act as a refuge for the insects, ensuring that the insect population continually included a high proportion of non-resistant individuals. As all the δ-endotoxin resistance pheno-types so far encountered are recessive, heterozygotes arising from a mating between a susceptible insect and a resistant partner would themselves be susceptible, contin-ually diluting the proportion of resistant insects in the population. Trials have been carried out and theoretical models examined to identify the most effective mixed plant-ing strategies. In practice, success or failure would depend to a very large extent on the farmers who grow the crops, as they would have to adhere to a precise planting strategy dictated by the scientists, despite the resulting loss in productivity due to damage suffered by the non-GM plants. Again, this introduces an element of risk. Clearly, the success of GM projects with plants depends on much more that the skills of the genetic engineers.

15.1.2 Herbicide-resistant crops

Although δ-endotoxin production has been engineered in crops as diverse as maize, cotton, rice, potato and tomato, these plants are not the most widespread GM crops grown today. In commercial terms the most important transgenic plants are those that have been engineered to withstand the herbicide glyphosate. This herbicide, which is widely used by farmers and horticulturists, is environmentally friendly as it is non-toxic to insects and to animals and has a short residence time in soils, breaking down over a period of a few days into harmless products. However, glyphosate kills all plants, both weeds and crop species, and so has to be applied to fields very carefully in order to prevent the growth of weeds without harming the crop itself. GM crops that are able to withstand the effects of glyphosate are therefore desirable as they would enable a less rigorous and hence less expensive herbicide application regime to be followed.

'Roundup Ready' crops

The first crops to be engineered for glyphosate resistance were called 'Roundup Ready', reflecting the trade name of the herbicide. These plants contain a modified gene for the enzyme enolpyruvylshikimate-3-phosphate synthase (EPSPS), which converts shikimate and phosphoenol pyruvate (PEP) into enolpyruvylshikimate-3-phosphate, an essential precursor for synthesis of the aromatic amino acids tryptophan, tyrosine and phenylalanine (Figure 15.6). Glyphosate competes with PEP for binding to the enzyme surface, thereby inhibiting synthesis of enolpyruvylshikimate-3-phosphate and preventing the plant from making the three amino acids. Without these amino acids, the plant quickly dies.

Initially, genetic engineering was used to generate plants that made greater than normal amounts of EPSPS, in the expectation that these would be able to withstand higher doses of glyphosate than non-engineered plants. However, this approach was unsuccessful because, although engineered plants that made up to 80-fold the normal amount of

Figure 15.6

Glyphosate competes with phosphoenol pyruvate in the EPSPS-catalyzed synthesis of enolpyruvylshikimate-3-phosphate, and hence inhibits synthesis of tryptophan, tyrosine, and phenylalanine.

EPSPS were obtained, the resulting increase in glyphosate tolerance was not sufficient to protect these plants from herbicide application in the field.

A search was therefore carried out for an organism whose EPSPS enzyme is resistant to glyphosate inhibition and whose EPSPS gene might therefore be used to confer resistance on a crop plant. After testing the genes from various bacteria, as well as mutant forms of *Petunia* that displayed glyphosate resistance, the EPSPS gene from *Agrobacterium* strain CP4 was chosen, because of its combination of high catalytic activity and high resistance to the herbicide. EPSPS is located in the plant chloroplasts, so the *Agrobacterium* EPSPS gene was cloned in a Ti vector as a fusion protein with a leader sequence that would direct the enzyme across the chloroplast membrane and into the organelle. Biolistics was used to introduce the recombinant vector into soybean callus culture. After regeneration, the GM plants were found to have a threefold increase in herbicide resistance.

A new generation of glyphosate-resistant crops

Roundup Ready versions of a variety of crops have been produced during recent years, and several of these – in particular, soybean and maize – are grown routinely in the USA and other parts of the world. However, these plants do not actually destroy glyphosate, which means that the herbicide can accumulate in the plant tissues. As glyphosate is not thought to be poisonous to humans or other animals, the use of such plants as food or forage should not be a concern, but accumulation of the herbicide can interfere with reproduction of the plant. The degree of resistance displayed by Roundup Ready crops has also been found to be too low to provide a major economic benefit with some crops, notably wheat.

Until recently, there have been only a few scattered reports of organisms capable of actively degrading glyphosate. However, searches of microbial collections have revealed that this property is relatively common among bacteria of the genus *Bacillus*, which possess an enzyme – glyphosate N-acetyltransferase (GLYAT) – which detoxifies glyphosate by attaching an acetyl group to the herbicide molecule (Figure 15.7a). The most active detoxifier known is a strain of *B. licheniformis*, but even this bacterium detoxifies glyphosate at rates that are too low to be of value if transferred to a GM crop.

(a) Detoxification of glyphosate by GLYAT

(b) Directed evolution to produce a highly active GLYAT enzyme

3 native GLYAT genes

11 rounds of multigene shuffling

New GLYAT gene

Enzyme is 10 000 times more active

Figure 15.7

Use of glyphosate *N*-acetyltransferase to generate plants that detoxify glyphosate. (a) GLYAT detoxifies glyphosate by adding an acetyl group (shown in blue). (b) Creation of a highly active GLYAT enzyme by multigene shuffling.

Is it possible to increase the activity of the GLYAT synthesized by *B. licheniformis*? The discovery that the bacterium possesses three related genes for this enzyme pointed a way forward. A type of directed evolution called multigene shuffling was used that involves taking segments of each member of a multigene family and reassembling the segments to create new gene variants. At each stage of the process, the most active genes are identified by cloning all variants in *E. coli* and assaying the recombinant colonies for GLYAT activity. The most active genes are then used as the substrates for the next round of shuffling. After 11 rounds, a gene specifying a GLYAT with 10 000-fold the activity of the enzymes present in the original *B. licheniformis* strain was obtained (Figure 15.7b). This gene was introduced into maize, and the resulting GM plants were found to tolerate levels of glyphosate sixfold higher than the amount normally used by farmers to control weeds, without any reduction in the productivity of the plant. This new way of engineering glyphosate resistance is being exploited in the development of herbicide-resistant soybean and canola.

15.1.3 Other gene addition projects

GM crops that synthesize δ-endotoxins or glyphosate resistance enzymes are by no means the only examples of plants engineered by gene addition. Examples of other gene addition projects are listed in Table 15.2.

These projects include an alternative means of conferring insect resistance, using genes coding for proteinase inhibitors, small polypeptides that disrupt the activities of enzymes in the insect gut, thus preventing or slowing growth. Proteinase inhibitors are produced naturally by several types of plant, notably legumes such as cowpeas and common beans, and their genes have been successfully transferred to other crops which do not normally make significant amounts of these proteins. The inhibitors are particularly effective against beetle larvae that feed on seeds, and so may be a better alternative than δ-endotoxin for plants whose seeds are stored for long periods.

Other projects are exploring the use of genetic modification to improve the nutritional quality of crop plants. The β-carotene (vitamin A) content of cassava has been increased 20-fold by the addition of genes for phytoene synthase from the bacterium *Pantoea agglomerans* (formerly called *Erwinia herbicola*) and deoxyxylulose-5-phosphate synthase from *Arabidopsis thaliana*. The iron and protein contents of cassava have also been improved, the former by introducing a metal transporter protein from *Chlamydomonas reinhardtii*, and the latter with a protein called sporazein. The sporazein gene is a fusion between sporamin, a protein from sweet potato tubers, and zein from maize kernels, designed to combine the best nutritional qualities of the sweet potato and maize proteins. Cassava is a staple crop in many parts of sub-Saharan Africa, but dependency on unmodified cassava leads to vitamin A and iron deficiencies, especially among children. The development of a biofortified version of cassava could therefore have a significant impact on nutrition and health in parts of Africa. In an analogous project, biofortified maize with improved β-carotene, ascorbate (vitamin C) and folate (vitamin B9) contents has been engineered by transfer of genes from rice and the bacteria *Escherichia coli* and *Pantoea ananatis*.

Finally, in a different sphere of commercial activity, ornamental plants with unusual flower colours are being produced by transferring genes for enzymes involved in pigment production from one species to another.

Table 15.2
Examples of gene addition projects with plants.

GENE FOR	SOURCE ORGANISM
Insect resistance	
δ-Endotoxin	*B. thuringiensis*
Proteinase inhibitors	Various legumes
Fungal resistance	
Chitinase	Rice
Glucanase	Alfalfa
Ribosome-inactivating protein	Barley
Bacterial resistance	
Ornithine carbamyltransferase	*Pseudomonas syringae*
Virus resistance	
2′-5′-Oligoadenylate synthetase	Rat
RNA polymerase, helicase	Potato leafroll luteovirus
Satellite RNAs	Various viruses
Virus coat proteins	Various viruses
Herbicide tolerance	
Acetolactate synthase	*Nicotiana tabacum*
Enolpyruvylshikimate-3-phosphate synthase	*Agrobacterium* spp.
Glyphosate oxidoreductase	*Ochrobactrum anthropi*
Glyphosate *N*-acetyltransferase	*B. licheniformis*
Nitrilase	*Klebsiella ozaenae*
Phosphinothricin acetyltransferase	*Streptomyces* spp.
Drought tolerance	
Acyl-coenzyme A binding protein	*Arabidopsis thaliana*
Dehydration-responsive element-binding protein	Soybean
Phosphatidylinositol-specific phospholipase C	Maize
Vacuolar pyrophosphatase	*A. thaliana*
Improved nutritional quality	
Acyl carrier protein thioesterase (fat/oil content)	*Umbellularia californica*
Carotene desaturase (β-carotene content)	*Pantoea ananatis*
Dehydroascorbate reductase (ascorbate content)	Rice
Delta-12 desaturase (fat/oil content)	*Glycine max*
Deoxyxylulose-5-phosphate synthase (β-carotene content)	*A. thaliana*
FEA1 metal transporter (iron content)	*Chlamydomonas reinhardtii*
GTP cyclohydrolase (folate content)	*Escherichia coli*
Methionine-rich protein (sulphur content)	Brazil nuts
Monellin, thaumatin (sweetness)	*Thaumatococcus danielli*
Phytoene synthase (β-carotene content)	*Pantoea agglomerans*
Sporazein (protein content)	Maize and sweet potato
Male sterility	
Barnase ribonuclease inhibitor	*Bacillus amyloliquefaciens*
DNA adenine methylase	*E. coli*
Fruit ripening	
S-Adenosylmethionine hydrolase	Bacteriophage T3
1-Aminocyclopropane-1-carboxylic acid deaminase	Various
Flower colour	
Dihydroflavanol reductase	Various flowering plants
Flavonoid hydroxylase	Various flowering plants

15.2 Gene subtraction

The second way of changing the genotype of a plant is by gene subtraction. This term is a misnomer, as the modification does not involve the actual removal of a gene, merely its inactivation. There are several possible strategies for inactivating a single, chosen gene in a living plant, the most successful so far in practical terms being the use of antisense RNA (p. 288).

15.2.1 Antisense RNA and the engineering of fruit ripening in tomato

To illustrate how antisense RNA has been used in plant genetic engineering, we will examine how tomatoes with delayed ripening have been produced. This is an important example of plant genetic modification as it resulted in one of the first GM foodstuffs to be approved for sale to the general public.

Commercially grown tomatoes and other soft fruits are usually picked before they are completely ripe, to allow time for the fruits to be transported to the marketplace before they begin to spoil. This is essential if the process is to be economically viable, but there is a problem in that most immature fruits do not develop their full flavour if they are removed from the plant before they are fully ripe. The result is that mass-produced tomatoes often have a bland taste, which makes them less attractive to the consumer. Antisense technology has been used in two ways to genetically engineer tomato plants so that the fruit-ripening process is slowed down. This enables the grower to leave the fruits on the plant until they ripen to the stage where the flavour has fully developed, there still being time to transport and market the crop before spoilage sets in.

Using antisense RNA to inactivate the polygalacturonase gene

The timescale for development of a fruit is measured as the number of days or weeks after flowering. In tomato, this process takes approximately eight weeks from start to finish, with the colour and flavour changes associated with ripening beginning after about six weeks. At this time a number of genes involved in the later stages of ripening are switched on, including one coding for the polygalacturonase enzyme (Figure 15.8). This enzyme slowly breaks down the polygalacturonic acid component of the cell walls

Figure 15.8

The increase in polygalacturonase gene expression seen during the later stages of tomato fruit ripening.

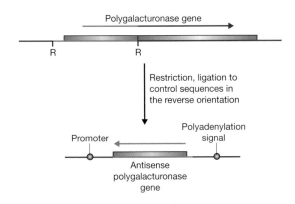

Figure 15.9

Construction of an antisense polygalacturonase gene.

in the fruit pericarp, resulting in a gradual softening that makes the fruit palatable. However, if taken too far this results in a squashy, spoilt tomato that is attractive only to students with limited financial resources.

A partial inactivation of the polygalacturonase gene should increase the time between flavour development and spoilage of the fruit. To test this hypothesis, a 730-bp restriction fragment was obtained from the 5′ region of the normal polygalacturonase gene, representing just under half of the coding sequence (Figure 15.9). The orientation of the fragment was reversed, a cauliflower mosaic virus (CaMV) promoter was ligated to the start of the sequence, and a plant polyadenylation signal attached to the end. The construction was then inserted into the Ti plasmid vector pBIN19 (p. 123). Once inside the plant, transcription from the CaMV promoter should result in the synthesis of an antisense RNA complementary to the first half of the polygalacturonase mRNA. Previous experiments with antisense RNA had suggested that this would be sufficient to reduce, or even prevent, translation of the target mRNA.

Transformation was carried out by introducing the recombinant pBIN19 molecules into *Agrobacterium tumefaciens* bacteria and then allowing the bacteria to infect tomato stem segments. Small amounts of callus material collected from the surfaces of these segments were tested for their ability to grow on an agar medium containing kanamycin (remember that pBIN19 carries a gene for kanamycin resistance; see Figure 7.14). Resistant transformants were identified and allowed to develop into mature plants.

The effect of antisense RNA synthesis on the amount of polygalacturonase mRNA in the cells of ripening fruit was determined by northern hybridization with a single-stranded DNA probe specific for the sense mRNA. These experiments showed that ripening fruit from transformed plants contained less polygalacturonase mRNA than the fruits from normal plants. The amounts of polygalacturonase enzyme produced in the ripening fruits of transformed plants were then estimated from the intensities of the relevant bands after the separation of fruit proteins by polyacrylamide gel electrophoresis, and by directly measuring the enzyme activities in the fruits. The results showed that less enzyme was synthesized in transformed fruits (Figure 15.10). Most importantly, the transformed fruits, although undergoing a gradual softening, could be stored for a prolonged period before beginning to spoil. This indicated that the antisense RNA had not completely inactivated the polygalacturonase gene, but had nonetheless produced a sufficient reduction in gene expression to delay the ripening process as desired. The GM tomatoes – marketed under the trade name 'FlavrSavr' – were one of the first genetically engineered plants to be approved for sale to the public, first appearing in supermarkets in 1994.

Figure 15.10

The differences in polygalacturonase activity in normal tomato fruits and in fruits expressing the antisense polygalacturonase gene.

Using antisense RNA to inactivate ethylene synthesis

The main trigger that switches on the genes involved in the later stages of tomato ripening is ethylene which, despite being a gas, acts as a hormone in many plants. A second way of delaying fruit ripening would therefore be to engineer plants so that they do not synthesize ethylene. Fruits on these plants would develop as normal for the first six weeks, but would be unable to complete the ripening process. The unripe fruit could therefore be transported to the marketplace without any danger of the crop spoiling. Before selling to the consumer, or conversion into paste or some other product, artificial ripening would be induced by spraying the tomatoes with ethylene.

The penultimate step in the ethylene synthesis pathway is the conversion of S-adenosylmethionine to 1-aminocyclopropane-1-carboxylic acid (ACC), which is the immediate precursor for ethylene. This step is catalysed by an enzyme called ACC synthase. As with polygalacturonase, ACC synthase inactivation was achieved by cloning into tomato a truncated version of the normal ACC synthase gene, inserted into the cloning vector in the reverse orientation, so that the construct would direct synthesis of an antisense version of the ACC synthase mRNA. After regeneration, the engineered plants were grown to the fruiting stage and found to make only 2% of the amount of ethylene produced by non-engineered plants. This reduction was more than sufficient to prevent the fruit from completing the ripening process. These tomatoes have been marketed as the 'Endless Summer' variety.

15.2.2 Other examples of the use of antisense RNA in plant genetic engineering

In general terms, the applications of gene subtraction in plant genetic engineering are less broad than those of gene addition. It is easier to think of useful characteristics that a plant lacks and which might be introduced by gene addition, than it is to identify disadvantageous traits that the plant already possesses and which could be removed by gene subtraction. There are, however, a growing number of plant biotechnology projects based on gene subtraction (Table 15.3), and the approach is increasing in importance as research groups explore the use of RNA interference (RNAi) as a second means of silencing selected genes. RNAi is one of the plant's natural processes for regulating gene expression, through the synthesis of microRNAs (miRNAs) and small interfering

Table 15.3
Examples of gene subtraction projects with plants.

TARGET GENE	MODIFIED CHARACTERISTIC
1-Aminocyclopropane-1-carboxylic acid synthase	Modified fruit ripening in tomato
1D-*myo*-inositol 3-phosphate synthase	Reduction of indigestible phosphorus content of rice grains
Chalcone synthase	Modification of flower colour in various decorative plants
Delta-12 desaturase	High oleic acid content in soybean
DET1 photomorphogenesis regulatory gene	Improved carotenoid and flavonoid content in tomato
Lycopene epsilon cyclase	Enhanced carotenoid content of rapeseed
5-Methylcytosine DNA glycosylase	Reduced immunogenicity of wheat seed proteins
Polygalacturonase	Delay of fruit spoilage in tomato
Polyphenol oxidase	Prevention of discoloration in fruits and vegetables
Starch synthase	Reduction of starch content in vegetables

RNAs (siRNAs) that bind to and inactivate their target mRNAs. The cloning of genes for artificial miRNAs and siRNAs could therefore be used to switch off specific genes and hence bring about gene subtraction.

15.3 Problems with genetically modified plants

Ripening-delayed tomatoes produced by gene subtraction were among the first GM whole foods to be approved for marketing. Partly because of this, plant genetic engineering has provided the battleground on which biotechnologists and other interested parties have fought over the safety and ethical issues that arise from our ability to alter the genetic make-up of living organisms. A number of the most important questions do not directly concern genes and the expertise needed to answer them will not be found in this book. For example, we cannot discuss in an authoritative fashion the possible impact, good or otherwise, that GM crops might have on local farming practices in the developing world. However, we can, and should, look at the biological issues.

15.3.1 Safety concerns with selectable markers

One of the main areas of concern to emerge from the debate over GM tomatoes is the possible harmful effects of the marker genes used with plant cloning vectors. Most plant vectors carry a copy of a gene for kanamycin resistance, enabling transformed plants to be identified during the cloning process. The kanamycin resistance gene is bacterial in origin and codes for the enzyme neomycin phosphotransferase II. This gene and its enzyme product are present in all cells of an engineered plant. Although the fear that neomycin phosphotransferase might be toxic to humans has been allayed by tests with animal models, two other safety issues remain:

- Could the kanamycin resistance gene contained in a genetically modified foodstuff be passed to bacteria in the human gut, making these bacteria resistant to kanamycin and related antibiotics?
- Could the kanamycin resistance gene be passed to other organisms in the environment, and would this result in damage to the ecosystem?

Figure 15.11

DNA excision by the Cre recombinase enzyme.

Neither question can be fully answered with our current knowledge. It can be argued that digestive processes would destroy all of the kanamycin resistance genes in a GM food before they could reach the bacterial flora of the gut and that, even if a gene did avoid destruction, the chances of it being transferred to a bacterium would be very small. Nevertheless, the risk factor is not zero. Similarly, although experiments suggest that the growth of GM plants would have a negligible effect on the environment, as kanamycin resistance genes are already common in natural ecosystems, the future occurrence of some unforeseen and damaging event cannot be considered an absolute impossibility.

The fears surrounding the use of kanamycin resistance and other marker genes have prompted biotechnologists to devise ways of removing these genes from plant DNA after the transformation event has been verified. One of these strategies makes use of an enzyme from bacteriophage P1, called Cre, which catalyses a recombination event that excises DNA fragments flanked by specific 34-bp recognition sequences (Figure 15.11). To use this system the plant is transformed with two cloning vectors. The first vector carries the gene being added to the plant along with its kanamycin resistance marker gene, the latter being flanked by the Cre target sequences, and the second vector carries the Cre gene. After transformation, expression of the Cre gene results in excision of the kanamycin resistance gene from the plant DNA.

But what if the Cre gene is itself hazardous in some way? This is immaterial as the two vectors used in the transformation would probably integrate their DNA fragments into different chromosomes, so random segregation during sexual reproduction would result in plants that contained one integrated fragment but not the other. A plant that contains neither the Cre gene nor the kanamycin resistance marker, but does contain the important gene that we wished to add to the plant's genome, can therefore be obtained.

15.3.2 The terminator technology

The Cre recombination system also underlies one of the most controversial aspects of plant genetic engineering, the so-called terminator technology. This is one of the processes by which companies that market GM crops attempt to protect their financial investment by ensuring that farmers must buy new seed every year, rather than simply collecting seed from the crop and sowing this second-generation seed the following year. In reality, even with conventional crops, mechanisms have been devised to ensure that second-generation seed cannot be grown by farmers, but the general controversies surrounding GM crops have placed the terminator technology in the public eye.

The terminator technology centres on the gene for the ribosome inactivating protein (RIP). The ribosome inactivating protein blocks protein synthesis by cutting one of the

(a) The RIP gene

(b) The terminator system

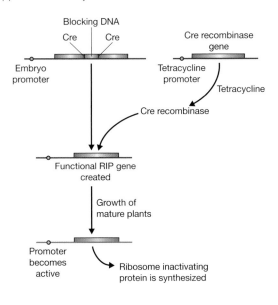

Figure 15.12

The terminator technology. (a) The RIP gene codes for a protein that blocks protein synthesis. (b) The system used to allow the production of first-generation seeds.

ribosomal RNA molecules into two segments (Figure 15.12a). Any cell in which the RIP is active will quickly die. In GM plants that utilize the terminator system, the RIP gene is placed under control of a promoter that is active only during embryo development. The plants therefore grow normally but the seeds that they produce are sterile.

So, how are the first-generation seeds that are sold to farmers obtained? To begin with, the RIP gene is non-functional because it is disrupted by a segment of non-RIP DNA (Figure 15.12b). However, this DNA is flanked by the 34-bp recognition sequences for the Cre recombinase. In these plants the gene for the Cre recombinase is placed under control of a promoter that is switched on by tetracycline. Once seeds have been obtained, the supplier activates the Cre recombinase by placing the seeds in a tetracycline solution. This removes the blocking DNA from the RIP gene, which becomes functional but remains silent until its own promoter becomes active during embryogenesis.

15.3.3 The possibility of harmful effects on the environment

A second area of concern regarding GM plants is that their new gene combinations might harm the environment in some way. These concerns have to be addressed individually for each type of GM crop, as different engineered genes might have different impacts. We will examine the investigations that have been carried out to assess whether it is possible that herbicide-resistant plants (one of the two examples of gene addition examined earlier in this chapter) can have a harmful effect. As these are the most widely grown GM crops, they have been subject to some of the most comprehensive environmental studies. In particular, in 1999, the UK Government commissioned an independent investigation into how herbicide-resistant crops – the growth of which was not permitted at that time in the UK – might affect the abundance and diversity of farmland wildlife.

After delays due to activists attempting to prevent the studies from being carried out, the UK research team reported their findings in 2003. The study involved 273 field trials conducted throughout England, Wales and Scotland, and included glyphosate-resistant sugar beet as well as maize and spring rape engineered for resistance to a second herbicide, glufosinate-ammonium. The results, as summarized in the official report (see Burke (2003) in Further Reading), were as follows:

"The team found that there were differences in the abundance of wildlife between GM crop fields and conventional crop fields. Growing conventional beet and spring rape was better for many groups of wildlife than growing GM beet and spring rape. There were more insects, such as butterflies and bees, in and around the conventional crops because there were more weeds to provide food and cover. There were also more weed seeds in conventional beet and spring rape crops than in their GM counterparts. Such seeds are important in the diets of some animals, particularly some birds. In contrast, growing GM maize was better for many groups of wildlife than conventional maize. There were more weeds in and around the GM crops, more butterflies and bees around at certain times of the year, and more weed seeds. The researchers stress that the differences they found do not arise just because the crops have been genetically modified. They arise because these GM crops give farmers new options for weed control. That is, they use different herbicides and apply them differently. The results of this study suggest that growing such GM crops could have implications for wider farmland biodiversity. However, other issues will affect the medium- and long-term impacts, such as the areas and distribution of land involved, how the land is cultivated and how crop rotations are managed. These make it hard for researchers to predict the medium- and large-scale effects of GM cropping with any certainty. In addition, other management decisions taken by farmers growing conventional crops will continue to impact on wildlife."

Further reading

Burke, M. (2003) GM crops: effects on farmland wildlife. Produced by the Farmscale Evaluations Research Team and the Scientific Steering Committee. ISBN: 0-85521-035-4. [For a more detailed description of this work see *Philosophical Transactions of the Royal Society, Biological Sciences*, **358**, 1775–1889 (2003).]

Castle, L.A., Siehl, D.L., and Gorton, R. (2004) Discovery and directed evolution of a glyphosate tolerance gene. *Science*, **304**, 1151–1154. [Cloning the GLYAT gene in maize.]

De Cosa, B., Moar, W., Lee, S.B., *et al.* (2001) Overexpression of the Bt *cry2Aa2* operon in chloroplasts leads to formation of insecticidal crystals. *Nature Biotechnology*, **19**, 71–74.

Feitelson, J.S., Payne, J., and Kim, L. (1992) *Bacillus thuringiensis*: insects and beyond. *Biotechnology*, **10**, 271–275. [Details of δ-endotoxins and their potential as conventional insecticides and in genetic engineering.]

Fischhoff, D.A., Bowdish, K.S., Perlak, F.J., *et al.* (1987) Insect-tolerant transgenic tomato plants. *Biotechnology*, **5**, 807–813. [The first transfer of a δ-endotoxin gene to a plant.]

Groot, A.T. and Dicke, M. (2002) Insect-resistant transgenic plants in a multi-trophic context. *Plant Journal*, **31**, 387–406. [Examines the impact of d-endotoxin plants on the ecological food chain.]

Koch, A. and Kogel, K.-H. (2014) New wind in the sails: improving the agronomic value of crop plants through RNAi-mediated gene silencing. *Plant Biotechnology Journal*, **12**, 821–831.

Koziel, M.G., Beland, G.L., Bowman, C., *et al.* (1993) Field performance of elite transgenic maize plants expressing an insecticidal protein derived from *Bacillus thuringiensis*. *Biotechnology*, **11**, 194–200. [Cloning a δ-endotoxin gene in maize.]

Matas, A.J., Gapper, N.E., Chung, M.-Y., Giovannoni, J.J., and Rose, J.K.C. (2009) Biology and genetic engineering of fruit maturation for enhanced quality and shelf-life. *Current Opinion in Biotechnology*, **20**, 197–203.

Miki, B. and McHugh, S. (2003) Selectable marker genes in transgenic plants: applications, alternatives and biosafety. *Journal of Biotechnology*, **107**, 193–232.

Naqvi, S., Zhu, C., Farre, G., *et al.* (2009) Transgenic multivitamin corn through biofortification of endosperm with three vitamins representing three distinct metabolic pathways. *Proceedings of the National Academy of Sciences of the United States of America*, **106**, 7762–7767.

Pollegioni, L., Schonbrunn, E., and Siehl, D. (2011) Molecular basis of glyphosate resistance – different approaches through protein engineering. *FEBS Journal*, **278**, 2753–2766.

Sayre, R., Beeching, J.R., Cahoon, E.B., *et al.* (2011) The BioCassava plus program: biofortification of cassava for sub-Saharan Africa. *Annual Review of Plant Biology*, **62**, 251-272.

Shelton, A.M., Zhao, J.Z., and Roush, R.T. (2002) Economic, ecological, food safety, and social consequences of the deployment of Bt transgenic plants. *Annual Review of Entomology*, **47**, 845–881. [Various issues relating to δ-endotoxin plants.]

Smith, C.J.S., Watson, C.F., Ray, J., *et al.* (1988) Antisense RNA inhibition of polygalacturonase gene expression in transgenic tomatoes. *Nature*, **334**, 724–726.

Tabashnik, B.E., Gassmann, A.J., Crowder, D.W., and Carriére, Y. (2008) Insect resistance to Bt crops: evidence versus theory. *Nature Biotechnology*, **26**, 199–202. [Discusses the effectiveness of refuges in reducing insect resistance to δ-endotoxin crops.]

Chapter 16

Gene Cloning and DNA Analysis in Forensic Science and Archaeology

Forensic science is the final area of biotechnology that we will consider. Hardly a week goes by without a report in the national press of another high-profile crime that has been solved thanks to DNA analysis. The applications of molecular biology in forensics centre largely on the ability of DNA analysis to identify an individual from hairs, bloodstains and other items recovered from the crime scene. In the popular media, these techniques are called genetic fingerprinting, though the more accurate term for the procedures used today is DNA profiling. We begin this chapter by examining the methods used in genetic fingerprinting and DNA profiling, including their use both in the identification of individuals and in establishing if individuals are members of a single family. This will lead us into an exploration of the ways in which genetic techniques are being used in archaeology.

Gene Cloning and DNA Analysis: An Introduction, Seventh Edition. T.A. Brown.
© 2016 John Wiley & Sons, Ltd. Published 2016 by John Wiley & Sons, Ltd.
Companion Website: www.wiley.com/go/brown/cloning

16.1 DNA analysis in the identification of crime suspects

It is probably impossible for a person to commit a crime without leaving behind a trace of his or her DNA. Hairs, spots of blood, and even conventional fingerprints contain traces of DNA, enough to be studied using the polymerase chain reaction (PCR). The analysis does not have to be done immediately, and in recent years a number of past crimes – so-called 'cold cases' – have been solved and the criminal brought to justice because of DNA testing that has been carried out on archived material. So how do these powerful methods work?

The basis to genetic fingerprinting and DNA profiling is that identical twins are the only individuals who have identical copies of the human genome. Of course, the human genome is more or less the same in everybody – the same genes will be in the same order with the same stretches of intergenic DNA between them. But the human genome, as well as those of other organisms, contains many polymorphisms, positions where the nucleotide sequence is not the same in every member of the population. We have already encountered the most important of these polymorphic sites earlier, because these variable sequences are the same ones that are used as DNA markers in positional cloning of disease susceptibility genes (p. 283). They include restriction fragment length polymorphisms (RFLPs), short tandem repeats (STRs), and single nucleotide polymorphisms (SNPs), the latter being positions in the genome where either of two different nucleotides can occur. All three types of polymorphism can occur within genes as well as in intergenic regions, and altogether there are several million of these polymorphic sites in the human genome, with SNPs being the most common.

16.1.1 Genetic fingerprinting by hybridization probing

The first method for using DNA analysis to identify individuals was developed during the mid-1980s by Sir Alec Jeffreys of Leicester University. This technique was not based on any of the types of polymorphic site listed above, but on a different kind of variation in the human genome called a hypervariable dispersed repetitive sequence. As the name indicates, this is a repeated sequence that occurs at various positions ('dispersed') in the human genome. The key feature of these sequences is that they are located at different positions in the genomes of different people (Figure 16.1a).

The particular repeat that was initially used in genetic fingerprinting contains the sequence GGGCAGGANG (where N is any nucleotide). To prepare a fingerprint, a sample of genomic DNA is digested with a restriction endonuclease, the fragments separated using agarose gel electrophoresis, and a Southern blot prepared (p. 151). Hybridization to the blot of a labelled probe containing the repeat sequence reveals a series of bands, each one representing a restriction fragment that contains the repeat (Figure 16.1b). Because the insertion sites of the repeat sequence are variable, the same procedure carried out with a DNA sample from a second person will give a different pattern of bands. These are the genetic fingerprints for those two individuals.

16.1.2 DNA profiling by PCR of short tandem repeats

Strictly speaking, genetic fingerprinting refers only to hybridization analysis of hypervariable dispersed repetitive sequences. This technique has been valuable in

(a) Polymorphic repeat sequences in the human genome

Figure 16.1

Genetic fingerprinting. (a) The positions of polymorphic repeats, such as hypervariable dispersed repetitive sequences, in the genomes of two individuals. In the chromosome segment shown, the second person has an additional repeat sequence. (b) An autoradiograph showing the genetic fingerprints of two individuals.

(b) Two genetic fingerprints

Lanes 1 and 2: DNA from two individuals

forensic work but suffers from three limitations:

- A relatively large amount of DNA is needed because the technique depends on hybridization analysis. Fingerprinting cannot be used with the minute amounts of DNA in hair and bloodstains.
- Interpretation of the fingerprint can be difficult because of variations in the intensities of the hybridization signals. In a court of law, minor differences in band intensity between a test fingerprint and that of a suspect can be sufficient for the suspect to be acquitted.
- Although the insertion sites of the repeat sequences are hypervariable, there is a limit to this variability and therefore a small chance that two unrelated individuals could have the same, or at least very similar, fingerprints. Again, this consideration can lead to acquittal when a case is brought to court.

The more powerful technique of DNA profiling avoids these problems. Profiling makes use of the polymorphic sequences called STRs. As described on p. 284, an STR is a short-sequence, 1–13 nucleotides in length, which is repeated several times in a tandem array. In the human genome, the most common type of STR is the dinucleotide repeat $[CA]_n$, where 'n', the number of repeats, is usually between 5 and 20 (Figure 16.2a).

The number of repeats in a particular STR is variable because repeats can be added or, less frequently, removed by errors that occur during DNA replication. In the population as a whole, there might be as many as ten different versions of a particular STR, each of the alleles characterized by a different number of repeats. In DNA profiling, the alleles of a selected number of different STRs are identified. This can be achieved quickly and with very small amounts of DNA by PCRs with primers that anneal to the DNA sequences either side of a repeat (see Figure 14.13). After the PCR, the products can be examined by agarose gel electrophoresis. The size of the band or bands seen in the gel indicate

(a) Two alleles of an STR

....CACACACACA.... $n = 5$

....CACACACACACA.... $n = 6$

(b) The results of PCR

1   DNA size markers
2, 3   PCRs of a single STR
        in two individuals
4   Multiplex PCR of three
        STRs (A–C)

(c) Analysis of multiplex PCR results by capillary gel electrophoresis

C   B   A

100   150   200

Size (bp)

Figure 16.2

DNA profiling. (a) DNA profiling makes use of STRs which have variable repeat units. (b) A gel obtained after DNA profiling. In lanes 2 and 3 the same STR has been examined in two individuals. These two people have different profiles, but have a band in common. Lane 4 shows the result of a multiplex PCR in which three STRs have been typed in a single PCR. (c) Capillary gel electrophoresis can be used to determine the sizes of multiplex PCR products.

the allele or alleles present in the DNA sample that has been tested (Figure 16.2b). Two alleles of an STR can be present in a single DNA sample because there are two copies of each STR – one on the chromosome inherited from the mother, and one on the chromosome from the father.

Because PCR is used, DNA profiling is very sensitive and enables results to be obtained with hairs and other specimens that contain only trace amounts of DNA. The results are unambiguous, and a match between DNA profiles is usually accepted as evidence in a trial. The current methodology, called CODIS (Combined DNA Index System), makes use of 13 STRs with sufficient variability to give only a one in 10^{15} chance that two individuals, other than identical twins, have the same profile. As the world population is around 6×10^9, the statistical likelihood of two individuals on the planet sharing the same profile is so low as to be considered implausible when DNA evidence is presented in a court of law. Each STR is typed by PCRs with primers that are fluorescently labelled and which anneal either side of the variable repeat region. The alleles present at the STR are then typed by determining the sizes of the amplicons by capillary gel electrophoresis. Two or more STRs can be typed together in a multiplex PCR if their product sizes do not overlap, or if the individual primer pairs are labelled with different fluorescent markers, enabling the products to be distinguished in the capillary gel (Figure 16.2c).

16.2 Studying kinship by DNA profiling

As well as the identification of criminals, DNA profiling can also be used to infer if two or more individuals are members of the same family. This type of study is called kinship analysis and its main day-to-day application is in paternity testing.

16.2.1 Related individuals have similar DNA profiles

Your DNA profile, like all other aspects of your genome, is inherited partly from your mother and partly from your father. Relationships within a family therefore become apparent when the alleles of a particular STR are marked on the family pedigree (Figure 16.3). In this example, we see that three of the four children have inherited the 12-repeat allele from the father. This observation in itself is not sufficient to deduce that these three children are siblings, though the statistical chance would be quite high if the 12-repeat allele was uncommon in the population as a whole. To increase the degree of certainty, more STRs would have to be typed but, as with identification of individuals, the analysis need not be endless, because a comparison of 13 STRs gives an acceptable probability that relationships that are observed are real.

16.2.2 DNA profiling and the remains of the Romanovs

An interesting example of the use of DNA profiling in a kinship study is provided by investigations carried out during the 1990s on the bones of the Romanovs, the last members of the Russian ruling family. The Romanovs and their descendants ruled Russia from the early 17th century until the time of the Russian Revolution, when Tsar Nicholas II was deposed and he and his wife, the Tsarina Alexandra, and their five children were imprisoned. On 17 July 1918, all seven of the family, along with their doctor and three servants, were murdered and their bodies disposed of in a shallow roadside grave near Yekaterinburg in the Urals. In 1991, after the fall of communism, the remains were recovered with the intention that they should be given a more fitting burial.

STR analysis of the Romanov bones
Although it was suspected that the bones recovered were indeed those of the Romanovs, the possibility that they belonged to some other unfortunate group of people could not be discounted. Nine skeletons had been found in the grave – six adults and three

Figure 16.3

Inheritance of STR alleles within a family.

children – and an examination of the bones suggested that four of the adults were male, two were female, and the three children were all female. If these were the remains of the Romanovs, then their son Alexei and one their daughters were, for some, reason, absent. The bodies showed signs of violence, consistent with reports of their treatment during their imprisonment and after death, and at least some of the remains were clearly aristocratic as their teeth were filled with porcelain, silver and gold – dentistry that was well beyond the means of the average Russian of the early 20th century.

DNA was extracted from the bones from each individual, and five STRs typed by PCR were used to test the hypothesis that the three children were siblings and that two of the adults were their parents, as would be the case if indeed these were the Romanovs. The results immediately showed that the three children could be siblings, as they had identical genotypes for the STRs called VWA/31 and FES/FPS and shared alleles at each of the three other loci (Figure 16.4). The THO1 data show that female adult 2 cannot be the mother of the children because she only possesses allele 6, which none of the children have. Female adult 1, however, has allele 8, which all three children have. Examination of the other STRs confirms that she could be the mother of each of the children, and so she is identified as the Tsarina. The THO1 data exclude male adult 4 as a possible father of the children, and the VWA/31 results exclude male adults 1 and 2. When all the STRs

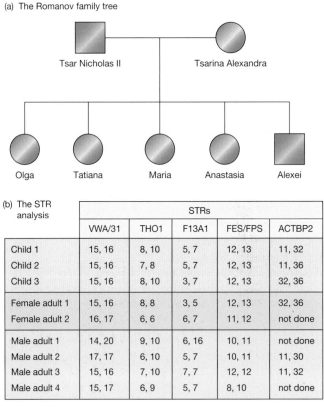

(a) The Romanov family tree

Tsar Nicholas II   Tsarina Alexandra

Olga   Tatiana   Maria   Anastasia   Alexei

(b) The STR analysis

	STRs				
	VWA/31	THO1	F13A1	FES/FPS	ACTBP2
Child 1	15, 16	8, 10	5, 7	12, 13	11, 32
Child 2	15, 16	7, 8	5, 7	12, 13	11, 36
Child 3	15, 16	8, 10	3, 7	12, 13	32, 36
Female adult 1	15, 16	8, 8	3, 5	12, 13	32, 36
Female adult 2	16, 17	6, 6	6, 7	11, 12	not done
Male adult 1	14, 20	9, 10	6, 16	10, 11	not done
Male adult 2	17, 17	6, 10	5, 7	10, 11	11, 30
Male adult 3	15, 16	7, 10	7, 7	12, 12	11, 32
Male adult 4	15, 17	6, 9	5, 7	8, 10	not done

Figure 16.4

Short tandem repeat (STR) analysis of the Romanov bones. (a) The Romanov family tree. (b) The results of the STR analysis. Data taken from Gill *et al.* (1994) (see Further Reading).

are taken into account, male adult 3 could be the father of the children and therefore is identified as the Tsar. Note that all these conclusions can be drawn simply from the THO1 and VWA/31 results: the other STR data simply provide corroborating evidence.

Mitochondrial DNA was used to link the Romanov skeletons with living relatives

The STR analysis showed that the skeletons included a family group as expected, if these were indeed the bones of the Romanovs. But could they be the remains of some other group of people? To address this problem, the DNA from the bones was compared with DNA samples from living relatives of the Romanovs. This work included studies of mitochondrial DNA, the small 16-kb circles of DNA contained in the energy-generating mitochondria of cells. Mitochondrial DNA contains polymorphisms that can be used to infer relationships between individuals, but the degree of variability is not as great as displayed by STRs, so mitochondrial DNA is rarely used for kinship studies among closely related people such as those of a single family group. But mitochondrial DNA has the important property of being inherited solely through the female line, the father's mitochondrial DNA being lost during fertilization and not contributing to the son or daughter's DNA content. This maternal inheritance pattern makes it easier to distinguish relationships when the individuals being compared are more distantly related, as was the case with the living relatives of the Romanovs.

Comparisons were therefore made between mitochondrial DNA sequences obtained from the skeletons and that of Prince Philip, the Duke of Edinburgh, whose maternal grandmother was Princess Victoria of Hesse, the Tsarina Alexandra's sister (Figure 16.5). The mitochondrial DNA sequences from four of the female skeletons – the three children and the adult female identified as the Tsarina – were exactly the same as that of Prince Philip, providing strong evidence that the four females were members of the same lineage. Comparisons were also made with two living matrilineal descendants of Tsar Nicholas's grandmother, Louise of Hesse-Cassel. This analysis was more complicated, as two sequences were present among the clones of the PCR product obtained from the adult male thought to be the Tsar. These sequences differed at a single position which was either a C or a T, the former being fourfold more frequent than the latter. This could indicate that the sample was contaminated with somebody else's DNA, but instead was interpreted as showing that the Tsar's mitochondrial DNA was heteroplasmic, an infrequent situation where two different mitochondrial DNAs coexist within the

Figure 16.5

Family tree showing the matrilineal relationship between Prince Philip, Duke of Edinburgh and Princess Victoria of Hesse, the Tsarina's sister. Males are shown as blue squares, and females as red circles.

same cells. The two descendants of the Tsar's grandmother both had the mitochondrial DNA version with a T at this position, suggesting that the mutation producing the C variant had occurred very recently in the Tsar's lineage. Support for this hypothesis was subsequently provided by analysis of DNA from the Tsar's brother, Grand Duke George Alexandrovich, who died in 1899, which showed that he also displayed heteroplasmy at the same position in his mitochondrial DNA. On balance, the evidence suggested that the Tsar's remains had been correctly identified.

The missing children

Only three children were found in the Romanovs' grave. Alexei, the only boy, and one of the four girls were missing. During the middle decades of the 20th century, several women claimed to be a Romanov princess, because even before the bones were recovered there had been rumours that one of the girls, Anastasia, had escaped the clutches of the Bolsheviks and fled to the West. One of the most famous of these claimants was Anna Anderson, whose case was first widely publicised in the 1920s. Anna Anderson died in 1984 but she left an archived tissue sample whose mitochondrial DNA does not match the Tsarina's and suggests a woman of Polish origin. There have also been various people claiming to be descended from Tsarevich Alexei. But these stories are almost certainly romances, as the partially cremated bodies of two other children found near Yekaterinburg in 2007 have now been shown to have mitochondrial DNA sequences that suggest that they are missing Romanov children.

16.3 Sex identification by DNA analysis

DNA analysis can also be used to identify the sex of an individual. The genetic difference between the sexes is the possession of a Y chromosome by males, so detection of DNA specific for the Y chromosome would enable males and females to be distinguished. Forensic scientists occasionally have to deal with bodies that are so badly damaged that DNA analysis is the only way of identifying their sex.

DNA tests can also be used to identify the sex of an unborn child. Finding out if a foetus is a boy or a girl is usually delayed until the anatomical differences have developed and the sex can be identified by scanning, but under some circumstances an earlier indication of sex is desirable. An example is when the pedigree of the family indicates that an unborn male might suffer from an inherited disease and the parents wish to make an early decision about whether to continue the pregnancy.

A third application of DNA-based sex identification, and the one that has been responsible for many of the developments in this field, is in the analysis of archaeological specimens. Male and female skeletons can be distinguished if key bones such as the skull or the pelvis are intact, but with fragmentary remains, or those of young children, there are not enough sex-specific anatomical differences for a confident identification to be made. If ancient DNA is preserved in the bones, a DNA-based method can tell the archaeologists if they are dealing with a male or a female.

16.3.1 PCRs directed at Y chromosome-specific sequences

The simplest way to use DNA analysis to identify sex is to design a PCR specific for a region of the Y chromosome. The PCR has to be designed with care, because the X

Figure 16.6

Sex identification by PCR of a Y-specific DNA sequence. Male DNA gives a PCR product (lane 2), but female DNA does not (lane 3). The problem is that a failed PCR (lane 4) gives the same result as female DNA.

1 DNA markers
2 Male DNA
3 Female DNA
4 Failed PCR

and Y chromosomes are not completely different, some segments being shared between the two. But there are many unique regions within the Y chromosome. In particular, there are several repeated sequences that are only located in the Y chromosome, these repeated sequences acting as multiple targets for the PCR and hence giving greater sensitivity – an important consideration if you are dealing with a badly damaged body or an ancient bone.

A PCR directed at Y-specific DNA sequences would give a product with male DNA but no band if the sample comes from a female (Figure 16.6). This is a clear distinction between the two alternatives and hence a perfectly satisfactory system for most applications. But what if the sample did not contain any DNA, or if the DNA was too degraded to work in the PCR, or if the sample also contained inhibitors of *Taq* polymerase that prevented the enzyme from carrying out the PCR? All of these possibilities could occur with archaeological specimens, especially those that have been buried in the ground and become contaminated with humic acids and other compounds known to inhibit many of the enzymes used in molecular biology research. Now, the test becomes ambiguous because a specimen that is unable to give a PCR product for one of these reasons could mistakenly be identified as female. The result would be exactly the same, there would be no band on the gel.

16.3.2 PCR of the amelogenin gene

The lack of discrimination between 'female' and 'failed PCR' that occurs when Y-specific sequences are studied has led to the development of more sophisticated DNA tests for sex identification, ones that give unambiguous results for both males and females. The most widely used of these involves PCRs that amplify the amelogenin gene.

The amelogenin gene codes for a protein found in tooth enamel. It is one of the few genes that are present on the Y chromosome and, like many of these genes, there is also a copy on the X chromosome. The two copies are far from identical, however, and when the nucleotide sequences are aligned a number of indels – positions where a segment of DNA has either been inserted into one sequence or deleted from the other sequence – are seen (Figure 16.7a). If the primers for a PCR anneal either side of an indel, the products obtained from the X and Y chromosomes would have different sizes. Female DNA would give a single band when the products are examined, because females only have the X chromosome, whereas males would give two bands, one from the X chromosome and one from the Y (Figure 16.7b). If the sample contains no DNA, or the PCR fails

Figure 16.7

Sex identification by PCR of part of the amelogenin gene. (a) An indel in the amelogenin gene. (b) The results of PCRs spanning the indel. Male DNA gives two PCR products, of 106 and 112 bp in the standard system used in forensics and biomolecular archaeology. Female DNA gives just the smaller product. A failed PCR gives no products and so is clearly distinguishable from the two types of positive result.

(a) Part of the amelogenin gene

Y chromosome

X chromosome

6 bp deletion in the X chromosome sequence

(b) Results of PCR

1   2   3   4

1 DNA markers
2 Male DNA
3 Female DNA
4 Failed PCR

for some other reason, then no bands will be obtained. There is no confusion between a failure and a male or female result.

The development of the amelogenin system for sex identification is having an important impact in archaeology. No longer is it necessary to assign sex to buried bones on the basis of vague differences in the structures of the bones. The greater confidence that DNA-based sex testing allows is resulting in some unexpected discoveries. In particular, archaeologists are now reviewing their preconceptions about the meaning of the objects buried in a grave along with the body. It was thought that if a body was accompanied by a sword then it must be male, or if the grave contained beads then the body was female. DNA testing has shown that these stereotypes are not always correct, and that archaeologists must take a broader view of the link between grave goods and sex.

16.4 Archaeogenetics: Using DNA to study human prehistory

Sex identification and kinship studies are not the only ways in which gene cloning and DNA analysis are being applied in archaeology. By examining DNA sequences in living and dead humans, archaeologists have begun to understand the evolutionary origins of modern humans (members of *Homo sapiens*), and the routes followed by prehistoric people as they colonized the planet. This area of research is called archaeogenetics.

16.4.1 The origins of modern humans

Palaeontologists believe that humans originated in Africa because it is here that all of the oldest pre-human fossils have been found. The fossil evidence reveals that humans first migrated out of Africa over one million years ago, but these were not members of our own species. Instead, they were an earlier type called *Homo erectus*, who were the

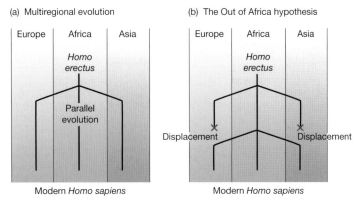

Figure 16.8

Two hypotheses for the origins of modern humans. (a) Multiregional evolution. (b) The Out of Africa hypothesis.

first humans to become geographically dispersed, eventually spreading to most parts of the Old World.

The events that followed the dispersal of *H. erectus* are controversial. From studies of fossils, many palaeontologists believe that the *H. erectus* populations which became located in different parts of the Old World gave rise to the modern *Homo sapiens* populations found in those areas today (Figure 16.8a). This process is called multiregional evolution. There may have been a certain amount of interbreeding between humans from different geographical regions but, to a large extent, these various populations remained separate throughout their evolutionary history.

DNA analysis has challenged the multiregional hypothesis

Doubts about the multiregional hypothesis were raised in 1987 when geneticists first started using DNA analysis to ask questions about human evolution. In one of the very first archaeogenetics projects, RFLPs were typed in mitochondrial DNA samples taken from 147 humans, from all parts of the world. The resulting data were then used to construct a phylogenetic tree showing the evolutionary relationships between different human populations. From this tree, various deductions were made:

- The root of the tree represents a woman (remember, mitochondrial DNA is inherited only through the female line) whose mitochondrial genome is ancestral to all the 147 modern mitochondrial DNAs that were tested. This woman has been called mitochondrial Eve. Of course, she was not equivalent to the Biblical character and was by no means the only woman alive at the time. She simply was the person who carried the ancestral mitochondrial DNA that gave rise to all the mitochondrial DNAs in existence today.
- Mitochondrial Eve lived in Africa. This was deduced because the ancestral sequence split the tree into two segments, one of which was composed solely of African mitochondrial DNAs. Because of this split, it was inferred that the ancestor was also located in Africa.
- Mitochondrial Eve lived between 140 000 and 290 000 years ago. This conclusion was drawn by applying the molecular clock to the phylogenetic tree. The molecular clock is a measure of the speed at which evolutionary change occurs in

mitochondrial DNA sequences, and is calibrated from the rate at which mutations are known to accumulate in mitochondrial DNA. By comparing the sequence inferred for Eve's mitochondrial DNA with the sequences of the 147 modern DNAs, the number of years needed for all of the necessary evolutionary changes to take place was calculated.

The key finding that mitochondrial Eve lived in Africa no earlier than 290 000 years ago does not agree with the suggestion that we are all descended from *H. erectus* populations that left Africa over a million years ago. A new hypothesis for human origins was therefore devised, called Out of Africa. According to this hypothesis, modern humans – *H. sapiens* – evolved specifically from those *H. erectus* populations that remained in Africa. Modern humans then moved into the rest of the Old World between 100 000 and 50 000 years ago, displacing the descendants of *H. erectus* that they encountered (Figure 16.8b).

At first, the mitochondrial Eve results were heavily criticized. It became apparent that the computer analysis used to construct the phylogenetic tree was flawed, mainly because the algorithms used to compare the RFLP data were not sufficiently robust to deal with this huge amount of information. However, the criticisms have now died away as the results of more extensive mitochondrial DNA studies, using actual DNA sequences rather than RFLPs and analysed using modern, powerful computers, have all confirmed the findings of the first project. To take one example, when the complete mitochondrial genome sequences of 53 people, again from all over the world, were compared, a date of 120 000–220 000 years for mitochondrial Eve was obtained. An interesting complement has been provided by studies of the Y chromosome which, of course, descends exclusively through the male line. This work has revealed that 'Y chromosome Adam' also lived in Africa between 120 000 and 338 000 years ago.

DNA analysis shows that Neanderthals are not the direct ancestors of modern Europeans

Neanderthals are an extinct type of human who lived in Europe between 200 000 and 30 000 years ago. According to the Out of Africa hypothesis, they were displaced when modern humans reached Europe about 50 000 years ago. Therefore, one prediction of the Out of Africa hypothesis is that Neanderthals are not the ancestors of modern Europeans. Analysis of ancient DNA from Neanderthal bones has been used to test this prediction.

The first Neanderthal specimen selected for study was the type specimen, which had been found in Germany in the 19th century. This fossil has not been precisely dated, but is between 30 000 and 100 000 years old. A short part of the mitochondrial DNA of this individual was sequenced by carrying out nine overlapping PCRs, each one amplifying less than 170 bp of DNA but together giving a total length of 377 bp. A phylogenetic tree was then constructed to compare the sequence obtained from the Neanderthal bone with the sequences of other mitochondrial DNA variants (called haplogroups) present in *H. sapiens*. The Neanderthal sequence was positioned on a branch of its own, connected to the root of the tree but not linked directly to any of the modern human sequences (Figure 16.9a). This was the first evidence suggesting that Neanderthals are not ancestral to modern Europeans.

Details of the first study of Neanderthal mitochondrial DNA were published in 1997. The results were confirmed in 2008 when a complete mitochondrial DNA sequence was obtained from three small pieces of Neanderthal bone, dating to 38 000 years ago, that

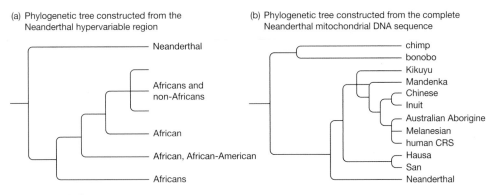

(a) Phylogenetic tree constructed from the Neanderthal hypervariable region

(b) Phylogenetic tree constructed from the complete Neanderthal mitochondrial DNA sequence

Figure 16.9

Phylogenetic analysis of ancient DNA suggests that Neanderthals are not directly related to modern humans. (a) The first comparison between Neanderthal and *H. sapiens* mitochondrial DNA. The *H. sapiens* mitochondrial sequences were obtained from various African and non-African groups. (b) Comparison based on a complete Neanderthal mitochondrial DNA sequence. The human sequences are from populations associated with different parts of the world. 'Human CRS' refers to the 'Cambridge Reference Sequence', which was the first human mitochondrial sequence to be obtained, and was from a person of European origin.

were found in a cave at Vindija in Croatia. The phylogenetic tree constructed from the complete DNA sequence again places Neanderthals on a branch of their own, separate from modern humans (Figure 16.9b). The Neanderthal mitochondrial genome therefore falls outside the *H. sapiens* range, which would be the expected result if Neanderthals are not the ancestors of modern Europeans. The results therefore provide an independent proof of the Out of Africa hypothesis and show that, at least for Europe, the multiregional model is incorrect.

The Neanderthal genome sequence suggests there was interbreeding with *H. sapiens*

One of the great achievements of next-generation sequencing (p. 182) has been the complete sequence of the Neanderthal nuclear genome, obtained from ancient DNA preserved in small pieces of a bone from a cave in the Altai mountains of Siberia. The genome sequence has enabled the intriguing possibility that Neanderthals and modern humans interbred to be addressed. *Homo sapiens* arrived in Europe about 45 000 years ago, and Neanderthals did not become extinct until about 30 000 years ago, so they must have lived in the same regions for several thousand years. Prior to the genome sequence becoming available, tentative evidence for interbreeding had been obtained from archaeological work. In 1998, the skeleton of a four-year-old child was found at Abrigo do Lagar Velho in Portugal, which seemed to have both Neanderthal and modern human features. Its status as a hybrid was doubtful, however, because the burial has been dated to 24 500 years ago, about 5000 years after Neanderthals are thought to have become extinct.

Much stronger evidence of interbreeding was obtained when comparisons were made between the Neanderthal genome and the genomes of *H. sapiens* from Europe and Africa. If there had been no interbreeding, then modern Europeans and Africans should be indistinguishable when compared with Neanderthals. However, the degree of divergence between Neanderthals and modern Europeans was slightly less than that between Neanderthals and modern Africans, which suggested that some Neanderthal DNA had

Figure 16.10

Extent of interbreeding between *H. sapiens* and extinct hominins. Arrows indicate the direction in which DNA has been transferred between the different types of human, and the percentages are the estimated magnitudes of the individual transfers.

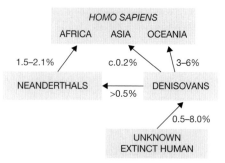

found its way into the genomes of modern Europeans. This indicates that there was a small amount of interbreeding between Neanderthals and *H. sapiens* during the 15 000 years or so that they were co-resident in Europe.

Ancient DNA sequencing has also identified a new type of human that was present in Eurasia around 40 000 years ago. The sequencing of ancient DNA from a finger bone from Denisova cave in Siberia revealed an unusual mitochondrial DNA variant whose features were unlike those of Neanderthals or modern humans, suggesting that the finger bone might have come from an unknown species who were called the 'Denisovans'. However, the complete Denisovan genome sequence shows a closer affinity with Neanderthals than indicated by the mitochondrial DNA, so possibly the Denisovans were an Asian version of Neanderthals. The Denisovan genome sequence gave additional evidence for interbreeding with *H. sapiens*, in this case specifically with the ancestors of modern inhabitants of Oceania. The most recent estimates are that 1.5–2.1% of the DNA of modern humans from outside of Africa is of Neanderthal origin, and 3.0–6.0% of the genomes of modern inhabitants of Oceania is derived from Denisovans (Figure 16.10). There is also evidence for interbreeding between Neanderthals and Denisovans, and between Denisovans and an unidentified extinct type of human.

16.4.2 DNA can also be used to study prehistoric human migrations

As well as establishing that our common ancestor lived in Africa relatively recently, DNA sequencing is also helping to trace and date the migrations by which our species colonized the rest of the planet. The earliest fossil remains of *H. sapiens* found outside of Africa were discovered in the Qafzeh and Skhul caves near Nazareth in modern Israel. These fossils are dated to 90 000–100 000 years ago, but they are unlikely to be the remains of members of the first wave of the advance of humans through the Old World. The caves were reoccupied by Neanderthals at a later date, and modern humans were not seen there again for some time. It seems that the Qafzeh and Skhul people either were part of a failed attempt at colonization, or merely represent an extension of the African population into Asia during a warm period.

Modern humans may have migrated from Ethiopia to Arabia

It now seems possible that the first modern human migration out of Africa left from Ethiopia, further south than the area of modern Suez that forms the physical link

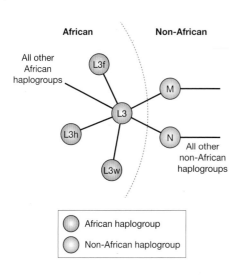

Figure 16.11

The two links between the African haplogroup L3 and haplogroups M and N in the non-African part of the human mitochondrial DNA network.

between Africa and Asia. This hypothesis is based partly on archaeological evidence and partly on analysis of mitochondrial DNA sequences. Over 100 mitochondrial DNA haplogroups are known in the modern human population, all of which can be linked together in a large network displaying their sequence relationships. Within this network, all of the haplogroups that are common today in Africa are clustered together, with just two links connecting them with the remainder of the network. These two links are between haplogroup L3 on the African side and M and N on the non-African side (Figure 16.11). Archaeogeneticists believe that these two links reveal that the first migration of *H. sapiens* out of Africa was from Ethiopia to southern Arabia. The argument is as follows:

- Haplogroups M and N are found mainly in Asia, suggesting that the initial migration was into Asia.
- The modern Africans whose L3 mitochondrial DNAs have the greatest similarity to M and N live in or close to Ethiopia, so the migration probably originated in the area we now call Ethiopia.
- The most direct route from Ethiopia to Asia is across the entrance of the Red Sea to the southern coast of Arabia (Figure 16.12).

Applying the molecular clock suggests that haplogroups M and N both originated 50 000–70 000 years ago, which suggests that the migration from Africa began around that time. The archaeological record indicates that modern humans reached Australia by 50 000–60 000 years ago, so the hypothesis is that the descendants of the first migration out of Africa moved along the southern Asian coast and then island-hopped all the way to Australia. The rate of migration would have been about 0.7–4.0 km per year.

Colonization of the New World

On their way to Australia, there would have been side branches that took some of the migrants into the interior of Asia. One of these side branches eventually gave rise to the first humans to arrive in the Americas – there is no evidence for the presence of *H. erectus* anywhere in the New World. The most likely point of entry is across the Bering

Figure 16.12

The route from Ethiopia to southern Arabia, thought to be the trajectory for the first migration of *H. sapiens* from Africa to Asia.

Strait, which separates Siberia from Alaska (Figure 16.13). The Bering Strait is quite shallow, and if the sea level dropped by 50 metres it would be possible to walk across from one continent to the other. It is believed that this 'Beringian land bridge' was the route taken by the first humans to venture into the New World.

The sea was 50 metres or more below its current level for most of the last Ice Age, between about 60 000 and 11 000 years ago, but for a large part of this time the route

Figure 16.13

The Bering Strait between modern Siberia and Alaska.

Figure 16.14

The ice-free corridor leading from Beringia to central North America.

would have been impassable because of the build-up of ice, not on the land bridge itself but in the areas that are now Alaska and northwest Canada. In addition, the glacier-free parts of northern America would have been arctic during much of this period, providing few game animals for the migrants to hunt and very little wood with which they could make fires. Only for a brief period around 14 000–12 000 years ago was the Beringian land bridge open, at a time when the climate was warming and the glaciers had receded to the extent that there was an ice-free corridor leading from Beringia to central North America (Figure 16.14). The implication is that the first modern humans might have reached the Americas by migrating through this corridor about 13 000 years ago.

The vast majority of Native Americans have one of four mitochondrial DNA haplogroups, namely A, B, C, and D, all of which are common in east Asia, suggesting that east Asia is the source of the populations that colonized the Americas. The New World versions of these four haplogroups originated between 24 000 and 14 000 years ago, implying that the initial migration into the New World occurred during this period. This is not quite what would be expected if the migration was through the ice-free corridor between 14 000 and 12 000 years ago, as the dates do not quite match up.

Some archaeologists believe that there is evidence of human occupation in the Americas earlier than 13 000 years ago. Ancient DNA sequencing has added to this evidence. Coprolites (fossilized excrement) dating to more than 14 000 years ago were discovered in a cave in Oregon. Judging by their size, shape and colour, these coprolites could be human in origin, though they may be from wild dogs. Excrement contains DNA from the animal or person that deposited the material, and coprolites have previously been shown to be a good source of ancient DNA, so DNA was sequenced from these specimens to determine whether they were human or canine in origin. The sequences were clearly human mitochondrial DNA, belonging to haplogroups A and B. This discovery pushes the date for the first migration into the Americas back to 15 000 years or so.

Archaeogenetics truly illustrates how broad ranging an impact gene cloning and DNA analysis have had on science.

Further reading

Brown, T.A. (2010) Stranger from Siberia. *Nature*, **464**, 838–839. [The discovery of the Denisovans.]

Cann, R.L., Stoneking, M., and Wilson, A.C. (1987) Mitochondrial DNA and human evolution. *Nature*, **325**, 31–36. [The first paper to propose the mitochondrial Eve hypothesis.]

Coble, M.D., Loreille, O.M., Wadhams, M.J., *et al.* (2009) Mystery solved: the identification of the two missing Romanov children using DNA analysis. *PLoS ONE*, **4**(3), e4838.

Gilbert, M.T.P., Jenkins, D.L., Götherstrom, A., *et al.* (2008) DNA from Pre-Clovis human coprolites in Oregon, North America. *Science*, **320**, 786–789.

Gill, P., Ivanov, P.L., Kimpton, C., *et al.* (1994) Identification of the remains of the Romanov family by DNA analysis. *Nature Genetics*, **6**, 130–135.

Green, R.E., Malaspinas, A.S., Krause, J., *et al.* (2008) A complete Neandertal mitochondrial genome sequence determined by high-throughput sequencing. *Cell*, **134**, 416–426.

Jeffreys, A.J., Wilson, V., and Thein, L.S. (1985) Individual-specific fingerprints of human DNA. *Nature*, **314**, 67–73. [Genetic fingerprinting by analysis of dispersed repeats.]

Jobling, M.A. and Gill, P. (2004) Encoded evidence: DNA in forensic analysis. *Nature Reviews Genetics*, **5**, 739–751.

Krings, M., Stone, A., Schmitz, R.W., *et al.* (1997) Neandertal DNA sequences and the origin of modern humans. *Cell*, **90**, 19–30.

Mellars, P., Gori, K.C., Carr, M., *et al.* (2013) Genetic and archaeological perspectives on the initial modern human colonization of southern Asia. *Proceedings of the National Academy of Sciences of the United States of America*, **110**, 10699–10704.

Nakahori, Y., Hamano, K., Iwaya, M., and Nakagome, Y. (1991) Sex identification by polymerase chain reaction using X–Y homologous primer. *American Journal of Medical Genetics*, **39**, 472–473. [The amelogenin method.]

Prüfer, K., Patterson, N., Jay, F., *et al.* (2014) The complete genome sequence of a Neanderthal from the Altai Mountains. *Nature*, **505**, 43–49.

Glossary

GLOSSARY

2 µm plasmid A plasmid found in the yeast *Saccharomyces cerevisiae* and used as the basis for a series of cloning vectors.

3′ terminus One of the two ends of a polynucleotide: that which carries the hydroxyl group attached to the 3′ position of the sugar.

5′ terminus One of the two ends of a polynucleotide: that which carries the phosphate group attached to the 5′ position of the sugar.

454 sequencing A next-generation sequencing method utilizing pyrosequencing of fragments immobilized on beads.

Adaptor A synthetic, double-stranded oligonucleotide used to attach sticky ends to a blunt-ended molecule.

Adeno-associated virus (AAV) A virus that is unrelated to adenovirus but which is often found in the same infected tissues, because AAV makes use of some of the proteins synthesized by adenovirus in order to complete its replication cycle.

Adenovirus An animal virus, derivatives of which have been used to clone genes in mammalian cells.

Affinity chromatography A chromatography method that makes use of a ligand that binds a specific protein and which can therefore be used to aid purification of that protein.

Agrobacterium tumefaciens The soil bacterium which, when containing the Ti plasmid, is able to form crown galls on a number of dicotyledonous plant species.

Ancient DNA Preserved DNA from an archaeological or fossil specimen.

Annealing Attachment of an oligonucleotide to a single-stranded DNA molecule by hybridization.

Antisense RNA An RNA molecule that is the reverse complement of a naturally occurring mRNA, and which can be used to prevent translation of that mRNA in a transformed cell.

Archaeogenetics The use of DNA analysis to study the human past.

Artificial gene synthesis Construction of an artificial gene from a series of overlapping oligonucleotides.

Autoradiography A method of detecting radioactively labelled molecules through exposure of an X-ray-sensitive photographic film.

Auxotroph A mutant microorganism that grows only when supplied with a nutrient not required by the wild-type.

Avidin A protein that has a high affinity for biotin and is used in a detection system for biotinylated probes.

BacMam vector A modified baculovirus that carries a mammalian promoter and so is able to express a cloned gene directly in a mammalian cell.

Gene Cloning and DNA Analysis: An Introduction, Seventh Edition. T.A. Brown.
© 2016 John Wiley & Sons, Ltd. Published 2016 by John Wiley & Sons, Ltd.
Companion Website: www.wiley.com/go/brown/cloning

Bacterial artificial chromosome (BAC) A cloning vector based on the F plasmid, used for cloning relatively large fragments of DNA in *E. coli*.

Bacteriophage or phage A virus whose host is a bacterium. Bacteriophage DNA molecules are often used as cloning vectors.

Baculovirus A virus that has been used as a cloning vector for the production of recombinant protein in insect cells.

Baits The oligonucleotides used to capture DNA fragments during target enrichment, prior to preparation of a next-generation DNA sequencing library.

Batch culture Growth of bacteria in a fixed volume of liquid medium in a closed vessel, with no additions or removals made during the period of incubation.

Bioinformatics The use of computer methods in studies of genomes.

Biolistics A means of introducing DNA into cells that involves bombardment with high-velocity microprojectiles coated with DNA.

Biological containment One of the precautionary measures taken to prevent the replication of recombinant DNA molecules in microorganisms in the natural environment. Biological containment involves the use of vectors and host organisms that have been modified so that they will not survive outside the laboratory.

Biotechnology The use of biological processes in industry and technology.

Biotin A molecule that can be incorporated into dUTP and used as a non-radioactive label for a DNA probe.

BLAST An algorithm frequently used in homology searching.

Blunt end or flush end An end of a DNA molecule at which both strands terminate at the same nucleotide position with no single-stranded extension.

Broad host range plasmid A plasmid that can replicate in a variety of host species.

Broth culture Growth of microorganisms in a liquid medium.

Buoyant density The density possessed by a molecule or particle when suspended in an aqueous salt or sugar solution.

Candidate gene A gene, identified by positional cloning, that might be a disease-causing or disease-susceptibility gene.

Capsid The protein coat that encloses the DNA or RNA molecule of a bacteriophage or virus.

Cassette A DNA sequence consisting of promoter–ribosome-binding site–unique restriction site–terminator (or for a eukaryotic host, promoter–unique restriction site–polyadenylation sequence) carried by certain types of expression vector. A foreign gene inserted into the unique restriction site is placed under control of the expression signals.

Cauliflower mosaic virus (CaMV) The best studied of the caulimoviruses, used in the past as a cloning vector for some species of higher plant. Cauliflower mosaic virus is the source of strong promoters used in other types of plant cloning vector.

Caulimoviruses One of the two groups of DNA viruses to infect plants, the members of which have potential as cloning vectors for some species of higher plant.

Cell extract A preparation consisting of a large number of broken cells and their released contents.

Cell-free translation system A cell extract containing all the components required for protein synthesis (i.e., ribosomal subunits, tRNAs, amino acids, enzymes and cofactors) and able to translate added mRNA molecules.

Chaotropic agent A chemical that interferes with the hydrogen bonding that normally holds water molecules together

Chimera (1) A recombinant DNA molecule made up of DNA fragments from more than one organism, named after the mythological beast. (2) The initial product of cloning using embryonic stem cells: an animal made up of a mixture of cells with different genotypes.

ChIP-seq A method for identifying the positions where individual DNA-binding proteins attach to a genome.

Chromatin The DNA–protein complex found in the nuclei of eukaryotic cells.

Chromatin immunoprecipitation sequencing (ChiP-seq) A method for identifying the positions where individual DNA-binding proteins attach to a genome.

Chromatography A group of techniques which achieve separation of substances by virtue on differential partitioning between mobile and stationary phases.

Chromosome One of the DNA–protein structures that contains part of the nuclear genome of a eukaryote. Less accurately, the DNA molecule(s) that contains a prokaryotic genome.

Chromosome walking A technique that can be used to construct a contiguous series of clones by identifying overlapping fragments of cloned DNA.

Cleared lysate A cell extract that has been centrifuged to remove cell debris, subcellular particles and possibly chromosomal DNA.

Clone A population of identical cells, generally those containing identical recombinant DNA molecules.

Clone fingerprinting Any one of a variety of techniques that compares cloned DNA fragments in order to identify ones that overlap.

Codon bias The fact that not all codons are used equally frequently in the genes of a particular organism.

Combinatorial screening A technique that reduces the number of PCRs or other analyses that must be performed by combining samples in an ordered fashion, so that a sample giving a particular result can be identified even though that sample is not individually examined.

Comparative genomics A research strategy that uses information obtained from the study of one genome to make inferences about the map positions and functions of genes in a second genome.

Compatibility The ability of two different types of plasmid to coexist in the same cell.

Competent A culture of bacteria that has been treated to enhance their ability to take up DNA molecules.

Complementary Two polynucleotides that can base pair to form a double-stranded molecule.

Complementary DNA (cDNA) cloning A cloning technique involving conversion of purified mRNA to DNA before insertion into a vector.

Conformation The spatial organization of a molecule. Linear and circular are two possible conformations of a polynucleotide.

Conjugation Physical contact between two bacteria, usually associated with transfer of DNA from one cell to the other.

Consensus sequence A nucleotide sequence used to describe a large number of related though non-identical sequences. Each position of the consensus sequence represents the nucleotide most often found at that position in the real sequences.

Contig A contiguous DNA sequence made up of a series of overlapping sequence reads.

Continuous culture The culture of microorganisms in liquid medium under controlled conditions, with additions to and removals from the medium over a lengthy period of time.

Contour-clamped homogeneous electric fields (CHEF) An electrophoresis technique for the separation of large DNA molecules.

Copy number The number of molecules of a plasmid contained in a single cell.

Cosmid A cloning vector consisting of the λ *cos* site inserted into a plasmid, used to clone DNA fragments up to 40 kb in size.

cos **site** One of the cohesive, single-stranded extensions present at the ends of the DNA molecules of certain strains of λ phage.

Covalently closed-circular DNA (cccDNA) A completely double-stranded circular DNA molecule, with no nicks or discontinuities, usually with a supercoiled conformation.

Coverage The average number of reads that cover each nucleotide position in a DNA sequence obtained by a next-generation method.

CpG island A GC-rich DNA region located upstream of approximately 56% of the genes in the human genome.

Defined medium A bacterial growth medium in which all the components are known.

Deletion analysis The identification of control sequences for a gene by determining the effects on gene expression of specific deletions in the upstream region.

Deletion cassette A segment of DNA that is transferred to a yeast chromosome by homologous recombination in order to create a deleted version of a target gene, in order to inactivate that gene and identify its function.

Denaturation Of nucleic acid molecules: breakdown by chemical or physical means of the hydrogen bonds involved in base pairing.

De novo **sequencing** A strategy in which a genome sequence is assembled solely by finding overlaps between individual sequence reads.

Density gradient centrifugation Separation of molecules and particles on the basis of buoyant density, by centrifugation in a concentrated sucrose or caesium chloride solution.

Deoxyribonuclease An enzyme that degrades DNA.

Dideoxynucleotide A modified nucleotide that lacks the 3′ hydroxyl group and so prevents further chain elongation when incorporated into a growing polynucleotide.

Directed evolution A set of experimental techniques that is used to obtain novel genes with improved products.

Direct gene transfer A cloning process that involves transfer of a gene into a chromosome without the use of a cloning vector able to replicate in the host organism.

Disarmed plasmid A Ti plasmid that has had some or all of the T-DNA genes removed, so it is no longer able to promote cancerous growth of plant cells.

DNA chip A wafer of silicon carrying a high-density array of oligonucleotides used in transcriptome and other studies.

DNA ladder A mixture of DNA fragments, whose sizes are multiples of 100 bp or of 1 kb, used as size markers during gel electrophoresis.

DNA ligase An enzyme that, in the cell, repairs single-stranded discontinuities in double-stranded DNA molecules. Purified DNA ligase is used in gene cloning to join DNA molecules together.

DNA marker A DNA sequence that exists as two or more alleles and which can therefore be used in genetic mapping.

DNA polymerase An enzyme that synthesizes DNA on a DNA or RNA template.

DNA profiling A PCR technique that determines the alleles present at different STR loci within a genome in order to use DNA information to identify individuals.

DNA sequencing Determination of the order of nucleotides in a DNA molecule.

Double digestion Cleavage of a DNA molecule with two different restriction endonucleases, either concurrently or consecutively.

Electrophoresis Separation of molecules on the basis of their charge-to-mass ratio.

Electroporation A method for increasing DNA uptake by protoplasts through prior exposure to a high voltage, which results in the temporary formation of small pores in the cell membrane.

Elution The unbinding of a molecule from a chromatography column.

Embryonic stem (ES) cell A totipotent cell from the embryo of a mouse or other organism, used in construction of a transgenic animal such as a knockout mouse.

End filling Conversion of a sticky end to a blunt end by enzymatic synthesis of the complement to the single-stranded extension.

Endonuclease An enzyme that breaks phosphodiester bonds within a nucleic acid molecule.

Episome A plasmid capable of integration into the host cell's chromosome.

Ethanol precipitation Precipitation of nucleic acid molecules by ethanol plus salt, used primarily as a means of concentrating DNA.

Ethidium bromide A fluorescent chemical that intercalates between base pairs in a double-stranded DNA molecule, used in the detection of DNA.

Exonuclease An enzyme that sequentially removes nucleotides from the ends of a nucleic acid molecule.

Expression vector A cloning vector designed so that a foreign gene inserted into the vector is expressed in the host organism.

Fermenter A vessel used for the large-scale culture of microorganisms.

Field inversion gel electrophoresis (FIGE) An electrophoresis technique for the separation of large DNA molecules.

Footprinting The identification of a protein binding site on a DNA molecule by determining which phosphodiester bonds are protected from cleavage by DNase I.

Forward genetics The strategy by which the genes responsible for a phenotype are identified by determining which genes are inactivated in organisms that display a mutant version of that phenotype.

Forward sequence One of the two directions in which a double-stranded DNA molecule can be sequenced.

Functional genomics Studies aimed at identifying all the genes in a genome and determining their expression patterns and functions.

Fusion protein A recombinant protein that carries a short peptide from the host organism at its amino or, less commonly, carboxyl terminus.

Gel electrophoresis Electrophoresis performed in a gel matrix so that molecules of similar electric charge can be separated on the basis of size.

Gel retardation A technique that identifies a DNA fragment that has a bound protein by virtue of its decreased mobility during gel electrophoresis.

Geminivirus One of the two groups of DNA viruses that infect plants, the members of which have potential as cloning vectors for some species of higher plants.

Gene A segment of DNA that codes for an RNA and/or polypeptide molecule.

Gene addition A genetic engineering strategy that involves the introduction of a new gene or group of genes into an organism.

Gene cloning Insertion of a fragment of DNA, carrying a gene, into a cloning vector, and subsequent propagation of the recombinant DNA molecule in a host organism. Also used to describe those techniques that achieve the same result without the use of a cloning vector (e.g., direct gene transfer).

Gene knockout A technique that results in inactivation of a gene, as a means of determining the function of that gene.

Gene mapping Determination of the relative positions of different genes on a DNA molecule.

Gene subtraction A genetic engineering strategy that involves the inactivation of one or more of an organism's genes.

Gene therapy A clinical procedure in which a gene or other DNA sequence is used to treat a disease.

Genetic engineering The use of experimental techniques to produce DNA molecules containing new genes or new combinations of genes.

Genetic fingerprinting A hybridization technique that determines the genomic distribution of a hypervariable dispersed repetitive sequence and results in a banding pattern that is specific for each individual.

Genetics The branch of biology devoted to the study of genes.

Genome The complete set of genes of an organism.

Genome annotation The process by which the genes, control sequences and other interesting features are identified in a genome sequence.

Genomic DNA Consists of all the DNA present in a single cell or group of cells.

Genomic library A collection of clones sufficient in number to include all the genes of a particular organism.

Genomics The study of a genome, in particular the complete sequencing of a genome.

Germline therapy A type of gene therapy in which a fertilized egg is provided with a copy of the correct version of the defective gene and reimplanted into the mother.

GM (genetically modified) crop A crop plant that has been engineered by gene addition or gene subtraction.

Haplogroup One of the major sequence classes of mitochondrial DNA present in the human population.

Harvesting The removal of microorganisms from a culture, usually by centrifugation.

Helper phage A phage that is introduced into a host cell in conjunction with a related cloning vector, in order to provide enzymes required for replication of the cloning vector.

Heterologous probing The use of a labelled nucleic acid molecule to identify related molecules by hybridization probing.

Heteroplasmic Possessing two haplotypes of the mitochondrial genome.

Hierarchical shotgun sequencing A DNA sequencing strategy which involves a pre-sequencing phase during which the genome is broken into large fragments, which are cloned and each sequenced individually by the shotgun method.

Homologous recombination Recombination between two homologous double-stranded DNA molecules; that is, ones which share extensive nucleotide sequence similarity.

Homology Refers to two genes from different organisms that have evolved from the same ancestral gene. Two homologous genes are usually sufficiently similar in sequence for one to be used as a hybridization probe for the other.

Homology search A technique in which genes with sequences similar to that of an unknown gene are sought, in order to confirm a gene identification or to understand the function of the unknown gene.

Homopolymer tailing The attachment of a sequence of identical nucleotides (e.g., AAAAA) to the end of a nucleic acid molecule, usually referring to the synthesis of single-stranded homopolymer extensions on the ends of a double-stranded DNA molecule.

Horseradish peroxidase An enzyme that can be complexed to DNA and which is used in a non-radioactive procedure for DNA labelling.

Host-controlled restriction A mechanism by which some bacteria prevent phage attack through the synthesis of a restriction endonuclease that cleaves the non-bacterial DNA.

Hybrid-arrest translation (HART) A method used to identify the polypeptide coded by a cloned gene.

Hybridization probe A labelled nucleic acid molecule that can be used to identify complementary or homologous molecules through the formation of stable, base-paired hybrids.

Hybrid-release translation (HRT) A method used to identify the polypeptide coded by a cloned gene.

Hypervariable dispersed repetitive sequence The type of human repetitive DNA sequence used in genetic fingerprinting.

Illumina sequencing A next-generation sequencing method utilizing reversible terminator sequencing of fragments immobilized on a slide.

Immunological screening The use of an antibody to detect a polypeptide synthesized by a cloned gene.

***In situ* hybridization** A technique for gene mapping involving hybridization of a labelled sample of a cloned gene to a large DNA molecule, usually a chromosome.

***In vitro* mutagenesis** Any one of several techniques used to produce a specified mutation at a predetermined position in a DNA molecule.

***In vitro* packaging** Synthesis of infective λ particles from a preparation of λ capsid proteins and a catenane of DNA molecules separated by *cos* sites.

Inclusion body A crystalline or paracrystalline deposit within a cell, often containing substantial quantities of insoluble protein.

Incompatibility group Comprises a number of different types of plasmid, often related to each other, that are unable to coexist in the same cell.

Indel A position where a DNA sequence has been inserted into or deleted from a genome, so-called because it is impossible from comparison of two sequences to determine which alternative has occurred, insertion into one genome or deletion from the other.

Induction (1) Of a gene: the switching on of the expression of a gene or group of genes in response to a chemical or other stimulus. (2) Of λ phage: the excision of the integrated form of λ and accompanying switch to the lytic mode of infection, in response to a chemical or other stimulus.

Insertion vector A λ vector constructed by deleting a segment of non-essential DNA.

Insertional inactivation A cloning strategy whereby insertion of a new piece of DNA into a vector inactivates a gene carried by the vector.

Ion-exchange chromatography A method for separating molecules according to how tightly they bind to electrically charged particles present in a chromatographic matrix.

Isoelectric focusing Separation of proteins in a gel that contains chemicals which establish a pH gradient when the electrical charge is applied.

Isoelectric point The position in a pH gradient where the net charge of a protein is zero.

Isotope-coded affinity tag (ICAT) Markers, containing normal hydrogen and deuterium atoms, used to label individual proteomes.

Kinship analysis An examination of DNA profiles or other information to determine if two individuals are related.

Klenow fragment (of DNA polymerase I) A DNA polymerase enzyme, obtained by chemical modification of *E. coli* DNA polymerase I.

Knockout mouse A mouse that has been engineered so that it carries an inactivated gene.

Labelling The incorporation of a marker nucleotide into a nucleic acid molecule. The marker is often, but not always, a radioactive or fluorescent label.

Lac selection A means of identifying recombinant bacteria containing vectors that carry the *lacZ'* gene. The bacteria are plated on a medium that contains an analogue of lactose that gives a blue colour in the presence of β-galactosidase activity.

Lambda (λ) A bacteriophage that infects *E. coli*, derivatives of which are used as cloning vectors.

Library (1) A collections of clones, e.g., a genomic library. (2) A collection of DNA fragments prepared for sequencing by a next-generation method.

Ligase (DNA ligase) An enzyme that, in the cell, repairs single-stranded discontinuities in double-stranded DNA molecules. Purified DNA ligase is used in gene cloning to join DNA molecules together.

Linkage analysis A technique for mapping the chromosomal position of a gene by comparing its inheritance pattern with that of genes and other loci whose map positions are already known.

Linker A synthetic, double-stranded oligonucleotide used to attach sticky ends to a blunt-ended molecule.

Liposome A lipid vesicle sometimes used to introduce DNA into an animal or plant cell.

Lysogen A bacterium that harbours a prophage.

Lysogenic infection cycle The pattern of phage infection that involves integration of the phage DNA into the host chromosome.

Lysozyme An enzyme that weakens the cell walls of certain types of bacteria.

Lytic infection cycle The pattern of infection displayed by a phage that replicates and lyses the host cell immediately after the initial infection. Integration of the phage DNA molecule into the bacterial chromosome does not occur.

M13 A bacteriophage that infects *E. coli*, derivatives of which are used as cloning vectors.

Massively parallel A high-throughput sequencing strategy in which many individual sequences are generated in parallel.

Mass spectrometry An analytical technique in which ions are separated according to their mass-to-charge ratios.

Matrix-assisted laser desorption ionization time-of-flight (MALDI-TOF) A type of mass spectrometry used in proteomics.

Melting temperature (T_m) The temperature at which a double-stranded DNA or DNA–RNA molecule denatures.

Messenger RNA (mRNA) The transcript of a protein-coding gene.

Microarray A set of genes or cDNAs immobilized on a glass slide and used in transcriptome studies.

Microinjection A method of introducing new DNA into a cell by injecting it directly into the nucleus.

Microsatellite A polymorphism comprising tandem copies of, usually, two-, three-, four-, or five-nucleotide repeat units. Also called a short tandem repeat (STR).

Minimal medium A defined medium that provides only the minimum number of different nutrients needed for growth of a particular bacterium.

Mitochondrial DNA The DNA molecules present in the mitochondria of eukaryotes.

Mitochondrial Eve The woman who lived in Africa between 140 000 and 290 000 years ago and who carried the ancestral mitochondrial DNA that gave rise to all the mitochondrial DNAs in existence today.

Modification interference assay A technique that uses chemical modification to identify nucleotides involved in interactions with a DNA-binding protein.

Molecular clock An analysis based on the inferred mutation rate that enables times to be assigned to the branch points in a phylogenetic tree.

Multicopy plasmid A plasmid with a high copy number.

Multigene family A number of identical or related genes present in the same organism, usually coding for a family of related polypeptides.

Multigene shuffling A directed evolution strategy that involves taking parts of each member of a multigene family and reassembling these parts to create new gene variants.

Multiplex PCR A PCR carried out with more than one pair of primers and hence targeting two or more sites in the DNA being studied

Multiregional evolution A hypothesis that holds that modern humans in the Old World are descended from *Homo erectus* populations that left Africa over one million years ago.

N50 size A measure of the degree of completeness of a genome sequence.

Next-generation sequencing A collection of DNA sequencing methods, each involving a massively parallel strategy.

NG50 size A measure of the degree of completeness of a genome sequence.

Nick A single-strand break, involving the absence of one or more nucleotides, in a double-stranded DNA molecule.

Nick translation The repair of a nick with DNA polymerase I, usually to introduce labelled nucleotides into a DNA molecule.

Northern transfer A technique for transferring bands of RNA from an agarose gel to a nitrocellulose or nylon membrane.

Nuclear transfer A technique, used in the production of transgenic animals, that involves transfer of the nucleus of a somatic cell into an oocyte whose own nucleus has been removed.

Nucleic acid hybridization Formation of a double-stranded molecule by base pairing between complementary or homologous polynucleotides.

Oligonucleotide A short, synthetic, single-stranded DNA molecule, such as one used as a primer in DNA sequencing or PCR.

Oligonucleotide-directed mutagenesis An *in vitro* mutagenesis technique that involves the use of a synthetic oligonucleotide to introduce the predetermined nucleotide alteration into the gene to be mutated.

Open-circular DNA (ocDNA) The non-supercoiled conformation taken up by a circular double-stranded DNA molecule when one or both polynucleotides carry nicks.

Open reading frame (ORF) A series of codons that is, or could be, a gene.

ORF scanning Examination of a DNA sequence for open reading frames in order to locate the genes.

Origin of replication The specific position on a DNA molecule where DNA replication begins.

Orphan An open reading frame thought to be a functional gene but to which no function has yet been assigned.

Orthogonal field alternation gel electrophoresis (OFAGE) A gel electrophoresis technique that employs a pulsed electric field to achieve separation of very large molecules of DNA.

Out of Africa hypothesis A hypothesis that holds that modern humans evolved in Africa, moving to the rest of the Old World between 100 000 and 50 000 years ago, displacing the descendants of *Homo erectus* that they encountered.

P element A transposon from *Drosophila melanogaster* used as the basis of a cloning vector for that organism.

P1 A bacteriophage that infects *E. coli*, derivatives of which are used as cloning vectors.

P1-derived artificial chromosome (PAC) A cloning vector based on the P1 bacteriophage, used for cloning relatively large fragments of DNA in *E. coli*.

Papillomaviruses A group of mammalian viruses, derivatives of which have been used as cloning vectors.

Partial digestion Treatment of a DNA molecule with a restriction endonuclease under such conditions that only a fraction of all the recognition sites are cleaved.

Pedigree analysis The use of a human family tree to analyse the inheritance of a genetic or DNA marker.

Peptide mass fingerprinting Identification of a protein by examination of the mass spectrometric properties of peptides generated by treatment with a sequence-specific protease.

Personalized medicine The use individual genome sequences to make accurate diagnoses of a person's risk of developing a disease, and the use that person's genetic characteristics to plan effective therapies and treatment regimes.

Phage display A technique involving cloning in M13 that is used to identify proteins that interact with one another.

Phage display library A collection of M13 clones carrying different DNA fragments, used in phage display.

Phagemid A double-stranded plasmid vector that possesses an origin of replication from a filamentous phage and hence can be used to synthesize a single-stranded version of a cloned gene.

Pharming Genetic modification of a farm animal so that the animal synthesizes a recombinant pharmaceutical protein, often in its milk.

Pilus One of the structures present on the surface of a bacterium containing a conjugative plasmid, through which DNA is assumed to pass during conjugation.

Plaque A zone of clearing on a lawn of bacteria caused by lysis of the cells by infecting phage particles.

Plasmid A usually circular piece of DNA, primarily independent of the host chromosome, often found in bacteria and some other types of cells.

Plasmid amplification A method involving incubation with an inhibitor of protein synthesis aimed at increasing the copy number of certain types of plasmid in a bacterial culture.

Polyethylene glycol A polymeric compound used to precipitate macromolecules and molecular aggregates.

Polylinker A synthetic double-stranded oligonucleotide carrying a number of restriction sites.

Polymerase chain reaction (PCR) A technique that enables multiple copies of a DNA molecule to be generated by enzymatic amplification of a target DNA sequence.

Polymorphism Refers to a locus that is present as a number of different alleles or other variations in the population as a whole.

Positional cloning A procedure that uses information on the map position of a gene to obtain a clone of that gene.

Positional effect Refers to the variations in expression levels observed for genes inserted at different positions in a genome.

Post-genomics Studies aimed at identifying all the genes in a genome and determining their expression patterns and functions.

Primer A short, single-stranded oligonucleotide which, when attached by base pairing to a single-stranded template molecule, acts as the start point for complementary strand synthesis directed by a DNA polymerase enzyme.

Primer extension A method of transcript analysis in which the $5'$ end of an RNA is mapped by annealing and extending an oligonucleotide primer.

Processivity Refers to the amount of DNA synthesis that is carried out by a DNA polymerase before dissociation from the template.

Productive A virus infection cycle that is able to proceed to completion and result in synthesis and release of new virus particles.

Promoter The nucleotide sequence, upstream of a gene, that acts as a signal for RNA polymerase binding.

Proofreading The $3'$ to $5'$ exonuclease activity possessed by some DNA polymerases which enables the enzyme to replace a misincorporated nucleotide.

Prophage The integrated form of the DNA molecule of a lysogenic phage.

Protease An enzyme that degrades protein.

Protein A A protein from the bacterium *Staphylococcus aureus* that binds specifically to immunoglobulin G (i.e., antibody) molecules.

Protein electrophoresis Separation of proteins in an electrophoresis gel.

Protein engineering A collection of techniques, including but not exclusively gene mutagenesis, that result in directed alterations being made to protein molecules, often to improve the properties of enzymes used in industrial processes.

Protein profiling The methodology used to identify the proteins in a proteome.

Proteome The entire protein content of a cell or tissue.

Proteomics The collection of techniques used to study the proteome

Protoplast A cell from which the cell wall has been completely removed.

Pyrosequencing A DNA sequencing method in which addition of a nucleotide to the end of a growing polynucleotide is detected directly by conversion of the released pyrophosphate into a flash of chemiluminescence.

Quantitative PCR A method for quantifying the amount of product synthesized during a test PCR by comparison with the amounts synthesized during PCRs with known amounts of starting DNA.

RACE (rapid amplification of cDNA ends) A PCR-based technique for mapping the end of an RNA molecule.

Radioactive marker A radioactive atom used in the detection of a larger molecule into which it has been incorporated.

Random priming A method for DNA labelling that utilizes random DNA hexamers, which anneal to single-stranded DNA and act as primers for complementary strand synthesis by a suitable enzyme.

Read A single sequence from the output of a next-generation sequencing run.

Reading frame One of the six overlapping sequences of triplet codons, three on each polynucleotide, contained in a segment of a DNA double helix.

Real-time PCR A modification of the standard PCR technique in which synthesis of the product is measured as the PCR proceeds through its series of cycles.

Recombinant A transformed cell that contains a recombinant DNA molecule.

Recombinant DNA molecule A DNA molecule created in the test tube by ligating together pieces of DNA that are not normally contiguous.

Recombinant DNA technology All of the techniques involved in the construction, study and use of recombinant DNA molecules.

Recombinant protein A polypeptide that is synthesized in a recombinant cell as the result of expression of a cloned gene.

Recombination The exchange of DNA sequences between different molecules, occurring either naturally or as a result of DNA manipulation.

Reference genome An existing genome sequence that is used to aid assembly of the reads obtained by next-generation sequencing of a related genome.

Relaxed (1) Refers to a plasmid with a high copy number of perhaps 50 or more per cell. (2) The non-supercoiled conformation of open-circular DNA.

Replacement vector A λ vector designed so that insertion of new DNA is by replacement of part of the non-essential region of the λ DNA molecule.

Replica plating A technique whereby the colonies on an agar plate are transferred *en masse* to a new plate, on which the colonies grow in the same relative positions as before.

Replicative form (RF) of M13 The double-stranded form of the M13 DNA molecule found within infected *E. coli* cells.

Reporter gene A gene whose phenotype can be assayed in a transformed organism, and which is used in, for example, deletion analyses of regulatory regions.

Reporter probe A short oligonucleotide that gives a fluorescent signal when it hybridizes with a target DNA.

Repression The switching off of the expression of a gene or a group of genes in response to a chemical or other stimulus.

Resin A chromatography matrix.

Restriction analysis Determination of the number and sizes of the DNA fragments produced when a particular DNA molecule is cut with a particular restriction endonuclease.

Restriction endonuclease An endonuclease that cuts DNA molecules only at a limited number of specific nucleotide sequences.

Restriction fragment length polymorphism (RFLP) A mutation that results in alteration of a restriction site and hence a change in the pattern of fragments obtained when a DNA molecule is cut with a restriction endonuclease.

Restriction map A map showing the positions of different restriction sites in a DNA molecule.

Retrovirus A virus with an RNA genome, able to insert into a host chromosome, derivatives of which have been used to clone genes in mammalian cells.

Reverse genetics The strategy by which the function of a gene is identified by mutating that gene and identifying the phenotypic change that results.

Reverse sequence One of the two directions in which a double-stranded DNA molecule can be sequenced.

Reverse transcriptase An RNA-dependent DNA polymerase, able to synthesize a complementary DNA molecule on a template of single-stranded RNA.

Reverse transcription–PCR A PCR technique in which the starting material is RNA. The first step in the procedure is conversion of the RNA to cDNA with reverse transcriptase.

Reversible terminator sequencing A DNA sequencing method in which the sequence is read by detection of the fluorescent label attached to each nucleotide that is added to a growing polynucleotide.

RFLP linkage analysis A technique that uses a closely linked RFLP as a marker for the presence of a particular allele in a DNA sample, often as a means of screening individuals for a defective gene responsible for a genetic disease.

Ribonuclease An enzyme that degrades RNA.

Ribosome-binding site The short nucleotide sequence upstream of a gene, which after transcription forms the site on the mRNA molecule to which the ribosome binds.

Ri plasmid An *Agrobacterium rhizogenes* plasmid, similar to the Ti plasmid, used to clone genes in higher plants.

RNA interference (RNAi) One of the natural processes for regulating gene expression in plants and other organisms, which can be utilized as a means of inactivating a plant gene and hence bringing about gene subtraction.

RNA-seq Next-generation sequencing of RNA.

S1 nuclease mapping A method for RNA transcript mapping.

Selectable marker A gene carried by a vector and conferring a recognizable characteristic on a cell containing the vector or a recombinant DNA molecule derived from the vector.

Selection A means of obtaining a clone containing a desired recombinant DNA molecule.

Sequence assembly Assembly of the many short reads obtained by next-generation sequencing into a contiguous DNA sequence.

Sequence depth The average number of reads that cover each nucleotide position in a DNA sequence obtained by a next-generation method.

Sequence tagged site (STS) A DNA sequence whose position has been mapped in a genome.

Serial analysis of gene expression (SAGE) A method for studying the composition of a transcriptome.

Short tandem repeat (STR) A polymorphism comprising tandem copies of, usually, two-, three-, four-, or five-nucleotide repeat units. Also called a microsatellite.

Shotgun approach A genome sequencing strategy in which the molecules to be sequenced are randomly broken into fragments which are then individually sequenced.

Shotgun cloning A cloning strategy that involves the insertion of random fragments of a large DNA molecule into a vector, resulting in a large number of different recombinant DNA molecules.

Shuttle vector A vector that can replicate in the cells of more than one organism (e.g., in *E. coli* and in yeast).

Simian virus 40 (SV40) A mammalian virus that has been used as the basis for a cloning vector.

Single molecule real-time sequencing A third-generation DNA sequencing method which uses an advanced optical system to observe the addition of individual nucleotides to a growing polynucleotide.

Single nucleotide polymorphism (SNP) A point mutation that is carried by some individuals of a population.

Somatic cell therapy A type of gene therapy in which the correct version of a gene is introduced into a somatic cell.

Sonication A procedure that uses ultrasound to cause random breaks in DNA molecules.

Southern transfer A technique for transferring bands of DNA from an agarose gel to a nitrocellulose or nylon membrane.

Sphaeroplast A cell with a partially degraded cell wall.

Spin column A method for accelerating ion-exchange chromatography by centrifuging the chromatography column.

Star activity The relaxation of the specificity of a restriction endonuclease that sometimes occurs at when the reaction is carried out under non-optimal conditions, e.g., in a low-salt buffer.

Stem–loop A hairpin structure, consisting of a base-paired stem and a non-base-paired loop, that may form in a polynucleotide.

Sticky end An end of a double-stranded DNA molecule where there is a single-stranded extension.

Stringent Refers to a plasmid with a low copy number of perhaps just one or two per cell.

Strong promoter An efficient promoter that can direct synthesis of RNA transcripts at a relatively fast rate.

Structural bioinformatics Computer methods that attempt to predict the structure and/or function of a protein or protein domain from its amino acid sequence.

Stuffer fragment The part of a λ replacement vector that is removed during insertion of new DNA.

Suicide gene therapy A gene therapy approach that makes use of a gene that selectively kills cancer cells or promotes their destruction by drugs administered in a conventional fashion

Supercoiled The conformation of a covalently closed-circular DNA molecule, which is coiled by torsional strain into the shape taken by a wound-up elastic band.

Synteny Refers to a pair of genomes in which at least some of the genes are located at similar map positions.

Target enrichment A method for enriching a next-generation DNA sequencing library for fragments derived from particular genes of interest.

***Taq* DNA polymerase** The thermostable DNA polymerase that is used in PCR.

T-DNA The portion of the Ti plasmid transferred to the plant DNA.

Temperature-sensitive mutation A mutation that results in a gene product that is functional within a certain temperature range (e.g., at less than 30 °C), but non-functional at different temperatures (e.g., above 30 °C).

Template A single-stranded polynucleotide (or region of a polynucleotide) that directs synthesis of a complementary polynucleotide.

Terminator The short nucleotide sequence, downstream of a bacterial gene, that acts as a signal for termination of transcription.

Terminator technology A recombinant DNA process which results in synthesis of the ribosome inactivating protein in plant embryos, used to prevent GM crops from producing viable seeds.

Thermal cycle sequencing A DNA sequencing method that uses PCR to generate chain-terminated polynucleotides.

Third-generation sequencing Methods in which DNA sequencing is carried out in real time.

Ti plasmid The large plasmid found in those *Agrobacterium tumefaciens* cells able to direct crown gall formation on certain species of plants.

Topoisomerase An enzyme that introduces or removes turns from the double helix by breakage and reunion of one or both polynucleotides.

Total cell DNA Consists of all the DNA present in a single cell or group of cells.

Totipotent Refers to a cell that is not committed to a single developmental pathway and can hence give rise to all types of differentiated cell.

Transcript analysis A type of experiment aimed at determining which portions of a DNA molecule are transcribed into RNA.

Transcriptome The entire mRNA content of a cell or tissue.

Transfection The introduction of purified virus DNA molecules into any living cell.

Transformation The introduction of any DNA molecule into any living cell.

Transformation frequency A measure of the proportion of cells in a population that are transformed in a single experiment.

Transgenic Refers to an animal or plant that contains a cloned gene in all of its cells.

Transgenic animal An animal that possesses a cloned gene in all of its cells.

Transposable element or transposon A DNA sequence that is able to move from place to place within a genome.

Two-dimensional gel electrophoresis A method for separation of proteins used especially in studies of the proteome.

Undefined medium A growth medium in which not all the components have been identified.

Universal primer A sequencing primer that is complementary to the part of the vector DNA immediately adjacent to the point into which new DNA is ligated.

UV absorbance spectrophotometry A method for measuring the concentration of a compound by determining the amount of ultraviolet radiation absorbed by a sample.

Vector A DNA molecule, capable of replication in a host organism, into which a gene is inserted to construct a recombinant DNA molecule.

Vehicle Sometimes used as a substitute for the word 'vector', emphasizing that the vector transports the inserted gene through the cloning experiment.

Virus chromosome The DNA or RNA molecule(s) contained within a virus capsid and carrying the viral genes.

Virus-induced gene silencing (VIGS) A technique involving a geminivirus vector used to study the function of a plant gene.

Watson–Crick rules The base-pairing rules that underlie gene structure and expression. A pairs with T, and G pairs with C.

Weak promoter An inefficient promoter that directs synthesis of RNA transcripts at a relatively low rate.

Western transfer A technique for transferring bands of protein from an electrophoresis gel to a membrane support.

Yeast artificial chromosome (YAC) A cloning vector comprising the structural components of a yeast chromosome and able to clone very large pieces of DNA.

Yeast episomal plasmid (YEp) A yeast vector carrying the 2 μm plasmid origin of replication.

Yeast integrative plasmid (YIp) A yeast vector that relies on integration into the host chromosome for replication.

Yeast replicative plasmid (YRp) A yeast vector that carries a chromosomal origin of replication.

Yeast two-hybrid system A technique involving cloning in *Saccharomyces cerevisiae* that is used to identify proteins that interact with one another.

Zero-mode waveguide A nanostructure that enables individual molecules to be observed.

Zoo blot A nitrocellulose or nylon membrane carrying immobilized DNAs of several species, used to determine if a gene from one species has homologues in the other species.

Index

Gene Cloning and DNA Analysis: An Introduction, Seventh Edition. T.A. Brown.
© 2016 John Wiley & Sons, Ltd. Published 2016 by John Wiley & Sons, Ltd.
Companion Website: www.wiley.com/go/brown/cloning